Microbes and People
An A–Z of Microorganisms in Our Lives

by Neeraja Sankaran

Oryx Press
2000

1/2002

The rare Arabian Oryx is believed to have inspired the myth of the unicorn. This desert antelope became virtually extinct in the early 1960s. At that time, several groups of international conservationists arranged to have nine animals sent to the Phoenix Zoo to be the nucleus of a captive breeding herd. Today, the Oryx population is over 1,000, and over 500 have been returned to the Middle East.

© 2000 by Neeraja Sankaran
Published by The Oryx Press
4041 North Central at Indian School Road
Phoenix, Arizona 85012-3397
www.oryxpress.com

Published simultaneously in Canada
Printed and bound in the United States of America

∞ The paper used in this publication meets the minimum requirements of American National Standard for Information Science—Permanence of Paper for Printed Library Materials, ANSI Z39.48, 1984.

Library of Congress Cataloging-in-Publication Data

Sankaran, Neeraja.
  Microbes and people : an A-Z of microorganisms in our lives / Neeraja
Sankaran.
    p. cm.
Includes bibliographical references and index.
  ISBN 1-57356-217-3 (alk. paper)
  1. Microbiology—Encyclopedias. I. Title.
  QR9 .S26 2000
  579'.03—dc21

                                        00-010117
                                            CIP

This book is dedicated to
Appa, for all the journeys you took to help me make this one
and
Poornette, who I hope will find it useful someday.

CONTENTS

List of Illustrations   vii

List of Tables and Figures   ix

Preface   xi

Acknowledgments   xiv

An A–Z of Microorganisms in Our Lives   1

Appendix 1. A Chronology of Epidemics in History   271

Appendix 2. Large Infectious Disease Outbreaks   275

Appendix 3. Leading Causes of Death in the World   276

Appendix 4. Leading Infectious Disease Killers   277

Appendix 5. Death Rates from Leading Causes of Death in the United States   278

Bibliography   279

Index   281

LIST OF ILLUSTRATIONS

Actinomyces colony on agar.   5

Sensitivity disc: penicillin showing zone of inhibition.   18

Athlete's foot infection with prominent lesions between the toes.   23

Walk-in autoclave.   24

Micrograph of *Streptococcus pneumo–niae.*   30

Micrograph of *Streptococcus mutans.*   30

Micrograph of *Bacillus brevis.*   30

Micrograph of *Pseudomonas aeru–ginosa.*   30

Micrograph of *Leptospira interrogans.*   30

Algal bloom in Lake Michigan, September 1999.   39

Assorted cheeses.   57

A colony of streptococci in goat milk feta cheese.   57

Micrograph of coronavirus OC43.   71

Skin ulcer due to cutaneous leishmania-sis.   75

Micrograph of *Treponema pallidum,* darkfield preparation.   79

Micrograph of dental plaque shows corn-cob configurations, an example of bacterial coaggregation. The central cores of the cobs are filamentous bacteria surrounded by streptococcal "kernels."   83

Ringworm lesion on the arm of a sheep-worker.   83

A painting showing inoculated horses being bled for their serum, which contains diphtheria antitoxin.   85

Electron microscope.   91

Micrograph of *Escherichia coli* O157:H7.   100

Micrograph of Marburg virus, one of the filoviruses.   104

Culture of *Histoplasma capsulatum* show-ing the typical fuzzy appearance of a mold colony.   110

Micrograph of *Giardia lamblia* trophozo-ite.   116

Micrograph of *Giardia lamblia* cyst.   116

Streptococcus pyogenes beta hemolysis on sheep blood agar.   124

Herpes simplex lesion on a lip.   129

Hooke's drawing of cork under magnifica-tion.   131

T cells infected with HIV virus.   132

Normal T cells.   132

Micrograph of *Escherichia coli* bacteria inhabiting the intestinal villi.   140

Technician uses a sterile inoculating loop to pick a single colony from a petri dish containing bacterial cultures.   142

Drawing of a sculpture by Monteverde showing Edward Jenner vaccinating his son.   144

Bacteriologist Robert Koch.   147

Drawings of "animalcules" from Leeu–wenhoek's September 1683 letter to the Royal Society in London.   151

Blood-feeding *Anopheles gambiae* mos-quito.   159

Laboratory worker using an optical micro-scope.   182

Bacteriologist and chemist Louis Pasteur. 189

Micrograph of *Plasmodium vivax,* immature schizont, in blood smear. 197

Micrograph of *Plasmodium vivax* trophozoite in blood smear. 197

Micrograph of *Plasmodium vivax,* mature schizont, 24 merozoites. 197

Micrograph of *Plasmodium ovale,* including a growing trophozoite with a "ring" nucleus. 197

Micrograph of poliovirus type 1. 200

Quarantine sign posted on a house. 208

Infant with congenital rubella and "blueberry muffin" skin lesions. 221

Sign promoting condom use in Helsinki, Finland. 227

Smallpox lesions on abdomen. 228

The last known person in the world to have smallpox, 1977. 228

Ultra thin section of *Escherichia coli* forming spheroplasts in a medium containing penicillin. 229

Trypanosoma forms in blood smear from patient with African trypanosomiasis. 245

Radiograph showing tuberculosis in the left lung. 246

Centrifuging. 251

Child receiving a vaccination. 252

Ixodes scapularis, tick vector for Lyme disease. 253

Micrograph of *Vibrio cholerae;* note the single terminal flagella. 254

Micrograph of *Cryptosporidium* and photograph of *Cryptosporidrium*-contaminated water. 262

LIST OF TABLES AND FIGURES

FIGURE 1. AIDS Cases in the United States, by Exposure Category and Year of Report, 1985-1998   2

FIGURE 2. Incidence of AIDS Cases and Deaths in the United States, 1985-1998   3

FIGURE 3. Percentage of Nosocomial Enterococci Reported as Resistant to Vancomycin in Intensive Care Units (ICUs) and Non-ICUs, 1989-1994   17

FIGURE 4. Rapid Increase of Dengue Fever, 1955-1998   81

TABLE 1. Antiviral Agents   19

TABLE 2. A Comparison of the Properties of Chlamydiae, Rickettsiae, and Viruses   60

TABLE 3. New and Emerging Infectious Diseases and Pathogens   93

TABLE 4. Arboviral Encephalitides   95

TABLE 5. Etiology and Distribution of Viral Hemorrhagic Fevers   125

TABLE 6. Recommended Childhood Immunization Schedule, United States, January-December 2000   138

TABLE 7. Infectious Diseases Designated as Notifiable at the National Level, United States, 1997   178

TABLE 8. Opportunistic Infections by Indigenous Microflora   181

TABLE 9. Rickettsial Diseases in Humans   218

TABLE 10. Classification of the Virus Families of Human/Public Health Relevance   258–59

PREFACE

Microbes and People: An A–Z of Microorganisms in Our Lives is a compilation of more than 750 entries about microorganisms and diseases designed to help readers navigate through the vast and often bewildering world of living things too small to be discerned with the naked eye. The book is aimed primarily at those with an interest in microbiology but not necessarily an extensive background in the subject. Microbes, long relegated to classrooms and laboratories at universities, research institutions, and hospitals, are increasingly gaining prominence in the public eye. But although we hear in the news about such exotic and alarming creatures and diseases as AIDS, Ebola viruses, mad-cow disease, and "flesh-eating" bacteria, most of us know little about where they came from or how they spread. Most written material about microorganisms is either too specialized (such as college-level text books and official reports) or too simplistic (news articles, for example). One of the main purposes in writing *Microbes and People* was to attempt to bridge the gap between the two extremes and provide younger and less technical audiences with information they can process and still find practical without sacrificing either scientific content or accuracy. Thus, a high school or beginning college student embarking on a research project about a specific microbial disease should find this guide useful for looking up key facts about a disease or its causative agent, and the information obtained can serve as a springboard for expanding on the topic. A more casual reader will find it useful for helping to put in perspective something they read in the news—the immediate implications of a reported outbreak of a "new" virus, for instance.

The Microbial World—A Working Definition

Before plunging into the world of microorganisms, it would be wise to define how this term is used in *Microbes and People.* Simply characterizing them, as we did earlier, as living beings too small to discern with the human eye is too loose a definition, especially when we consider that the smallest creatures that we can see unaided are about a millimeter (mm) in diameter or thickness. Unfortunately, this limit still allows for size differences of exponential orders of magnitude—from creatures whose diameter can be measured in nanometers (1 nm is only a millionth of 1 mm) to those that are approaching a millimeter. Furthermore, one must realize that if size alone can be such a variable, then there is a tremendous potential for other, more complex differences among the microbes.

Indeed, there is a very diverse collection of creatures—bacteria, fungi, protozoa, and nematodes, to name but a few—that fit the above description. Over the past century or so, the discipline of microbiology has come to focus largely upon bacteria and viruses, and, accordingly, these entities are the subject of the bulk of this book. But there are some other organisms, notably protozoan parasites and certain fungi, that should not be ignored. In the interests of providing as broad a coverage of the microbial world as possible under the size constraints of this publication, we have set our boundaries to include the prokaryotic and single-celled eukaryotic microorganisms. (Definitions of these terms may be found in the body of this volume).

Public Health: A Practical Context

If microorganisms are the main subject of *Microbes and People,* then public health is the principle context for their discussion. One of the main reasons for this choice is perhaps the familiarity aspect; the impact of microbes upon human disease is arguably the best recognized thing about them. For well over a century, people have known about "germs" and their ill effects upon health. One has only to look as far as the nearest television and wait for a commercial break to encounter an advertisement for "antibacterial" hand soap or disinfectant cleaners to realize the extent to which the fear of microbial diseases pervades our daily lives. But diseases are by no means the only or even the most important effects of microbes. Indeed, their good deeds, so to speak, may be considered to far outweigh their harmful effects—the world without its microbial population would not be one inhabitable by the human race. Such mundane considerations as our supply of air for breathing or our ability to digest certain types of food are also functions of microbial activity. Therefore, this book takes the broadest possible view of public health—the public's health, if one will—rather than confines its discussions to infectious diseases alone. Every attempt has been made to sample the various spheres of human life that are influenced by microbes, and there are entries covering various aspects of environmental, industrial, and food microbiology in addition to those pertaining to the microbiology of health and disease.

Criteria for Inclusion

In defining the arenas of microbiology and public health, I have, for the most part, explained the criteria for the inclusion and exclusion of the book's content. Interspersed among these essays are brief entries on the people who made important contributions to our modern understanding of microbes and their roles in public health and disease. In addition, I have included a number of entries that I consider as "background" or conceptual definitions and essays. These contain the sort of information that is necessary for the reader to gain more than the most perfunctory facts about an organism and its impact on humans. They deal with basic features of living things in general, and microbes in particular, that enable one to understand how a microbe lives and why it may cause a disease, and are designed to give the reader a vocabulary to help them navigate through the book. Concepts such as basic biochemistry—DNA and RNA for instance—and details of microbial structure, biochemistry, and metabolism (e.g., fermentation and photosynthesis) are examples of such entries.

Arrangement and Structure of Entries

Given the diversity of the subject matter, it was not possible to adhere to a single formula in composing the entries, nor is that strategy necessarily the most effective method of packaging the information in the most accessible way. The terms are arranged alphabetically to enable readers to find them with ease. Wherever possible, I have tried to orient the reader to the organism or disease by calling upon the most familiar or famous facts about it, and then going on to describe details. Each entry is cross-referenced to other pertinent entries, without reiterating the obvious connections. Thus, for example, every bacterial genus will not be cross-referenced to the general entry on "bacteria," although readers may find it helpful to refer to general entries such as those on bacteria and viruses either before or during their perusal of a specific organism. The reader is also likely to run into some of the basic concept terms in the descriptions of individual microbes or diseases, e.g., words such as "heterotrophic" or "aerobic" in the description of a bacterium or parasite. Every attempt has been made to mention these terms in context, so that readers will not have to spend time jumping around the pages every time they see an unfamiliar term, although some degree of page-flipping is unavoidable, especially when one first begins to use this book. The length of individual entries varies, and depends on the amount of available information about a subject and its relative importance.

Additional Features

The 750 or so entries in this book are supplemented with photographs as well as tables and maps to enhance the accessibility of the material. The placement of these materials has been chosen so that they convey the maximum information when read in conjunction with the appropriate entries, although they would not be devoid of meaning when viewed on their own. The classification of viruses for example, is placed alongside the general entry on viruses, so that readers might get a feel for the layout of the viral world in a single glance, and read about the details in the text. Of course, readers might also find the table a useful reference point when reading about a specific virus. The appendixes are virtually stand-alone sections that provide some interesting facts and figures about infectious diseases and epidemics at a glance.

The scientific names of all organisms, save the viruses—which as one will find out do not really qualify as "true" living beings—appear in italics, with the capitalized generic name followed by the species name in lower case, as is the convention. Within each entry, the full name of an organism is spelled out only once, after which it is abbreviated by reducing the genus name to its first letter. Thus *Escherichia* (genus) *coli* (species) may be represented as *E. coli*. The years following the names of personalities refer to the years of their births and where applicable, their deaths. Except in the names of the organisms, all spellings follow the American convention, e.g., color rather than "colour."

ACKNOWLEDGMENTS

It is with great pleasure that I thank my friends and family, teachers and editors who helped me in a myriad ways to bring this book about. First, my love and gratitude to my father, for his support and help through all of this, and for taking me to such a wonderful place to do the bulk of my writing. Iram, for putting me up and putting up with me during the earliest iterations of this volume. Liz Welsh, my editor at Oryx, for bearing with me and my many phone calls. A big thanks to the Mathematics Department at the University of Hawaii for allowing me to work there, and to Dr. Karl Dovermann and Dr. Dale Myers for their help in sorting out computer problems. Also, Adam, the Ramanathans, and my cousin Krishna, whose computers were used with gratitude when the need arose. My teachers from undergrad onwards—especially Dr. Campbell, and Dr. Roy who showed me the world of microorganisms, Dr. John Wilkes who taught me to show it to others through writing, and to Geoff Montgomery for opening new doors when several were closing. Thanks also to Professor Larry Holmes for encouraging me to take time off from school to devote my entire attention to completing the project. Sudha and Rick for such prompt help when asked. Other overworked writer friends, Sean and Jonny, who empathize as no one else can, about the frustrations of writer's block and the (guilty) pleasures of procrastination. For moral support and good company through the days when I had my nose to the grindstone—Monika, Sean, Kaine, Jonny, Ron and Margaret Brown, and Fran and Winston Ota in Hawaii; and my housemates, Adam, Art, Chris, and Manish in New Haven. Art, Chris, Shipra, and Tara for helping me with proofreading. Leila Caleb and Larry O'Hanlon, whose faith in my ability "to see things through" kept me doggedly writing and rewriting even when I felt like giving up in frustration. To my family back in India—my mother, grandparents, and Vidya for their love and long-distance support and my brother Mahesh who thinks it's cool to have an author in the family even if the only thing the book has to do with computers is that it was written on one.

A

Acetobacter

Bacteria often found in association with fermenting or spoiling fruit, characterized by—and named for—the ability to convert alcohol (usually ethanol) to acetic acid. These organisms have not been found associated with human diseases, but they have applications in the food industry and in the environment. For instance, the most common species, *Acetobacter aceti*, is used commercially in the production of edible vinegar from wine or cider. Another species, *A. xylinum,* is used in the fermentation of tea leaves, which is a necessary step in converting raw, fresh leaves into an edible product. A recently discovered species called *A. diazotrophicus* is capable of nitrogen fixation and thus plays a role in the earth's cycling of this element. *Acetobacter* species are often implicated in the spoilage of wine, beer, and other alcoholic beverages. Viewed under the microscope, the bacteria appear as small, gram-negative rods. They are obligate aerobes and produce acetic acid by the oxidation of ethanol. Because these organisms do not form spores, they may be easily destroyed by such methods as pasteurization. *See also* FERMENTATION; NITROGEN CYCLE; NITROGEN FIXATION; VINEGAR.

acid-fast stain

A type of stain used specifically for the visualization and identification of bacteria that have the property of acid fastness. Acid fast-ness is the ability of bacteria to remain impermeable to mixtures of acid and alcohol, even in the presence of heat, and it is a function of the presence of certain acid- and alcohol-resistant compounds in the cell walls. Examples of acid-fast bacteria include the mycobacteria, causative agents of tuberculosis and leprosy, whose cell walls contain special lipids called mycolic acids. The staining procedure is a two-step process in which the bacterial cells are first stained with a dye called carbolfuchsin (which colors the cells a deep red) and then washed with a mixture of warm acid and ethanol. Most bacteria—except acid-fast organisms—are decolorized in this step. These non-acid-fast organisms may be visualized by staining with a contrasting dye such as malachite green. Like the Gram stain, the acid-fast stain, which is also called the Ziehl-Neelsen stain, is a differential stain that may be used to distinguish among different organisms in a single sample. *See also* DIFFERENTIAL STAINING; *Mycobacterium*; STAINING.

Acinetobacter

While not as familiar to most of us as the disease-causing germs, members of the genus *Acinetobacter* are nevertheless common in our daily environment, as residents of soil and water and as part of the normal inhabitants (i.e., indigenous microflora) of many animals. In humans, these bacteria may be found as part of the normal microbial population of the urino-genital areas, and they do not cause any significant damage to the host under most circumstances. However, they are capable of causing infections under conditions of weakened host resistance or when inadvertently transferred into the bloodstream via cuts or during surgical procedures. *Acinetobacter* is frequently used in sewage treatment plants, along with other aerobic bacteria, to remove organic material from wastewater. They are gram negative rods with simple nutritional requirements and can use a variety of simple compounds such as ethanol, acetic acid, or lactic acid as their primary carbon source. *A. calcoaceticus* is the best known species. *See also* INDIGENOUS MICROFLORA; OPPORTUNISTIC PATHOGEN; SEWAGE TREATMENT.

Acquired Immunodeficiency Syndrome (AIDS)

Fifty or a hundred years from now, when people look back on the medical history of our times, they will, in all likelihood, label the final decades of the twentieth century as the era of AIDS. Certainly, no other disease has gained as much attention worldwide in as short a time. One has only to visit the nearest bookstore or library and peruse a few books to see that virtually all contemporary subjects are discussed in the context of AIDS.

AIDS is actually a constellation of many possible diseases or symptoms whose common, underlying cause is the breakdown of the cell-mediated arm—that is, the T cell component—of the immune system. Most medical texts state that the immune system breakdown is caused by the action of a retrovirus called the human immunodeficiency virus (HIV), although there is still a significant voice of dissent on this matter. Aspects on the debate about HIV/AIDS are discussed in the entry for human immunodeficiency virus (HIV). This entry deals with the less contentious aspects of the disease such as clinical features, treatment, and epidemiology.

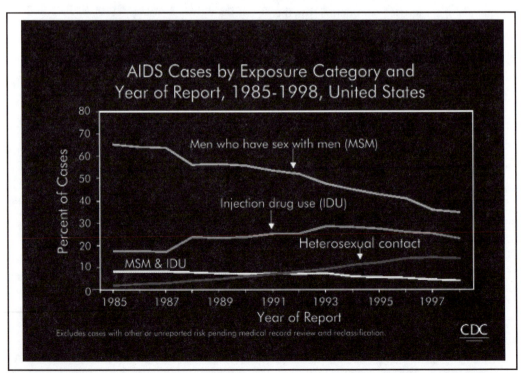

Figure 1. AIDS Cases in the United States, by Exposure Category and Year of Report, 1985–1998. *Centers for Disease Control.*

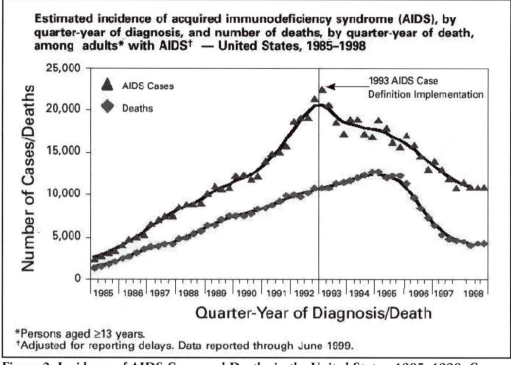

Figure 2. Incidence of AIDS Cases and Deaths in the United States, 1985–1998. *Centers for Disease Control.*

Historically, the syndrome we now recognize as AIDS first gained publicity during the early 1980s as a group of seemingly unrelated infections and tumors that occurred predominantly, as it seemed at the outset, in homosexual men. The disease always had a fatal outcome, marked toward the final stages by painful symptoms of brain damage such as dementia. The infections seen in the affected populations were unusual in that the causative agents were opportunistic organisms that did not affect people with normal immune systems, e.g., pneumonia with *Pneumocystis carinii*, chronic infection with *Cryptosporidium*, *Toxoplasma* infections of the central nervous system, and candidiasis of the lower respiratory tract or esophagus. Similarly, the cancers associated with these patients were relatively rare except in immuno-compromised individuals—examples included such conditions as Kaposi's sarcoma, B cell lymphoma of the brain, and a non-Hodgkin's type of lymphoma. The

common feature underlying all these diseases was an apparent breakdown in the immune system of the patients. The constellation of diseases was labeled as "acquired" immunodeficiency syndrome because the deficiency appeared to be transmissible from one individual to another.

Contrary to the early indications, this disease (which was briefly known as gay-related immunodeficiency or GRID) was not restricted to homosexuals but could be transferred from any individual to another via bodily fluids such as blood or semen. Hemophiliacs and other people receiving blood transfusions were also found to contract these diseases at a higher rate than the general public. Other high-risk groups included intravenous drug users, prostitutes, and people with multiple sexual partners. A search for the infectious agent that could cause the disease when introduced in nondiseased hosts led to the discovery and identification—independently by Luc Montagnier and Robert

Gallo—of a virus (later named the human immunodeficiency virus or HIV) that appeared to cause T cell destruction.

By 1987, the U.S. Centers for Disease Control (CDC) had revised its definition of AIDS to include diseases in which certain indicator diseases such as extrapulmonary tuberculosis, sensory neuropathy, or dementia were associated with the presence of HIV as detected by antibody assays. This definition was revised again in 1993 to describe the syndrome that is now recognized as AIDS within the United States. According to the currently accepted definition, a person is diagnosed with AIDS—regardless of the presence or absence of symptoms—if he or she

- tests positive for anti-HIV antibodies, and
- has a T-helper cell count of less than 200 cells per cubic millimeter, or a lymphocyte population of less than 14 percent of T cells, including helper-T cells.

In addition, the spectrum of indicator diseases was expanded to include pulmonary tuberculosis, recurrent pneumonia, and an invasive cervical cancer.

The time frame for the appearance of disease symptoms after initial infection with the virus is variable, and infected people may remain asymptomatic for several years. However, scientists have shown that this period of latency is not one of viral inactivity, but rather a period of rapid multiplication and proliferation and a gradual, almost insidious decline in the population of the immune cells. AIDS knows no geographical boundaries. At first, incidence seemed highest in the United States, where some 500,000 cases had been diagnosed by early 1995, but since then numbers have steadily risen in all parts of the world. According to the World Health Organization, in 1999 there were more than 33.5 million people infected and living with HIV worldwide. A significant proportion of this population—23.5 milllion—was in sub-Saharan Africa. In the two decades since it first appeared, AIDS had claimed nearly 16.5 million lives.

Risk factors for contracting AIDS include practices such as unprotected (condomless) sexual relations, especially with multiple partners, and the sharing of hypodermic needles. The transfer of body fluids of infected individuals is necessary for transmitting HIV infection; simple contact is *not* believed to be a factor. During the early years of its discovery, a number of cases were seen in hemophiliacs and other recipients of transfusions, due to blood donations from infected donors. However, with careful screening procedures now in place at blood banks and other facilities, the danger of acquiring HIV infections from this source has fallen considerably. Yet another method of HIV spread is the vertical transmission from an infected pregnant mother to her child before it is even born.

Death is the outcome in nearly all people with symptomatic AIDS. Treatment strategies against the disease must be twofold, simultaneously aimed at getting rid of HIV (and hence, controlling lymphocyte destruction), as well as alleviating or treating the secondary disease, e.g., the opportunistic infection or tumor, which is almost always the direct cause of death. One of the first drugs to become broadly available was Zidovudine (popularly known as AZT) but the virus has since mutated and developed resistance against this drug, thereby limiting its usefulness. In fact, the high rate of mutability of the retrovirus has placed serious obstacles in the way of developing effective, long-term therapies against AIDS. The development of "cocktail" therapy by David Ho, in which the patient is treated with multiple drugs at once, is aimed at more or less paralyzing the viruses with too much information, so that they are unable to adapt to any of the threats. This approach has proven to be one of the most effective treatments for HIV infections until now, although scientists advise caution against the overuse of these cocktails because the long-term consequences are as yet unknown. *See also* ANTIVIRAL AGENT; *CRYPTOSPORIDIUM*; GALLO, ROBERT; HO, DAVID; HUMAN IMMUNODEFICIENCY VIRUS; IMMUNE SYSTEM; KAPOSI'S SARCOMA; MONTAGNIER, LUC; OPPORTUNISTIC PATHOGEN;

Pneumocystis carinii; retrovirus; T cell; *Toxoplasma gondii*.

Actinobacillus

A relatively uncommon bacterium, *Actinobacillus* nevertheless deserves mention because of its involvement in some human infections, including skin abscesses and occasionally endocarditis. While seldom known to cause disease by itself, these bacteria have often been isolated from skin lesions containing the granular yellow discharge typical of *Actinomyces* infections. *Actinobacillus* species show marked morphological and biochemical differences from the actinomycetes, however, in that they are gram-negative in nature and do not exhibit any filamentous forms typical of the latter. In practical terms, these differences result in widely variant antibiotic sensitivities between the two groups of bacteria, which has implications for the treatment and monitoring of infections where both pathogens may be involved. The most commonly isolated human pathogen in this genus is *A. actinomycetemcomitans*, which is also a rare cause of endocarditis (inflammation of the heart muscular tissue). Perhaps the most widespread member of this genus is *A. lignieresii*, found mostly in veterinary diseases and occasionally implicated in granulomas and abscesses in humans. As a group, *Actinobacillus* species are rather difficult to cultivate in the laboratory. They grow slowly even in complex media containing blood or serum, and, consequently, they may be overlooked in diagnoses. Infections may be treated with antibiotics that are effective against gram-negative organisms or with broader-spectrum drugs. *See also* ACTINOMYCES; GRANULOMA.

Actinomyces

This large and diverse group of bacteria is a curiosity in the world of microbes, interesting to humans not only because of their association with different disease conditions but also due to their many unusual morphological and biochemical features. One has only to glance at *Bergey's Manual* to realize the importance of these organisms to scientists—of the four volumes that make up the latest edition, one entire volume is devoted to members of this and some closely related bacterial genera (collectively called the actinomycetes). In terms of public health, the actinomycetes are known for their involvement in the formation of dental plaque, which is the first step in the development of dental caries (tooth decay) and gum disease. In addition, when introduced to other locations in the body (via the bloodstream, for instance), *Actinomyces* species have been known to cause chronic skin lesions by triggering a delayed-type hypersensitivity response in the host organisms.

The actinomycetes are often mistakenly identified as fungi because they consist of branching, filamentous cells that resemble fungal hyphae. Nevertheless, they are considered to be true bacteria because of their prokaryotic cellular structure and the presence of peptidoglycan in their cell walls. They are gram-positive, non-motile, and non-spore-forming and are typically found in nature in association with animals. Individual organisms vary greatly in shape and size—from short single rods to filamentous forms or those with swollen, club-like ends. Smaller cells multiply by simple cell division, while longer filaments are capable of fragmenting into smaller entities, which are then capable of independent survival and multiplication.

The hypersensitivity reactions induced by various actinomycetes often result in the development of pus-filled lesions characterized

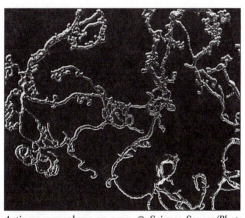

Actinomyces colony on agar. © *Science Source/Photo Researchers.*

by the presence of yellowish particles called sulfur granules. The presence of sulfur granules is an important diagnostic clue for identifying actinomycete infections, although the name is misleading because there is no evidence for the presence of sulfur in these leisons. Upon close scrutiny under the microscope, the granules are seen to contain filaments of the bacteria. A common feature of many of these infections is the presence of other organisms, such *Actinobacillus* or *Nocardia* in the lesions. These organisms are considered co-pathogens because the symptoms they induce are nonspecific, and no single bacterium can be conclusively identified as the primary pathogen. In addition to human infection, *Actinomyces* also causes serious infections of the jawbone of cattle, known as lumpy jaw. In other animals, such as dogs and pigs, *Actinomyces* tends to target soft tissue rather than bone.

Because these organisms induce a range of nonspecific effects in the host, a differential diagnosis—confirmed by the isolation and definitive identification of organisms from the infected patient—is important. Various species are susceptible to a spectrum of penicillin-type antibiotics, and infections may be checked by the use of the same once diagnosis is confirmed and the specific culprits identified. At least five distinct species of *Actinomyces*, differentiable on the basis of morphology as well as biochemical makeup, have been identified thus far: *A. bovis, A. odontolyticus, A. israelii, A. naeslundii,* and *A. viscosus*. When grown on solid media, these species present different types of colony morphology both microscopically and to the naked eye. They also have widely variant oxygen requirements. All species can ferment glucose to produce different end products such as acetic and lactic acid. In addition to these species, there are some additional organisms whose taxonomic position is not clear, e.g., *A. eriksonii*, whose cell wall composition and fermentation abilities are more similar to the genus *Bifidobacterium*; and, *A. humiferus*, which resembles other *Actinomyces* in morphology and biochemistry but is found free-living in the soil. This last species

is also unusual in that it, unlike most other gram-positive microbes, is sensitive to the enzyme lysozyme *See also* BACTERIA; *BERGEY'S MANUAL OF DETERMINATIVE BACTERIOLOGY*; DENTAL PLAQUE; FUNGI; GRAM STAIN; LUMPY JAW; LYSOZYME; MYCETOMA; PROKARYOTE.

activated sludge

Semi-solid mixture of mud-like consistency containing live, active microbes, used for the treatment of liquid wastes in sewage treatment plants. Organisms that are present in activated sludge typically include such aerobic, heterotrophic bacteria as *Acinetobacter, Alcaligenes*, and *Zoogloea*, as well as a number of amoebae and unicellular ciliate and flagellate protozoa. The sludge is used as a secondary means of treating sewage water already separated from solid wastes through either sedimentation (settling) or filtration. The liquid wastes thus obtained are mixed with the activated sludge and briskly aerated and agitated for several hours so that the bacteria and protozoa assimilate various organic molecules present in the sewage. After an appropriate time, the liquid is separated from the sludge—now greatly increased in bulk, due to the growth and multiplication of the various microbes—and treated with disinfectants to ensure the removal of pathogenic organisms. Chemically, these treated waters have greatly reduced organic content and may be used for irrigation and for various industrial processes, although not for drinking. At least some of the sedimented solid sludge is used as the starter material for subsequent batches of sewage treatment, while the rest is treated in the same manner as other solid wastes. *See also* SEWAGE TREATMENT.

adenosine triphosphate. *See* ATP.

adenovirus

These viruses were first isolated in 1953 from adenoid and tonsil tissues of people with respiratory infections in certain military populations, but they have since been found associated with several other conditions and organ systems. Although adenovirus infections are frequently asymptomatic, they can cause a variety of disorders such as pneumonia; conjunctivitis; acute febrile conjunctivitis, a whooping cough–like disease that is virtually indistinguishable from the bacterial disease (except for the cause); an epidemic keratoconjunctivitis; gastroenteritis; and acute hemorrhagic cystitis. The specific nature of the disease caused by an infection depends on several factors, including virus type, how the virus enters the body, the immune status of the infected individual, and the geographical environment. Infections are diagnosed by the isolation of viruses from the patient—these may be obtained from throat swabs, nasal washings, conjunctival swabs or scrapings, feces (for gastrointestinal disorders), etc. The virus seems to spread quickly and easily among people in crowded living situations. A vaccine has been developed and is advised for high-risk groups such as military personnel who live and move in large groups. However, because of the overall low risk of exposure to the adenoviruses, a routine vaccination is not recommended.

Physically, the adenoviruses are regular, geometrical particles, containing a double-stranded DNA genome. The characteristic "space satellite" type of shape that is revealed through an electron microscope is endowed by the protein coat or capsid of the virus. Emanating from the vertices of this shell are fibers called pentons, believed to play a role in the pathogenicity of the virus. Replication is a long and slow process, lasting from 32 to 36 hours, which takes place in distinct cytoplasmic and nuclear phases. The adenoviruses exhibit a fair degree of host specificity and are classified into two main categories—avian parasites and mammalian parasites—according to this trait. This host specificity appears to be maintained within the groups, at least to the extent that that viruses isolated from humans can only infect other humans. Individual adenoviruses of humans may be differentiated on the basis of such properties as their surface antigens (as detected by hemagglutination tests), ability to induce tumors in animals, and the poten-

tial to induce cellular transformation in tissue culture. Identification becomes an issue to epidemiologists when they are trying to trace the source or route of infectious outbreak or institute measures to contain infections. *See also* HEMAGGLUTININ; TRANS—FORMATION; WHOOPING COUGH.

aerobe

An organism that requires oxygen from the air for respiration. Some bacteria are strict or obligate aerobes—organisms that will die in the absence of oxygen and even experience trouble surviving conditions of low oxygen tension. Other bacteria are facultative and can survive the absence of oxygen for periods of time. A special subcategory of the aerobes are the microaerophilic organisms, which require oxygen but typically at much lower concentrations than is found in the atmosphere. *See also* ANAEROBE; FACULTATIVE AEROBE; FACULTATIVE ANAEROBE; RESPIRATION.

Aerococcus viridans

Part of the normal microflora of the intestinal lining in humans, *A. viridans* is rarely ever a primary cause of human infections, although it has been implicated as an opportunistic pathogen in various inflammatory conditions including septic arthritis, endocarditis, meningitis, and urinary tract diseases. A small, gram-positive coccus, it was long thought to be a member of the genus *Streptococcus* and only recently has been reclassified into a separate genus on the basis of fundamental differences in DNA content and sequence. *See also* OPPORTUNISTIC PATHOGEN.

Aeromonas

Often found residing in freshwater (nonmarine) environments, various species of these bacteria have been found associated with a number of diseases in both warm- and cold-blooded animals. One example of the former is *Aeromonas salmonicida*, a parasitic pathogen of fish. *A. hydrophila* and *A. caviae* are the names of species that cause human infections, usually due to the consumption of contaminated water or food. These infections typically manifest themselves in symptoms of gastroenteritis and diarrhea, similar to the disorder induced by other intestinal pathogens. The severity of the disease may vary from a mild chronic condition extending over several days to an acute dysentery-like disease with a high volume of watery stools, nausea, and vomiting. *Aeromonas* species have also been implicated in wound infections following some sort of trauma or the exposure of cuts and scratches to contaminated water. The diagnosis of various infections is achieved by the isolation of *Aeromonas* from clinical samples. Organisms are resistant to penicillin but susceptible to other antibiotics, and infections may be treated accordingly.

Currently, these gram-negative rods are classified in the same family as the genus *Vibrio* (best known as the causative agent of cholera—see entry on *Vibrio cholerae*), although biochemical characteristics and nutritional requirements indicate that they are less related than originally thought. *Aeromonas* also shares several biochemical properties with members of the Enter–bacteriaceae family—e.g., *E. coli*, *Salmonella*, *Klebsiella*—but is different in that it produces the enzyme oxidase. Such differential information about these intestinal pathogens becomes important to epidemiologists when they are called upon to identify the cause of a sudden disease outbreak, to trace the source of infection, and to institute adequate measures to control the spread of the organisms. *See also* VIBRIO; WATER-BORNE INFECTION.

aerosol

A fine suspension of liquid droplets in air, such as the one created by a sneeze or by a spray bottle. Aerosols may function as vehicles for the transport of large numbers of microorganisms from a reservoir to a new location—e.g., an uninfected person may breathe in the organisms in the aerosols cre-

ated by the sneezing or coughing of an infected individual. *See also* DROPLET INFECTION.

Aflatoxin

Toxic compounds produced by fungi such as *Aspergillus flavus* and *A. parasiticus*, which are associated with episodes of animal poisoning. These toxins were first discovered around 1960, when over 100,000 turkeys in England died as a result of consuming moldy peanut meal imported from South America and Africa. The compounds are aromatic chemicals that are only produced by the fungi under the right temperature and moisture conditions. For instance, the optimal temperature for aflatoxin production appears to be about 24–28°C, and fungi that grow at temperatures less than 15°C or at a relative humidity over 75 percent will not produce these poisons. But even under controlled laboratory studies, the production of aflatoxins has proven unpredictable, and it is possible that other factors such as special nutrients also determine their production. They have been found in numerous foods, typically from agricultural sources—for example, in nuts, grains, meats, cheese, and flour—held at ambient temperatures. They appear to act by binding to DNA, and inducing mutations, which in turn are carcinogenic. The liver appears to be especially susceptible to aflatoxins. Circumstantial evidence connects these toxins with human disease, but no firm links have been established. *See also* ASPERGILLUS

African sleeping sickness

Systemic disease caused by an infection with the protozoan *Trypanosoma brucei*, which is spread by the tsetse fly. This type of trypanosomiasis is seen in the central part of the African continent and is geographically limited by the natural habitat range of the flies. There are two forms of the disease caused by two distinct subspecies of the trypanosome, which predominate in different areas. The East African sleeping sickness, caused by *T. brucei rhodesiense*, is primarily a zoonotic infection because the tsetse fly appears to prefer animal blood to that of humans. This organism is also the cause of nagana, a rapidly fatal type of sleeping sickness in horses and other animals. The West African subspecies, *T. brucei gambiense,* infects only humans. The main difference in the two infections in humans is the rate of development of external symptoms, which are otherwise largely the same for both types. The *rhodesiense* disease takes a more acute form and is characterized by high numbers of parasites in the blood and low lymphatic involvement, while the *gambiense* trypanosomiasis tends to cause a chronic disease with fewer organisms in the blood and with prominent involvement of the regional lymph nodes.

Sleeping sickness occurs in stages. The first stage begins with the insect bite and is marked by the development of a hard nodule or chancre at the site within 4–10 days, during which time the organisms complete their transformation and begin to invade the bloodstream. From there they travel to the local lymph nodes, where they proliferate and reside, causing intermittent episodes of fever. Each episode may last anywhere from a day to a week, during which time the patient experiences symptoms of high fever, malaise, headaches, and night sweats. The spleen and liver may also become enlarged. The *gambiense* or Gambian form of the disease may remain at this blood-lymphatic phase for several years before developing into the full-fledged disease. The "sleeping" sickness that gave this disease its name occurs when the parasites migrate to the central nervous system (CNS) and begin to proliferate there. This phase is marked by visible changes in a person's behavior and personality. There is steady progressive meningitis, apathy, confusion, fatigue, loss of coordination, and other symptoms of CNS damage. In the classic, final period of this disease, the patient develops an uncontrollable desire to sleep, severe loss of motor control and malnutrition, and finally falls into a coma, which if left untreated leads to death.

Trypanosomiasis is diagnosed by the detection of the parasites in the circulating blood, preferably during the blood-lymphatic stage, before CNS infection sets in. The two subspecies are indistinguishable in morphological terms and are differentiated on the basis of occurrence and symptoms. Because the organisms are highly infectious, blood should be handled in specialized labs and only by trained professionals. Serological tests are also used for diagnosis. Treatment options vary according to the stage at which the disease is diagnosed. Early stages are treated with drugs called suramin or pentamidine to clear the parasites from the blood. Treatment is considerably more complicated after CNS invasion has occurred. While drugs such as melasoprol or tryparsamide are useful in stalling the progression of infection, they cannot restore lost or damaged nerve tissue. *See also* TRYPANSOMA BRUCEI.

agar

Material derived from seaweed, widely used in microbiology laboratories for the preparation of solid media used to grow a variety of bacteria and fungi and to isolate individual colonies of these organisms. Chemically, agar is a polysaccharide that is not degraded by most bacteria. In addition, it offers the advantage of maintaining solidity at 30-37°C, which is optimal for the growth of most microbes, especially human pathogens. Only organisms that grow at high temperatures (above 60°C)—at which point agar is liquid—may not be isolated on agar-containing media. *See also* COLONY; MEDIUM.

Agrobacterium

This bacterium is probably best known as a pathogen of plants, causing a variety of tumor-like growths in different kinds of plants. The best-characterized organism is *Agrobacterium tumifaciens*, the causative agent of a disease known as "crown gall" in a variety of herbaceous and woody-stemmed plants, including apples, pears, and stone fruits like peaches. The bacterium, which is motile, enters the plant via wounds in the roots or crowns, and delivers tumor-inducing genes to the plant. These genes, which are present in bacterial plasmids rather than the main chromosome, are delivered in the form of a DNA-protein complex called the T-complex. When introduced into a plant cell, the T-complex induces an uncontrolled growth of the cells in the infected region, which results in the formation of a gall. Beginning as a small, smooth outgrowth at the site of infection, the gall progresses over time to form larger, woody, misshapen tumors with gnarled, irregular surfaces. The formation of these galls will often attract other pathogens such as fungi and bacteria to the site. Meanwhile, the bacteria continue to grow and multiply within the gall and spread through the plant to the roots, where they may be released into the soil again. Good agricultural practices are required to keep these pathogens at bay. The starting stocks should be free of infection at the outset, and soils should be well drained.

In addition to crown galls, the related species *A. rubi* and *A. rhizogenes* cause plant diseases called cane gall and hairy root, respectively. Cane galls are large swellings that occur in long masses along the trunks or stems of trees and split the bark. In the case of hairy root disease, small wiry roots may be found emerging singly or in bunches from the main root. The basic mechanisms of pathogenesis and bacterial control are the same in all cases. Physically these bacteria are gram-negative rods with well-demarcated capsules. *See also* CROWN GALL DISEASE.

AIDS. *See* ACQUIRED IMMUNODEFICIENCY SYNDROME

akinete

Metabolically inactive form, or the resting stage, assumed by different species of cyanobacteria in response to unfavorable environmental conditions. This structure is equivalent to the spores formed by other bacteria. *See also* CYANOBACTERIA; SPORE.

Alcaligenes

Generally found inhabiting various soils and water sources, this bacterium is interesting to environmental microbiologists because its different metabolic activities may be harnessed in various industries. Because it has a rapid rate of metabolism, this organism is widely used in wastewater and sewage treatment plants to efficiently remove suspended organic materials from the water. The most widespread species, and the best-characterized, is *A. faecalis*, which is an opportunistic pathogen of humans—when consumed in large quantities via infected water, for instance—that can be distinguished from other species because it emits a strong fruity odor. Another interesting species is *Alcaligenes eutrophus*, which is capable of synthesizing a plastic-like material called polyhydroxybutyrate (PHB). PHB is becoming increasingly favored for use as a plastic because it is biodegradable and thus causes less harm to the environment than many other plastics in use today. PHB is a by-product of the metabolism of *A. eutrophus* and is produced as a means of storing excess chemical energy that these bacteria generate during the conversion of hydrogen to water. It is stored in the form of granules in the cytoplasm and is used as a reserve supply of energy when the organism is faced with sugar-deficient conditions. The morphology of *Alcaligenes* is fairly variable, and the bacteria exhibit a variety of shapes from nearly spherical forms to short rods when viewed under a microscope. Aside from being gram-negative, the taxonomic position of these bacteria is not completely clear. *See also* OPPORTUNISTIC PATHOGEN; SEWAGE TREATMENT.

alcoholic fermentation

Fermentation reaction typical of yeasts, such as *Saccharomyces cerevisae,* in which sugar is ultimately broken down to alcohol—primarily ethanol—and gas. This is the main metabolic activity at work in such commercial microbiological processes as beer-making and wine-making. The organisms contain an enzyme called alcohol dehydrogenase, which catalyzes the conversion of pyruvate (produced from the breakdown of sugars during glycolysis) to ethanol (alcohol) and carbon dioxide. Because this chemical reaction requires no oxygen, the process takes place under conditions of low aeration. *See also* BEER; FERMENTATION; GLYCOLYSIS; WINE.

algae

Although not implicated as human pathogens at a scale comparable to the bacteria, protozoa, and fungi, this group of living organisms are nevertheless important to humans in a multitude of ways. On the positive side, the algae—photosynthetic eukaryotes—play an essential role in harnessing the sun's energy into the biosphere, and consequently they serve as the first link in the food chain, particularly within marine and other aquatic ecosystems. On the negative side, algae also pose public health problems by fouling both drinking water sources and beaches.

Of all the groups of microbes discussed in this volume, the algae are perhaps the most anomalous because the microbial forms constitute only a fraction of the entire kingdom. Indeed the range of morphological forms of algae is vast—from microscopic unicellular organisms such as the diatoms to huge multicellular entities such as kelp, a single piece of which may be as large as several trees. In addition to the true multicellular organisms, some algal species are colonial in that they represent aggregates of single-celled organisms. The underlying features that tie these different life forms together into one cohesive group are the eukaryotic organization of the cells, the ability to perform photosynthesis, and the maintenance of a haploid nucleus for the majority of the life cycle. Indeed, it is this last property that distinguishes between algae and the so called "higher" plants that exist in a diploid state for most of their lifecycle. Individual algal cells have an external cell wall like plant cells. Unicellular algae are usually motile and possess flagella. Multicellular organisms are non-motile, although their reproductive cells—gametes—may also possess flagella similar to the types seen in the single-celled algae.

Algae are widespread on earth, showing a preference for various aquatic environments, although many species may also be found growing in soils and on wood and rocks where there is plenty of moisture. An overabundance of algal growth can significantly alter the properties of the water sources in which they grow, as well as the immediate environment, resulting in the fouling of the water source. These organisms are autotrophic and perform their photosynthetic functions with the help of a number of different pigments such as chlorophyll, carotenoids, and xanthophyll. The assimilated carbon is stored in the algal cells and used as required. Common examples of storage materials produced by algae include starch, mannitol, paranylon, and laminarin. Most organisms are free-living, although some species establish symbiotic relationships with fungi to form lichens. Nearly all the algae display some mode of sexual reproduction, during which process compatible organisms exchange genetic material. In multicellular algae, some cells differentiate to form simple sex organs that are distinguishable as male and female. In unicellular organisms, the vegetative cells function as gametes during the reproductive phase of the life cycle. *See also* CHLOROPHYLL; CYANOBACTERIA; BLOOM; EUKARYOTE; LICHEN; PHOTOSYNTHESIS.

allergen

Any substance—antigen—capable of eliciting an allergic reaction in an individual when brought into contact with the immune system, either directly in the bloodstream or by inhalation, ingestion, or contact via skin. Some common examples of allergens include pollen, animal hair, and food items such as nuts. *See also* ALLERGIC REACTION; ANTIBODY; ANTIGEN; HYPERSENSITIVITY; IMMUNE SYSTEM; INFLAMMATION.

allergic reaction

Often manifesting itself in various uncomfortable and often dramatic symptoms, including hives on the skin, wheezing and asthma, stuffy sinuses and runny noses, gastro-intestinal upsets, and diarrhea, an allergic reaction is actually a constellation of effects arising from a hypersensitive immune response, specifically mediated by antibodies that are produced in response to the offending substance or allergen. Because this type of reaction occurs from within a few minutes to as much as a day after exposure to the antigen, it is also known as an immediate or Type I hypersensitivity reaction. The accumulation of the antibody at the site of antigen exposure prompts the activity of a number of other immune cells, which not only attack the body's own cells at the site with the same vigor that they fend off the allergen, but also set into motion other processes that are ultimately responsible for the various systemic symptoms of an allergic reaction. Just as secondary immune responses are stronger than primary responses, second and subsequent exposures to an allergen will typically produce more violent episodes than the first exposure. *See also* ALLERGEN; ANTIBODY; ANTIGEN; HYPERSENSITIVITY; IMMUNE RESPONSE; IMMUNE SYSTEM; INFLAMMATION.

ameba

Group of intestinal protozoan parasites frequently associated with gastrointestinal infections in humans known as amebiasis. Different types of amoebae have been isolated from the intestinal tracts of a wide variety of host animals, where they either coexist with their hosts without affecting them in any noticeable manner or cause gastrointestinal diseases.

The amoebae are among the simplest known eukaryotic organisms, both in terms of life cycle and structure. The life cycle alternates between two phases: an active stage in the form of a trophozoite, and a relatively inactive cyst phase. The trophozoite, a single-celled structure with no regular shape, is the active, feeding stage of an ameba's life cycle. It cannot survive outside a living host, and thus represents the "parasitic" phase of this organism. In the pathogenic species, this is

the stage responsible for inducing disease symptoms. Trophozoites of amoebae are capable of independent motion with the help of protoplasmic extensions called pseudopodia or "false feet." They may either divide by a process of simple binary fission to produce two progeny cells, or change into the cyst stage of the parasite. The amoeboid cyst is a non-motile, single-celled structure consisting of a highly dense cytoplasm and nucleus encased in a thick, resistant cell wall. During this stage of the life cycle, the amoebae are metabolically inactive. While it is known that cyst formation occurs while the trophozoite is still in the intestine, the exact triggering factors are not completely understood. The cysts represent the infective stage of these organisms' life cycles, and unlike the trophozoites, they can survive outside the host, in soil and water, when passed with the feces. The ingestion of contaminated water or food, or the passive transfer from soil by flies or fomites, are various routes through which the cysts may gain entry into a new host. Once in the intestine, they undergo a process called excystation, whereupon the cell wall ruptures to release new trophozoites.

The structural features of the trophozoites and cysts of various amoebae are distinct and may be used as a means of differentiating between different genera and species. Trophozoites exhibit differences in their nuclear structure, the nature of inclusions in their cytoplasm, their size, and the way in which they move. Cysts of different organisms are distinguishable on the basis of nuclear structure, cyst shape and size, and the number of nuclei they contain. Some common genera are *Entamoeba, Endolimax, Blastocystis,* and *Acanthamoeba. See also* AMEBIASIS; CYST; CYTOPLASM; *ENTAMOEBA*; FOMITE; NUCLEUS; TROPHOZOITE.

amebiasis

Collective name for a spectrum of diseases caused by an intestinal infection with the protozoan parasite *Entamoeba histolytica.* Primary infection with this parasite is by ingestion of the cysts via contaminated food or water or via fomites. The release of trophozoites into the intestines may sometimes pass unnoticed, but more often triggers various symptoms of gastrointestinal upset. The most common manifestation of infection is amebic dysentery, typified by bloody and mucoid stools accompanied by fever and chills. In other cases, the disease may take the form of alternating bouts of mild diarrhea and constipation. In some instances, the constant irritation to a specific site on the intestinal wall due to the presence of the organism results in inflammation and the development of a lesion called an ameboma. Parasites may also gain entry into the bloodstream, through which they may travel to and induce lesions in remote organs such as the liver, and, less commonly, the lungs or brain.

Because the intestinal symptoms of *Entamoeba* infections are rather nonspecific, early diagnosis is important so as to be able to administer proper treatment. Diagnosis is achieved by the examination of stool smears under the microscope for the presence of trophozoites or cysts, as well as *E. histolytica*–specific immunodiagnostic tests. Depending on the type and severity of disease, the patient may be treated with the drug metronidazole. The treatment of asymptomatic carriers as a means of controlling outbreaks is still a matter of debate because the same species often has both pathogenic and nonpathogenic strains, and the undue exposure of the latter to drugs may induce drug resistance, which can be transferred easily to the pathogenic strains and species. *See also* AMEBA; ANTIBIOTIC RESISTANCE; CARRIER STATE; DRUG RESISTANCE; DYSENTRY; *ENTAMOEBA*.

American Type Culture Collection (ATCC)

Better known by its acronym, ATCC, this organization is a centralized repository for standardized cultures of various microorganisms—including bacteria, yeasts, algae, fungi, protozoa, and viruses—as well as various tissue cultures and recombinant DNA products. Formally established in 1925 by a committee of scientists from several academic and government institutions, ATCC is operated on a non-profit basis and is the standard supplier of culture materials to most of the laboratories in the world. The organization has its origins in a bacterial collection established in 1911 at the American Museum of Natural History in New York City by American microbiologist C. E. A. Winslow, who conceived of it as a "museum of living bacteria for the benefit of working laboratories all over the world." Over the years, the collection has expanded and today supplies some 60,000 authenticated cultures from its facility in Manassas, Virginia.

Ames test

A microbe-based laboratory procedure used to test the potential of unknown chemical substances to induce cancer. The basic assumption underlying this test is that most carcinogenic (cancer-promoting) agents are mutagens—i.e., they induce mutations in a gene by interacting with DNA molecules—and, in fact, the Ames test may be considered a test of mutagenicity rather than carcinogenicity. Specifically, the ability of the chemical to induce mutations is measured by its ability to revert an auxotrophic mutant to the wild type. The test is conducted by incubating mutant strains of *Salmonella typhimurium*, auxotrophic for the amino acid histidine in a mixture containing the test chemical, and a minimal medium without histidine. Only the samples containing test chemicals with a high mutagenic/carcinogenic potential will show abundant growth of the bacteria, indicating a high percentage of reversion to the wild type. *See also* AUXOTROPH; MEDIUM; MUTAGEN; MUTATION.

amino acid

A class of organic molecules that constitute the individual units or building blocks of protein molecules. Twenty different amino acids make up the proteins in living systems. Chemically, an amino acid consists of a carbon atom attached to an amino ($-NH_2$) group, a carboxyl ($-COOH$) group, and a side chain that confers specific properties on each amino acid. The amino and carboxyl groups are used to forge the links (called the peptide bonds) between adjacent amino acids in a protein. *See also* PROTEIN.

aminoglycosides

Group of broad-spectrum, bactericidal antibiotics that act by binding to the ribosomes of bacteria, thereby inhibiting protein synthesis and the growth and multiplication of these organisms. Examples include both natural and synthetic antibiotics such as kanamycin, gentamycin, streptomycin, and neomycin. *See also* ANTIBIOTIC; STREPTOMYCIN.

ammonia assimilation

Chemical pathway through which most microbes incorporate (assimilate) nitrogen from their surroundings into their cells. The reaction is one of fundamental importance because it is, in most cases, the first step toward providing nitrogen—one of the essential components of all living matter—to the rest of the living world. The nitrogen taken up by microbes in this manner travels up the food web in many different, complex ways.

Chemically, ammonia assimilation involves an enzyme-catalyzed reaction that incorporates nitrogen present in ammonia or ammonium compounds (furnished either by nitrogen fixation or by inorganic compounds in the environment) into an amino acid. The amino acids become the building blocks used to synthesize the various proteins the organism requires in its lifetime. Amino acids also serve as the nitrogen sources in the biosynthesis of the building blocks of nucleic acids—the purines and pyrimidines. Ammonia assimilation works through one of two basic reaction pathways catalyzed by specific enzymes. Most bacteria contain an enzyme called glutamic dehydrogenase, which catalyzes the incorporation of the amino group from ammonia into $\propto$-ketoglutaric acid (a Krebs Cycle intermediate) to form the amino acid glutamic acid. Glutamic acid can incorporate a second amino group in the presence of the enzyme glutamine synthetase to form glutamine. Glutamine is then used as a source

of the amino group in the synthesis of other amino acids. Enzymes called transaminases catalyze this last class of reactions. *See also* FOOD CHAIN/WEB; NITROGEN CYCLE.

ammonification

Final step in the breakdown of different types of nitrogenous organic compounds in living organisms, resulting in the release of ammonia (NH_3) or ammonium ions to the air, soil, or water. This reaction is seen in numerous organisms throughout the living world and occurs through a series of chemical reactions catalyzed by a diverse array of enzymes. Larger molecules such as proteins and nucleic acids are initially hydrolized (broken down in the presence of water) to smaller units by specific enzymes known as proteases and nucleases, respectively. Individual enzymes vary widely in their origins and in their chemical and molecular properties. The actual process of releasing ammonia takes place at the level of the smaller breakdown products of the macromolecules, usually the amino acids. A diverse battery of enzymes named dehydrogenases, oxidases, and carbon lyases participate in the process of ammonification. *See also* NITROGEN CYCLE; PUTREFACTION.

amoebiasis. *See* AMEBIASIS

anaerobe

An organism that does not use oxygen for respiration. The strict or obligate anaerobes cannot tolerate the presence of oxygen, while facultative anaerobes may survive under aerobic conditions without using oxygen. To humans, for whom oxygen is absolutely essential for survival, the notion of anaerobiosis (living without oxygen) might seem peculiar, yet this is the mode of life that evolved first on earth, eons before the aerobic mode of living developed. *See also* AEROBE; FACULTATIVE ANAEROBE; RESPIRATION.

anaerobic respiration

Biochemical process seen in certain groups of bacteria. Anaerobic respiration differs from normal respiration only in that energy is derived from various substrates with the help of external oxidizing agents *other* than oxygen. Common examples of these oxidizing sources include the following:

- nitrates—used by *E. coli* and *Clostridium* species, as well as denitrifying bacteria such as *Bacillus licheniformis* and *Pseudomonas stutzeri*

- sulfates—used by *Desulfovibrio* and *Desulfococcus* species

- sulfur—used by *Desulfuromonas*

- fumarate—used by strains of *E. coli*

Respiratory mechanisms such as these endow the microbes with the ability to live under very different chemical environments and hence occupy diverse ecological niches on earth. *See also* ARCHAEBACTERIA; NITRATE RESPIRATION; NITROGEN CYCLE; RESPIRATION.

anemia

Physiological condition of the body arising from an inability of the blood to carry normal amounts of oxygen, due either to a reduction in the number of circulating red blood cells or to decreased quantities or abnormal types of hemoglobin. It is a frequent symptom in many different types of infection, where the pathogens may destroy the blood cells by hemolysis or hemagglutination (for example, in viral infections) or invade and colonize the blood cells (for example, in parasitic infections, such as malaria) and thus use up the oxygen meant for the body.

animal virulence test

A diagnostic test to confirm the pathogenicity or virulence of an infectious agent. The test was developed as a means of diagnosing diphtheria and is primarily used for this purpose. Pure cultures of the organism isolated from clinical samples are introduced via sub-cutaneous injections into two guinea pigs, one of which has been inoculated with the diphtheria antitoxin. Virulent bacteria will produce a well-defined area of inflammation at the site of injection in 24–48 hours in the animal that did not previously receive the antitoxin, while untreated animals will not produce such a reaction and may succumb to the disease itself. *See also* ANTIBODY; ANTITOXIN; *CORYNEBACTERIUM*; DIPHTHERIA; IMMUNE REACTION.

anthrax

A zoonotic disease caused by infection with the bacterium *Bacillus anthracis*. The disease is most commonly seen among ruminants—cattle, sheep, and goats—but a host of other animals, including horses, rats, and birds, may also be infected. Anthrax is enzootic in parts of Asia, Africa, and South America, where it has a serious economic impact on the agricultural and dairy industries. Human incidence in these areas is also much higher compared with North America and Europe, no doubt due to agricultural exposure.

B. anthracis forms resistant spores that are capable of lying dormant in the soil or on plants for several decades. These spores may be ingested by animals along with food or may gain entry via open wounds or sores. On germination, the bacteria multiply and produce toxins that are responsible for the major symptoms of the disease. Anthrax in animals is characterized by a fulminant septicemia, which may or may not result in death. If an infected animal's carcass is not disposed of properly, it will serve as a fresh source of bacteria and spores to the environment.

Depending on the route of infection, anthrax can take three forms in humans. The most common type is cutaneous anthrax, which was commonly seen among farm workers who milked cows or handled animals regularly. This is usually a self-limiting form of the disease, beginning with the formation of a small red vesicle resembling an insect bite and progressing to necrosis and the development of a black, fluid-filled lesion called an eschar, which heals spontaneously most of the time. Pulmonary anthrax, which occurs as a result of inhaling a large quantity of spores, was first observed among English workers in woolen mills, as a result of which it was named "wool sorter's disease." Spores settle and germinate in pockets in the lungs, where they cause necrosis and lead to common respiratory symptoms. The alimentary canal, which is the most common portal of entry in various animals, is—surprisingly enough—the least-used route in humans. In part this is because meat from diseased animals is almost never consumed and infection of foods is likely only as a rare accident. Early symptoms of gastrointestinal anthrax include nonspecific signs such as abdominal pain and vomiting, followed by bloody stools and toxemia (the presence of anthrax toxin in the blood). Generalized systemic complications, including septicemia, shock, and even death (mortality is about 20 percent), will follow unless the infection is treated with appropriate antibiotics. *See also* BACILLUS ANTHRACIS; ZOONOSIS.

antibiotic

A substance which, even at low concentrations, can inhibit or kill certain microorganisms. The term was originally used specifically to describe antimicrobial substances produced by microbes—bacteria and fungi—but nowadays is used in a broader sense to include synthetic compounds or modified natural compounds with similar properties.

Some antibiotics are active against only a narrow range of microbial species, while others, called broad-spectrum antibiotics, can attack many organisms. Depending upon its concentration, mode of action, and spectrum of activity, an antibiotic may either be bactericidal (able to kill its target microbe) or bacteriostatic (able to inhibit its growth and multiplication). Typically, an antibiotic acts on a specific site in the cell such as the cell membrane, cell wall, or its synthetic machinery. Because the properties of these structures differ in prokaryotic and eukaryotic cells, most antibiotics do not exert their toxic activity as strongly in animals and humans as they do against bacteria. When two or more antibiotics act simultaneously on an organism, their activities can either negate or complement each other. Antibiotics with similar or the same targets tend to act synergistically. The negating phenomenon, known as antagonism, occurs when the two antibiotics have different targets. For instance, penicillin, which acts by inhibiting cell-wall synthesis in growing cells, would be antagonistic toward chloramphenicol, an antibiotic that inhibits cell growth.

The property of selective toxicity makes antibiotics ideal for treating various bacterial diseases. Although they proved useful in this regard during the earlier part of the twentieth century, the widespread and almost indiscriminate use of these compounds led to development of antibiotic resistance in various bacteria, which greatly decreased the efficacy of antibiotics in treating disease. *See also* AMINOGLYCOSIDE; ANTIBIOTIC RESISTANCE; ANTIBIOTIC SENSITIVITY TEST; CHLORAMPHENICOL; B-LACTAM ANTIBIOTICS; NALIDIXIC ACID; NOVOBIOCIN; PENICILLIN; POLYMYXIN, RIFAMYCIN; SULFONAMIDE ANTIBIOTICS; TETRACYCLINE.

antibiotic resistance

Property that enables microbes to withstand inactivation or destruction by an antibiotic. Resistance to an antibiotic may be either natural or acquired. For instance, an organism may be naturally resistant because it lacks the molecule, structure, or process that is specifically attacked by the antibiotic. A good example is the resistance of *Mycoplasma*, which has no cell walls, to penicillin and related antibiotics that specifically inhibit the formation of peptidoglycans, which are found exclusively in bacterial cell walls. Microbes may also be resistant to specific antibiotics due to physical barriers, such as an outer membrane (in gram-negative bacteria), or because they have specific enzymes that can inactivate the antibiotic, (for example, the b-lactamase enzymes of *Staphylococcus* makes these bacteria resistant to penicillin). Antibiotic resistance may be acquired through a process of mutation, in response to the continued presence of the offending agent over successive generations, or by the genetic

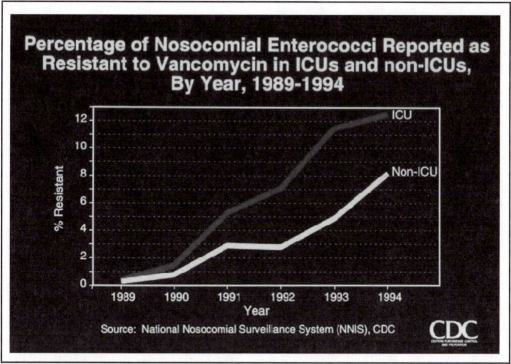

Figure 3. Percentage of Nosocomial Enterococci Reported as Resistant to Vancomycin in Intensive Care Units (ICUs) and Non-ICUs, 1989–1994. *National Nosocomial Surveillance System (NNIS), Centers for Disease Control.*

transfer of a resistance gene from one organism to another via a plasmid or through conjugation.

One of the serious problems facing society today is the emergence of drug-resistant species of many pathogens once considered conquered by medical science. At least part of the reason for the development of resistance was the overuse of antibiotics in the early decades of their discovery. Health professionals are advised to exercise caution in prescribing antibiotics to minimize the exposure of these drugs to microbes, which may not only acquire resistance themselves but may also transfer this property to other bacteria in nature.

Another area in which to exercise caution is in the use of hygiene products like soap and air fresheners that are labeled as "antibacterial." Two good questions to consider when encountered with such a label are, "Which bacteria are these soaps fighting, and why?" After all, bacterial organisms are all around us, in the air we breath, the ground

we walk on, and the water that gushes from our taps. Eliminating them all is simply impossible, and indeed not even desirable. Many are harmless and some are helpful—but most antimicrobial agents cannot tell the "good" bugs from the "bad." By unduly exposing these microorganisms to various antibacterial agents, we are encouraging them to develop resistance, which can be transferred to pathogens later. *See also* ANTIBIOTIC; DRUG RESISTANCE.

antibiotic sensitivity test

Laboratory procedure that tests the susceptibility of a microbial pathogen to different antibiotics. This type of test is useful to clinicians attempting to select optimal antibiotics to use for a particular therapy. The pattern of sensitivities and resistance of a given strain to a panel of antibiotics is called its antibiogram.

Two major protocols are used for determining the antibiotic sensitivities of an or-

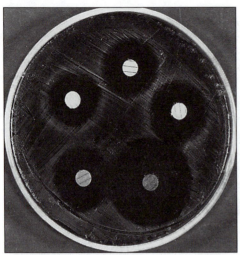

Sensitivity disc: penicillin showing zone of inhibition. ©
Science Source/Photo Researchers.

ganism. In the disc diffusion test, discs of absorbent paper containing different antibiotics are placed at intervals in a solid medium evenly spread with a suspension of the test organisms. The plate is incubated for a period sufficient to enable the organisms to grow and form colonies. During this time, the antibiotics will diffuse outward from the disc into the medium, along a concentration gradient. If the organism is susceptible to a particular antibiotic, its growth will be inhibited, and a clear zone will develop around the corresponding disc. A disc that contains an antibiotic to which the organism is resistant will not display any such clear zone. Organisms that have inducible enzymes conferring antibiotic resistance—e.g., the b-lactamase in *Staphylococcus aureus*—will form a narrow zone of inhibition close to the disc (where they were unable to synthesize adequate amounts of enzyme in time) but also will show growth at a much closer distance than genuinely sensitive strains as they begin to synthesize the enzyme.

A second type of antibiotic sensitivity test is a dilution test, in which organisms are inoculated into a set of media with serially diluted concentrations of antibiotic. The lowest concentration of antibiotic capable of inhibiting microbial growth is called the minimum inhibitory concentration (MIC) for that antibiotic/microbe pair and is a useful index for making dose recommendations in antibiotic therapies. *See also* ANTIBIOTIC.

antibody

Protein produced by the immune systems of higher animals (such as humans) in response to unfamiliar substances—antigens—encountered by immune cells in the bloodstream or in other parts of the body such as the respiratory tract or alimentary canal. Examples of antigen sources include bacteria, viruses, food products, and airborne particles such as dust and pollen. Antibodies combine with the antigens to neutralize their harmful effects or to change them in such a manner as to make it easier for the body to get rid of them. Chemically, the anibodies are globular proteins that are also known as immunoglobulins. The five main classes of immunoglobulins (abbreviated Ig) differ from one another with respect to their gross structure and the types of immune responses that they are involved in. The five classes of immunoglobulins are IgG, Ig M, IgA, IgD and Ig E. *See also* ANTIGEN; B CELL; IMMUNE RESPONSE; OPSONIN.

antigen

A substance that elicits a reaction by the immune system of the animal into whose system it is introduced. If we were to make an analogy between an immune response and a war, then antigens represent the enemy soldiers, entities that are identified as foreign and whose foreignness is a sign for the immune system to defend the body against an invasion. Antigens are present on the surface of various infectious organisms, in different foods items and inhalants, and on the surface of cancer cells. Chemically, proteins and protein-carbohydrate complexes (glycoproteins) are the most antigenic molecules. Different antigens can stimulate either a cellular response or the production of antibodies in the host organism. *See also* ALLERGEN; ANTIBODY; IMMUNE RESPONSE.

antigenic drift

Phenomenon exhibited by viruses such as the influenza virus, in which the antigenic structure of the surface glycoproteins—hemagglutinin and neuraminidase—change slightly over successive generations. This change allows the virus to disguise itself from the host just long enough to avoid detection and complete inactivation by antibodies that were produced in response to an earlier antigenic variant. Antigenic drift is a result of spontaneous mutations that occur in the viral genome during replication. This property, superimposed by the ability of the influenza virus to undergo periodic drastic antigenic shifts, is one of the key reasons for difficulties in curtailing influenza epidemics. *See also* ANTIGEN; ANTIGENIC SHIFT; INFLUENZA; INFLUENZA VIRUS.

antigenic shift

Phenomenon exhibited by viruses, most commonly the influenza virus, characterized by an abrupt change in the antigenic structure of its hemagglutinin protein. The end result of such a shift enables the virus to evade, or at least to delay, an attack by the immune system of the host, and is hence responsible for large epidemic outbreaks. The exact origins of this type of change are unknown. *See also* ANTIGEN; ANTIGENIC DRIFT; INFLUENZA; INFLUENZA VIRUS.

antisepsis

The destruction of microbes. Laboratory sterilization and disinfection procedures are examples of antiseptic techniques. *See also* ASEPTIC TECHNIQUE; ASEPSIS; STERILIZATION.

antiserum

The serum fraction of the blood from an animal that has been actively immunized against a specific antigen (either proteins or whole organisms may be used). The antiserum contains immunoglobulin (antibody) molecules against the antigen used in the immunization. *See also* ANTIBODY; ANTIGEN; IMMUNIZATION.

antitoxin

Antibody against a toxin, capable of binding to the poisonous substance and neutralizing its harmful effects. Antitoxins are used in therapies against toxin-mediated diseases such as diphtheria and botulism. *See also* ANTIBODY; ANTIGEN; TOXIN.

antiviral agent

Chemical agent capable of inactivating a virus in humans and therefore useful for treat-

Table 1. Antiviral Agents

Drug	Mechanism/Target	Infections Treated	Administration
Acyclovir	Nucleoside analog that prevents viral DNA replication by inhibiting action of viral DNA polymerase	Herpes simplex infections Chicken pox (at high doses of drug)	Oral, topical, or intravenous infusion
Ganciclovir	Derivative of acyclovir with similar DNA polymerase inhibition activity	Cytomegalovirus (CMV) infections in AIDS patients and transplant recipients[1]	Intravenous only
Interferon[2]	Appears to affect translation by interacting with mRNA.	Broad spectrum of viruses[3]	Intravenous only
Ribavirin	Nucleoside analog that inhibits both DNA and RNA synthesis	Severe respiratory syncytial virus (RSV) infections Lassa and Hanta viruses	Aerosols via a mask Oral or intravenous
Rimantidine	Prevents assembly of virus particles by blocking membrane ion channels	Influenza A virus	Oral or aerosol spray
Zidovudine/AZT	Nucleoside (TTP) analog that inhibits reverse transcriptase	HIV infections[4]	Oral or intravenous

Notes

[1]Ganciclovir is highly toxic and should be reserved for life-threatening infections in immuno-suppressed patients.

[2]Natural cellular products in response to viral infections, also produced in high quantities in laboratories for therapeutic purposes.

[3]Used primarily as a supplementary therapy in cancers.

[4]Usefulness reduced with the development of drug resistance on the part of viruses.

ing viral infections. Due to the unique status of viruses in the living world and their peculiar modes of replication and survival, the applicability of these agents is far more limited that those of antibiotics against cellular pathogens. *See also* VIRUS.

aphthovirus

This member of the picornavirus family is known to cause foot-and-mouth disease in cloven-hoofed animals such as swine and cattle. In addition to having a different host range, these viruses are physically distinct from the other members of the picornavirus family because they have a larger genome— 8.5 kb in comparison with the 7–7.5 kb of the others. Also, the antigenic sites of their surface glycoproteins are placed in a different manner from other picornaviruses. These differences have implications in the susceptibility of the viruses to various drugs and antibodies, and hence in determining treatment and disease-management strategies. *See also* FOOT AND MOUTH DISEASE; PICORNAVIRUS.

Arachnia

Named for the spider-like (or arachnid-like) appearance of its colonies on solid media, this bacterial species closely resembles *Actinomyces* both in physical appearance and in the clinical symptoms it induces, but it is different enough from the latter in terms of biochemistry to be classified in a separate genus. The differentiation of the genera has implications for diagnosis, treatment, and other public-health considerations. Like the actinomycetes, *Arachnia* is gram-positive and filamentous. It is a facultative anaerobe that grows best in low oxygen concentrations. A single species, *A. propionica,* has been identified to date. Among the main features that differentiate it from *Actinomyces* are the production of propionic acid (the acid responsible for the flavor of Swiss cheese) during fermentation and the chemical composition of its cell wall. Despite its fermentative abilities, *A. propionica* shows very low DNA homology with other propionic acid-producing

bacteria such as *Propionibacterium*. *See also* ACTINOMYCES; PROPIONIBACTERIUM.

arbovirus

Name given to a large group of RNA viruses that were originally classified together because they were found to cause human diseases that could be transmitted by insects or other arthropods, namely *arthropod-bo*rne viruses. The group contains some 400 species that are, at best, only loosely related to one another and which were more recently reclassified into distinct families of single-stranded RNA viruses: the Togavirus, Bunyavirus, Arenavirus, and Flavivirus families. Certain genera of double-stranded RNA viruses in the reovirus family were also considered among the arboviruses.

Archaea. *See* ARCHAEABACTERIA

Archaebacteria

Although they have seldom, if ever, been implicated in human disease, these organisms are important because they represent some of Earth's most ancient life forms. They are single-celled prokaryotic living organisms, fundamentally different from the true bacteria and with a distinctive evolutionary lineage. Many scientists have proposed that they be classified in a separate kingdom or domain— a new taxonomic division hierarchically one level above kingdoms—called Archaea.

The archaea differ from the bacteria with respect to the structure of their cell walls (when present) and other molecular features. They are divided into subcategories on the basis of their physiology, which enables them to survive in a variety of environments that are hostile to most other forms of life. For instance, one group (thermophiles) contains organisms that live at very high temperatures and reduce elemental sulfur. Organisms from this group have been isolated from such sites as geothermal vents and hot springs. A second group consists of the methanogens (methane producers), which convert CO_2 and simple organic molecules to methane and are found in swamps where they are responsible for producing marsh gas. Finally, there are the extreme halophiles that require very high concentrations of salts for their growth. *See also* BACTERIA; HALOPHILE; METHANOGEN; THERMOPHILE.

Archaeoglobus fulgidus

Isolated almost exclusively from deep-sea, hydrothermal vents, this member of the archaebacteria is an obligately anaerobic, spherical organism with polar flagella at one end for motility. The taxonomic position of this organism is somewhat puzzling because it possesses characteristics of all three subgroups within the Archaea domain. Not only is it methanogenic and halophilic (requiring at least 1 percent salinity), but it also has the high temperature requirements of the thermophilic group—its optimal growth temperature is about 83°C. In laboratory cultures, *A. fulgidus* forms greenish-black colonies that fluoresce under ultraviolet light. It is the only archaebacterial species known to have the ability to reduce sulfate, and it can derive energy from hydrogen, carbon dioxide, or organic compounds, depending on the availability of substrates. *See also* ARCHAEBACTERIA.

arenavirus

One of the families of arthropod-borne viruses found occurring naturally in rodent hosts and known to be associated with various human diseases, including meningitis and a number of hemorrhagic fevers. There are two distinct serological groups or complexes within the arenavirus family. The "Old World" group, usually found in Europe and Africa, but also in the Americas, includes the lymphocytic choriomeningitis virus (LCMV) of mice and the virus that causes Lassa fever, while the "New World" group predominates in South America and causes different South American hemorrhagic fevers. The LCM virus is widespread throughout the world in its natural host, the common mouse, and frequently infects other rodents such as hamsters, as well as immunodeficient laboratory mice, which leads to various problems in research laboratories. These viruses may be transmitted from the mice to humans, where they cause persistent but often asymptomatic or mild respiratory infections. Rarely—typically when the individual is severely immuno-compromised due to some other disease, drugs, or stress—the virus can cause more serious inflammations of central nervous system tissue, including aseptic meningitis and encephalitis.

The arenaviruses are seen as spherical enveloped particles, each of which contains circular segments of the nucleocapsid. Also found often within the envelope are ribosomes of the host cell—incorporated into the virion during budding—which appear like grains of sand under an electron microscope ("arena" means sand). The genome is composed of two segments of linear, single-stranded RNA molecules, both of which may form circular forms by hydrogen bonding at the ends. A distinctive feature of these RNA molecules is their mixed polarity. While most of the molecule in both segments is of the negative sense, tracts of sequence near one end in both molecules is of the positive sense and capable of functioning as a messenger RNA. Only one other group of viruses, namely the bunyaviruses (also a family of arboviruses), also exhibits this dual nature. The arenaviruses multiply in the cytoplasm of the host cell and typically persist in the cells for prolonged periods of time. They rarely, if ever, induce cell death. This characteristic is reflected in the mostly low-grade symptoms and slow progression of the diseases induced by arenaviruses. *See also* ARBOVIRUS; HEMORRHAGIC FEVERS.

asepsis

The prevention of microbial growth. This is different from antisepsis in that there is no active killing of the organisms. *See also* ANTISEPSIS.

aseptic technique

Method of working in a microbiology laboratory so as to prevent the contamination of instruments, media, and other materials with organisms from the surroundings. Because microbes are a constant presence in most environments on Earth, special precautions must be taken to ensure sterility of all materials both prior to and during an experiment. Aseptic technique involves the sterilization, before use, of all glassware, media, and other materials to be used in an experiment and minimal contact of these materials with the air during their use. The risk of contamination may be reduced by wiping all surfaces with an antiseptic solution or by irradiating surfaces to remove any living microbes. Containers of sterile media are kept plugged at all times and are opened fleetingly when access is required. All work is done close to a flame, and the openings of containers are heated at this flame to ensure the killing of any organisms that might have entered. In modern laboratories, sterile conditions are maintained by the use of laminar flow hoods specially designed for the handling of infection specimens. *See also* ASEPSIS; DISINFECTION; STERILIZATION.

Aspergillus

With over 600 species floating about in the air around us and resting on all possible surfaces, it is just as well that this fungus is, for the most part, a relatively harmless organism that causes disease only under conditions of extreme host susceptibility to infections and allergies. The diseases caused by *Aspergillus* species are collectively called the aspergilloses. Most often, these take the form of allergic reactions, as in the case of allergic bronchiopulmonary aspergillosis, caused by a hypersensitive response due to the exposure of lung tissue to fungal spores. Typical symptoms of this disease include wheezing coughs and the production of brownish plugs of mucus material containing spores and immune cells. In other instances, these fungi may invade the body, resulting in either localized or systemic infections. Finally, some strains of *A. flavus*, as well as another species called *A. parasiticus*, produce toxic metabolites called aflatoxins, which, when ingested along with food, cause various symptoms of food poisoning.

Structurally, *Aspergillus* species are filamentous fungus with septate hyphae and asexual spores that develop from specialized flask-shaped cells called conidia.

Aspergillus can be cultured easily on simple media at 25°C. Fluffy or velvety white colonies become visible within 48 hours. On further incubation, the colonies become powdery in texture and change in color corresponding to the spores they form. Different species, such as *A. fumigatus, A. flavus,* and *A. niger,* can be distinguished on the basis of the color of their spores—gray-green, reddish, and black, respectively. *See also* AFLATOXIN; ALLERGIC REACTION; HYPERSENSITIVITY.

assimilation

The incorporation of various elements and molecules from nonliving sources, such as soil and air, into living matter. Assimilation does not refer to any single chemical reaction, but is a general term to describe all such reactions. Thus, photosynthesis is a means of assimilating carbon from the atmosphere into living cells, and nitrate reduction is a means of assimilating nitrogen in soil into the biomass. Assimilation is a fundamental ability of all living creatures. *See also* AMMONIA ASSIMILATION; CARBON CYCLE; NITRATE REDUCTION; NITROGEN CYCLE.

astrovirus

Family of small RNA viruses known to be associated with human diarrhea. They are endemic in nearly all parts of the world. The disease they cause is a milder version of a rotavirus infection, with a watery diarrhea accompanied by abdominal discomfort and some vomiting that typically resolves within 4–8 days. Infection is spread via contaminated water and food and occurs primarily in young children whose immune systems have not yet built up resistance to these viruses. The immune response to astroviruses seems effective and long-lasting, and, consequently, infections seldom appear in adults. The astroviruses resemble the picornaviruses in many respects of structure and genome organization—i.e., they lack envelopes and have RNA genomes—but they are classified in a separate group because they have different proteins in their capsids. In addition, the virus particles have a highly characteristic appearance (apparent under an electron microscope) of a five- or six-pointed star with a central hollow core. Replication takes place in the cytoplasm of the host cells, where mature virions accumulate in crystalline arrays that may be detected microscopically. These arrays are useful as a diagnostic clue. *See also* DIARRHEA; ROTAVIRUS.

ATCC. *See* AMERICAN TYPE OF CULTURE COLLECTION

athlete's foot

Superficial fungal infection or derma–tophycosis in the region of the foot, also known as tinea pedia, caused by fungal species such as *Trichophyton rubrum, T. mentagrophytes,* and *Epidermophyton floccosum.* The disease is characterized by the scaling or cracking of skin on the feet and the production of infective fluid-filled blisters, especially in the regions between the toes. In severe cases, an allergic reaction to the fungi may manifest as a skin reaction in other parts of the body, but these lesions are not infective. The fungi are transmitted via their spores, which are shed at different locations, such as floors, shower stalls, and carpets, from the feet of an infected person. The oral antibiotic griseofulvin has proved useful in keeping such fungal infections un-

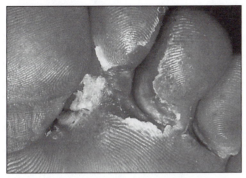

Athlete's foot infection with prominent lesions between the toes. © *Science Source/Photo Researchers.*

der control. Athlete's foot lesions may also be treated with topical antifungal agents. *See also* DERMATOPHYCOSIS.

ATP

One of the chemicals in which living things store energy to use for various metabolic reactions. ATP (adenosine triphosphate) is often regarded as one of the chief "currency molecules" in living systems because cells save and spend it in much the same way as we do money. Energy is stored in the terminal phosphate bond, i.e., between the second and third phosphate groups in the ATP molecule. When an organism requires energy for a biochemical reaction, e.g., breakdown of a sugar, it splits the ATP molecule into ADP (adenosine diphosphate) and a phosphate molecule, which releases the energy stored in the bond. Conversely, energy-generating reactions, such as photosynthesis, result in the net production of ATP by adding a phosphate (i.e., phosphorylating) group to an ADP molecule. In addition to its role in energy metabolism, ATP is also one of the building blocks for the synthesis of RNA molecules. *See also* BIOSYNTHESIS; METABOLISM; PHOSPHORYLATION.

attack rate

This is a proportion that measures the cumulative incidence of a disease; it is a particular epidemic, measured in terms of population, or a limited period of time. Also termed as case rate, this proportion is usually expressed in terms of a particular percentage, e.g., the number of cases per 100 in the group. A related concept is the secondary attack rate, which refers to the number of cases of the disease occurring due to the exposure of family members and other people in close contact to the primary case of infection. This concept is important for determining the epidemiology of a disease.

attenuated vaccine

Vaccine prepared with live organisms—or in case of viruses, active particles—which have been altered or "attenuated" in such a manner that they have lost their virulence or ability to cause disease without losing their specific antigenic character. Different ways to attenuate the organisms include treating them with chemicals or radiation to alter their chemistry or growing several generations in animal hosts or tissue cultures. *See also* MMR VACCINE; POLIOMYELITIS; SABIN, ALBERT B.; VACCINE.

Australia antigen

Original name for the antigen associated with the surface envelope of the hepatitis B virus, the discovery of which led to the eventual identification of the cause of this class of hepatitis (also called infectious hepatitis or serum hepatitis). This protein-lipid complex was so named because it was first discovered in 1964 in the blood of Australian Aboriginals by Baruch Blumberg. At first, scientists thought that this antigen was unique to Aboriginal people, but investigations of blood from various sources revealed that it was more common in people who had received transfusions. When blood work on one child showed a conversion from the negative to positive reactions for the antigen, scientists began to look for possible reasons for this conversion. Their search led to the conclusion that the presence of the Australia antigen correlated with hepatitis B, thus leading the way to the discovery of this virus, which until then had not been isolated. *See also* BLUMBERG, BARUCH; HEPATITIS B VIRUS.

autoclave

Instrument for heating materials in steam under pressure, widely used in hospitals and microbiology labs for sterilization. This is one of the most efficient methods of sterilization because it allows the heat to penetrate materials quickly, resulting in the rapid oxidization of various cellular components and the coagulation of cellular proteins, which kills microorganisms quickly. Autoclaving for 15–20 minutes at 115–120°C under 15–20 pounds of pressure is the standard method for sterilizing media, bottles, and instruments, as well as for killing bacteria in cultures and infected materials. *See also* STERILIZATION.

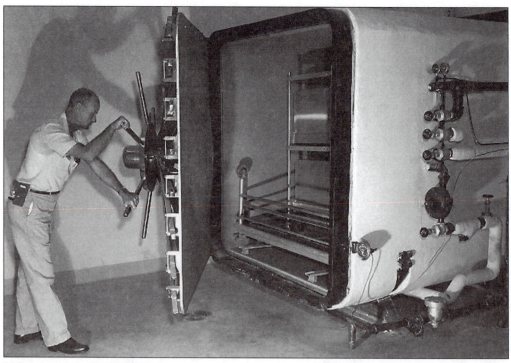

Walk-in autoclave. *CDC/W. Richter.*

autoradiography

A technique for imaging small entities such as microorganisms, trace elements, and contents of cells, which relies on the sensitivity of photographic emulsions to radioactivity. The basic technique involves overlaying a photographic film (or X-ray film) on a radioactive or radio-labeled specimen, i.e., a specimen that has been stained with a radioactive dye. The localization of the radioactivity will then appear as dark spots when the photographic plate is developed. The sharpness, as well as the level of resolution of images, depends mainly on the path length of the emitted radiation. Different elements used in the autoradiography of biological specimens include tritium (a high-molecular-weight form of hydrogen), and radioisotopes of carbon, sulfur, phosphorus, and nitrogen.

Autoradiography has wide applications in microbiology, biochemistry, and molecular biology. Perhaps the most familiar use is in nucleic acid sequencing techniques, in which radio-labeled nucleotides are used to produce radioactive pieces of DNA or RNA, which may then be separated according to size by electrophoresis. In DNA hybridization, a radioactive probe made of a specific sequence may be used to detect complementary or near-complementary sequences in a larger DNA pool. Autoradiography is also useful for localizing specific elements within a cell or for tracking the pathway of a specific element through various metabolic cycles. For instance, by feeding an organism with C^{14}-labeled sugars, one may be able to find out the exact pathway of sugar metabolism by taking autoradiographs of the organism at fixed time intervals. The use of radio-labeled nucleotides is also useful for localizing DNA and other intracellular structures too small to be observed using ordinary microscopy. *See also* DNA HYBRIDIZATION; ELECTROPHORESIS.

autotroph

Free-living organism capable of assimilating inorganic carbon from the environment in the presence of some energy source and reducing agent. The main carbon source is atmospheric carbon dioxide, although a few species of archaebacteria use methane. Different groups of organisms use either radiant or light energy (phototrophs) or chemical energy (chemotrophs). The most widely used reducing agent is water, although a few groups depend on other simple inorganic compounds (lithotrophs) or organic compoounds (organotrophs). *See also* CHEMOAUTOTROPH; CHEMOORGANOTROPH; CHEMOTROPH; LITHOTROPH; PHOTOLITHOTROPH; PHOTOORGANOTROPH; PHOTOSYNTHESIS; PHOTOTROPH.

auxotroph

A bacterial strain that is metabolically dependent upon a specific substance such as an amino acid, sugar, or growth factor, and is unable to grow in the absence of this compound. An organism can become auxotrophic for a specific substance via a mutation that induces an obstacle in the biochemical pathway through which the organism normally synthesizes that substance. *See also* BIOSYNTHESIS; METABOLISM; MUTANT.

Avery, Oswald, T. (1877–1955)

Lead investigator of the team that purified DNA and identified it as the material in genes through the analysis of bacterial transformation. The first report of DNA as the transforming principle was published in 1944 (Avery, MacLeod, and McCarty, *Journal of Experimental Medicine* 79:137–58). Two subsequent papers published in 1946 offered important evidence to confirm the evidence offered in the first paper. *See also* MACLEOD, COLIN; McCARTY, MACLYN; TRANSFORMATION (BACTERIAL).

avian sarcoma virus

Group of oncogenic (cancer-causing) retroviruses that induce the formation of tumors in different avian (bird) species. Like all retroviruses, these viruses are enveloped particles with surface glycoprotein spikes and a diploid, positive sense RNA genome.

In addition to the normal complement of viral genes, the avian sarcoma viruses contain some mutated version of a gene from the host cell. The expression of this gene is responsible for inducing tumors in the host. An example of an avian sarcoma virus is the Rous sarcoma virus (RSV), which was first discovered by Peyton Rous as the causative agent of a sarcoma (connective tissue tumor) in chickens. The mechanism of action of the virus was not worked out until several decades after the original discovery. *See also* CANCER; RETROVIRUS; ROUS, PEYTON.

axenic

Term used to describe conditions of living where only a single living species is present. Thus, an axenic culture is a pure culture uncontaminated with other species, and an axenic organism refers to one that harbors no commensals or parasites. Axenic conditions seldom, if ever, occur naturally and are usually produced and maintained for specific purposes in the laboratory. *See also* PURE CULTURE.

Azomonas

Found mainly in various aquatic regions in the world, these bacteria are important because of their role in fixing nitrogen into the biosphere. They are gram-negative, free-living aerobic rods that contain the nitrogen-fixing enzyme complex nitrogenase, encoded by the *Nif* genes. *Azomonas* are classified along with the more abundant *Azotobacter* in the family Azotobacteriaceae and differ from the latter mainly in the ability to form spores or cysts. *See also* AZOTOBACTER; *Nif* GENES; NITROGEN CYCLE; NITROGEN FIXATION; NITROGENASE.

Azotobacter

This bacterium is one of the most abundant free-living, nitrogen-fixing organisms on Earth and plays a significant role in the nitrogen cycle. Together with symbiotic species such as *Rhizobium*, *Azotobacter* species account for nearly 60 percent of the total nitrogen-fixing activity on earth. The bacterium is an obligate aerobe with the highest rate of respiration or oxygen uptake of any known living organism. This high oxidative rate probably plays a role in maintaining low oxygen tension within the cells, which is necessary to protect the nitrogen-fixing enzyme, nitrogenase. Physically, this genus consists of gram-negative rods measuring between 2–3 μm, most commonly found in neutral and alkaline soils in all parts of the world. Two main species have been identified: *A. vinelandii* and *A. chromococcum*. A characteristic feature of *Azotobacter* is the formation of cysts when carbon compounds such as butanol are present in the environment. These cysts resemble the spores of other bacteria in that they are resistant to desiccation and radiation damage, and they have little respiratory activity. Unlike spores, however, they are not completely dormant because they are able to oxidize a variety of substrates without transforming into the vegetative form of the cell. Also, they are not especially resistant to heat. It is believed that the formation of cysts might also be a strategy to avoid undue exposure of the nitrogenase enzyme to oxygen. *See also* *Nif* GENE; NITROGEN CYCLE; NITROGEN FIXATION; NITROGENASE.

B

B cell

A type of lymphocyte found circulating in the blood of humans and various other higher animals—e.g., mammals and birds—whose primary function is to defend the body against specific harmful pathogens. B cells function by producing specific antibodies (also called immunoglobulins) against different types of antigens. In addition, these cells may play the role of presenting antigens to T cells. At any given time, the B cells may constitute 5–20 percent of the total circulating lymphocytes in the body. *See also* ANTIBODY; ANTIGEN; IMMUNE SYSTEM; IMMUNE RESPONSE; LYMPHOCYTE; T CELL.

Babesia

A protozoan parasite that closely resembles the malaria-causing parasite in appearance as well as in disease symptoms, *Babesia* is a tick-borne (rather than mosquito-borne) organism that infects the red blood cells of many mammalian species. *B. microti* is the species most commonly associated with human infections. The disease, which is known as babesiosis, often occurs along with Lyme disease, because the organisms causing these diseases share reservoirs and vectors. Hence, public health measures instituted against one of these would also be effective against the other. However, other aspects of these two diseases, such as target tissues and organs,

clinical features, and treatment strategies are distinct.

The life cycle of *Babesia* is somewhat complex and has distinct phases in the mammalian and tick hosts. Cattle and rodents serve as reservoirs of the parasites that reside in their trophozoite form in the blood cells of these animals. The sexual phase of *Babesia* begins when the trophozoites differentiate to form gametes. When a tick takes a blood meal, it ingests the gametes, which penetrate the tick's stomach wall and undergo fertilization to form a diploid cell called a merozoite. The merozoite divides to give rise to haploid cells called sporozoites that are released into the insect's mouth and gut. The sporozoites are the infective forms of *Babesia*, and they are injected into the bloodstream of the mammalian host (either humans or animals) during the tick's blood meal. In the mammalian host, the sporozoites invade the blood cells and multiply to form pear-shaped trophozoites that generally lie in pairs or in tetrads (often in a characteristic Maltese cross form). They may also be present in a ring-shaped form similar to the trophozoites of *Plasmodium falciparum*.

The proliferation of the parasites in the blood following a tick bite induces a disease with symptoms that are similar to malaria—high fevers, chills, body aches, and malaise—but without the periodicity, since *Babesia* trophozoites cannot reinfect new blood cells after one cycle of reproduction. In some instances, babesiosis might also be spread via the transfusion of infected blood. Clinical symptoms appear about 2–3 weeks after exposure to the parasite. In addition to chills and fevers, the patient may develop anemia due to massive red cell destruction, as well as the symptoms of mild spleen and liver damage. Untreated, the disease can be fatal. Diagnosis is achieved by examination of thick smears of whole or capillary blood, where the appearance of parasite tetrads in the Maltese cross form serves as a good diagnostic clue. Chloroquine may provide symptomatic relief but does not clear the parasite from the blood. Thus far, the most promising treatment in clinical trials has been to use a combination of quinine (chloroquine) and the antibiotic clindamycin. *See also* LYME DISEASE; MALARIA; *PLASMODIUM*.

Bacillus

With over 40 known species, *Bacillus* is one of the largest and most heterogeneous bacterial genera, found widely distributed in soil and water and consisting of organisms whose only common traits are that they are gram-positive aerobic rods that form spores. So ubiquitous are they, the term "bacillus"—without capitals or italics—is used as a label for rod-shaped organisms in general. Of the numerous organisms classified within this genus, only two species, *B. anthracis* and *B. cereus*, are known for certain to be human pathogens, while various other species may infect other animals, insects, or plants. Many species are of human interest because their metabolic activities have useful applications. Species show a wide variation in DNA content, with GC ratios varying from 32–62 moles percent. *See also* GC RATIO; SPORE.

Bacillus anthracis

This species is the causative agent of anthrax in humans and several other animals. One of the major contributors to the pathogenicity of this bacterial species is a thick capsule made of poly-D-glutamic acid. Strains of *B. anthracis* that lack or lose their capsules are not virulent. A second pathogenic determinant is an exotoxin. The production of both the capsule and exotoxin is under the control of genes on plasmids, rather than the bacterial chromosome. Another distinctive property of *B. anthracis* is its ability to form highly resistant spores that can survive in soil and plants for several decades, which may result in dramatic unexpected outbreaks of anthrax.

During World War II, *B. anthracis* was tested for its potential as a weapon of biological warfare. In 1942, the British government conducted an experiment on Gruinard Island, off the coast of Scotland, in which they detonated "bombs" consisting of anthrax spores over herds of sheep. While this action was never repeated on human populations, the experiment certainly proved the deadliness of the organism—most of the sheep began to die within days of the explosion. Furthermore, the soil on the island remained heavily contaminated with anthrax spores for several decades and was only disinfected after a systematic cleanup was undertaken in 1986–87. *See also* ANTHRAX; BIOLOGICAL WARFARE; EXOTOXIN.

Bacillus cereus

This organism is a common culprit of food poisoning incidents, particularly those associated with food products such as cocoa, spices, and cereals, which are stored at ambient temperatures and provide an aerobic environment in which the bacteria can thrive. *B. cereus* is also associated with a wide variety of nonspecific infections such as septicemia, wound infections, pulmonary infections, and eye infections. It also causes bovine mastitis, i.e., an inflammation of the udder tissue in cows. The organism produces a number of different extracellular products that contribute to its pathogenicity, including two types of hemolysins, a lethal exotoxin that induces the necrosis of skin, and an enterotoxin. In addition, enzymes such as lecithinase, b-lactamase, proteases, and nucleases enhance its virulence. Finally, the ability to form spores that are then capable of withstanding many antimicrobial treatments—including, on occasion, pasteurization—makes these organisms effective agents of food poisoning. *See also* FOOD POISONING; HEMOLYSIN.

Bacillus israeliensis

This species of *Bacillus* is an insect pathogen that causes a lethal infection of the black fly *(Simulium),* which is the chief vector for river blindness, a crippling parasitic disease of humans seen in parts of Africa and South America. *B. israeliensis* is a good example of the success of biological pest control as a means of controlling insect vectors in an ecologically safe manner. During the 1980s, black flies in Africa began to develop resistance to various chemical insecticides, with the result that there was a rapid resurgence of river blindness, which had been brought under control by killing black flies in the area. By a fortunate coincidence, *B. israeliensis* was discovered in a lab in Israel at about the same time, and scientists soon began using it to control the insecticide-resistant flies. *See also* BIOLOGICAL PEST CONTROL.

Bacillus polymyxa

This species is the dominant member of a subgroup of the genus *Bacillus* that is capable of nitrogen fixation. It is most abundant in temperate and tundra regions, where it may account for up to 10 percent of the total nitrogen-fixing capacity of the soil. It has also been shown to establish symbiotic relationships with certain plants such as grasses and wheat, although it does not induce the formation of nodules like *Rhizobium* species do in legumes. *B. polymyxa* is also known because it was the first organism to be found producing the antibiotic polymyxin. A closely related species, *B. macerans,* is capable of symbiotic nitrogen fixation in the rumen (stomach compartment) of sheep. *See also* NITROGEN FIXATION; POLYMYXIN; SYMBIOSIS.

Bacillus stearothermophilus

Representative of the thermophilic (heat-loving) *Bacillus* species, this organism is notable because of its utility in processes such as com-

post degradation and water treatment. It is one of the few species of *Bacillus* that is facultatively anaerobic, and it has been found in a variety of settings, including naturally hot ecological niches such as thermal vents or hot springs, and also in such sites as compost heaps, hot infusions from sugar refineries, and sewage treatment plants. *See also* COMPOST; SEWAGE DISPOSAL; THERMOPHILE.

Bacillus thuringiensis

This species is distinguished from other *Bacillus* species by its ability to produce a powerful toxic substance (exotoxin) that is active against many species of caterpillars. In the past, the organism was packaged and sold to farmers as a control measure against these crop pests. With the advance of genetic engineering techniques, scientists have created plants containing the bacterial gene and are thus able to produce the toxin by themselves without the bacteria being present. *See also* EXOTOXIN.

bacteremia

Condition referring to the presence of bacteria in the bloodstream. Bacteremia is not a specific disease, but rather a condition that may ocur in virtually any bacterial infection. Depending upon the nature of the infecting organism and the status of the host's immune system, bacteremia may be transient (very short lived), intermittent, or continuous. Intermittent bacteremia arises when the host is unable to contain a localized infection within the primary lesion; it may also arise from a number of localized infections such as abscesses (skin or mucus membranes), pneumonia (lungs), or meningitis (brain). When the primary infection is in a vascular source, e.g., in the case of endocarditis (a heart infection) or an infected aneurysm, then the supply of bacteria to the blood is constant and bacteremia is continuous. Both intermittent and continuous bacteremias frequently give rise to symptoms of fever, chills, aches, septicemia, and shock. *See also* SEPTICEMIA; SEPTIC SHOCK.

bacteria

Group of microorganisms that share the fundamental characteristics of a prokaryotic, unicellular mode of existence. This description encompasses a wide variety of organisms that represent some of Earth's earliest living beings. The bacteria are a very diverse group of more than 2,000 species that differ from one another with respect to their shape, cellular composition, nutritional requirements, metabolic capabilities, and preferred habitats.

Morphologically, the bacteria average about 1–2 micrometers across and may be present as simple spheres (cocci), rods, curved forms, or spiral forms. All possess a cell wall, which lies outside the cell membrane and is responsible for giving the organism its

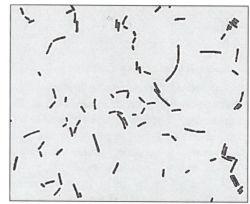

Micrograph of *Bacillus brevis. CDC/Dr. William A. Clark.*

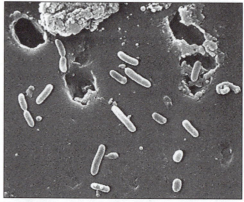

Micrograph of *Pseudomonas aeruginosa. CDC/Janice Carr.*

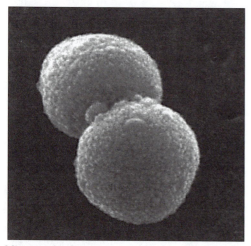

Micrograph of *Streptococcus pneumoniae. CDC/Dr. Richard Facklam and Janice Carr.*

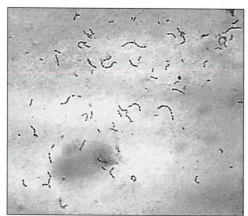

Micrograph of *Streptococcus mutans. CDC/Dr. Richard Facklam.*

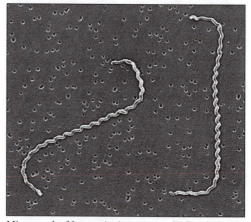

Micrograph of *Leptospira interrogans. CDC/NCID/HIP/ Janice Carr and Rob Weyant.*

characteristic shape. Within the membrane is enclosed the bacterial cytoplasm, which contains various substances, including ribosomes, but which lacks membrane-bound cytoplasmic organelles such as chloroplasts or mitochondria. As prokaryotes, these organisms lack a defined nucleus but contain a single, circular DNA molecule—also called the bacterial chromosome—which contains all the necessary genes. Some species may also contain a smaller piece of DNA called a plasmid that encodes non-essential functions. Bacteria may be either stationary or motile due to the presence of flagella. Organisms divide by a simple process of binary fission, whereby a single cell (the mother) doubles in size and quantity of different components before splitting into two roughly equal progeny cells.

Different bacteria possess a vast repertoire of metabolic enzymes, which enables them to live under a wide spectrum of environmental conditions and derive food and energy from a variety of sources, including sunlight (phototrophs) and organic and inorganic chemicals (lithotrophs). Bacteria obtain their nutrition by absorbing food molecules in various forms; they do not ingest large food particles or other organisms. One common feature among all bacteria, regardless of their ability to perform photosynthesis, is their inability to produce oxygen. Most bacteria are free-living in nature and, due to their diverse metabolic capabilities, occupy almost every environmental niche.

The bacteria perform many vital functions for human life—e.g., they produce vitamins in the gut and they ferment milk into yogurt and cheese. Yet they also pose some of the worst threats to our health and well-being, and are among the principal causes of human disease. *See also* BINARY FISSION; BIOSYNTHESIS; METABOLISM; PROKARYOTE; TAXONOMY.

bacteriocin

Bacterial protein produced by one strain of an organism that is toxic to closely related strains of the same species. These toxins are produced by one strain in an attempt to eliminate competition for nutrients by closely related strains, which would typically share most metabolic requirements. Some attempts have been made to harness bacteriocins for therapeutic purposes, although their application is somewhat limited due to the narrow spectrum of organisms that any single bacteriocin may inhibit. However bacteriocins are often used for "typing" or identifying and distinguishing among closely related strains of bacteria. Genes for different bacteriocins may be found either on the bacterial genome or in plasmids. *See also* COLICIN.

bacteriophage

Special group of viruses that specifically infect bacterial hosts rather than plants or animals. Phages are among the simplest viruses known, for which reason scientists have used them extensively in various kinds of experiments aimed at understanding different aspects of molecular biology and genetics, not just of viruses but also of all other forms of life. A bacteriophage consists of a single species of nucleic acid encased in a protein coat called a capsid. The genome may be composed of any configuration of nucleic acid, i.e., it may be either single- or double-stranded DNA or RNA. The outer coat may be polyhedral or filamentous, and some phages have a "tail" for attachment to the bacteria. In many instances, a bacteriophage is specific for a given bacterial genus, species, or strain.

There are two main classes of phages—virulent and temperate—based on their life cycle within the bacterial cell. Virulent or lytic phages multiply within the bacterial cell. When they have accumulated to an amount called their "burst size," the host cell undergoes lysis and the progeny viruses are released into the surroundings. An example of a lytic phage is the T4 phage of *E. coli*. In a few cases, the phage does not lyse the host cell immediately but releases progeny from an intact cell for a few generations. A temperate phage establishes a stable relationship with its host by integrating a copy of its own genome into the bacterial genome. This relationship is called lysogeny, and the host bacterium is said to be lysogenic. A lysogenic bacterium is typically immune from infection by phages related to the infecting bacteriophage. Examples of temperate phages include the λ-phage of *E. coli* and the M13 phage, which infects a large number of enterobacteria. *See also* PLAQUE; VIRUS.

Bacteroides

Residents of the intestinal tracts of many warm-blooded animals including humans, the bacteria of this genus constitute nearly half the fecal matter that is shed by these animals. They are, for the most part, an avirulent bacteria and are often used as indicators to test for the contamination of water. Under rare conditions, such as very severe trauma to the abdominal region, *Bacteroides* may cause opportunistic infections, usually resulting in the formation of an abscess at the site of injury. These abscesses consist of the bacteria, cellular debris, and immune cells, and may become walled in because of the accumulation of collagen around the site of bacterial invasion. Occasionally these walls disintegrate and cause the spread of the organisms into the neighboring tissues or the bloodstream. Systemic infections may give rise to symptoms of fever. Infections are treated with antibiotics such as metronidazole or clindamycin.

Bacteroides are gram-negative anaerobic rods with large capsules. They do not form spores but may give a misleading impression of doing so because of the presence of large vacuoles in the cell. Unlike other intestinal inhabitants, such as *E. coli*, these organisms produce no endotoxins, another reason for their low virulence. They ferment glucose and lactose, while some are capable of using peptones as their primary carbon source. Although anaerobic, they are able to tolerate oxygen and even produce the enzyme catalase. The most abundant organism of this genus is *B. fragilis*. *See also* OPPORTUNISTIC INFECTION.

baculovirus

These viruses have garnered a tremendous amount of interest from the scientific community because of their potential as agents of biological pest control against a variety of agricultural pests. The distinguishing characteristic of baculoviruses that makes them such attractive candidates in agriculture is a very narrow host range restricted to the arthropods—insects, arachnids (spiders), and crustaceans (crabs)—and the inability to infect higher plants or animals. Left to themselves, these viruses kill their hosts rather slowly, but it is possible to engineer their genomes to carry genes for insect-specific toxins and thus to create a pesticide that has no adverse effects on either the crops or humans. This has been accomplished, for example, by splicing the genes for scorpion toxins into the baculovirus genome, thus rendering these viruses lethal to pests such as caterpillars. In addition to their narrow host range, these viruses do not appear to have any relatives in plants or animals, which also reduces the long-term risks of releasing harmful genetically engineered products into the environment.

Physically, the baculoviruses are large, enveloped rod-shaped viruses consisting of circular double-stranded DNA genomes. They are capable of surviving outside the host cell in the form of particles called nuclear polyhedrosis viruses (NPV), which consist of a protective protein crystal that encases one to several nucleocapsid particles. NPV crystals are the infective form of baculoviruses and enter the host organism via ingestion. Once in the arthropod gut, the protein coat is broken down and the virion particles that are released infect the cells of the intestinal mucosa. Viral replication, which takes place in the nucleus, occurs in two phases. During the early phase, the virus forms nucleocapsids, which emerge from the host cell by budding to form enveloped, extracellular particles that may infect new host cells. During the later phase, enveloped nucleocapsids begin to accumulate in the cell, forming either granular inclusions called granulosis virus (GV) particles, or NPV, due to the synthesis of the crystalline occlusion protein. These are eventually released from the cell through lysis. *See also* BIOLOGICAL PEST CONTROL.

Balantidium coli

A large protozoan parasite known to cause colon infections resulting in diarrhea and dysentery in humans. Although the parasite is found worldwide, disease incidence is relatively low, with occasional outbreaks coincident with the contamination of water in areas where there are large populations of swine. In addition to pigs, rats and certain non-human primates may also carry *B. coli*. The life cycle of these organisms is relatively simple, consisting of infective, resistant cysts that are passed in the feces and reintroduced into animals (including humans) via water or food, and trophozoites, which are released by the excystation in the small intestine. The trophozoites of *B. coli* are motile due to the presence of cilia on the cell surface. They undergo a few cycles of multiplication by binary fission in the large intestine before developing a cyst wall, which surrounds the entire ciliated cell. Balantidiasis, as the protozoan disease is known, presents with rather nonspecific symptoms of intestinal upset: diarrhea, accompanied by abdominal pain (colic), nausea, and vomiting. Dysentery, when it does occur, is characterized by blood and mucus, but no pus. Diagnosis is by the identification of ciliated trophozoites or cysts in the feces and intestinal lining. Drugs used for treatment include tetracycline, iodoquinol, and metronidazole. *See also* CYST; DIARRHEA; DYSENTERY; TROPHOZOITE; WATER-BORNE INFECTION.

Baltimore, David (1938–)

An American scientist whose discovery of the enzyme reverse transcriptase—demonstrating that RNA molecules could act as templates for DNA synthesis—helped provide the experimental evidence for the mechanism of action of cancer-inducing RNA viruses such as the Rous sarcoma virus. This contribution gained him a share in the 1975 Nobel Prize in Physiology/Medicine along with virology pioneers Renato Dulbecco and Howard Temin for what the Nobel committee described as their "discoveries concerning the interaction between tumor viruses and the genetic material of the cell." *See also* REVERSE TRANSCRIPTASE.

Bartonella

Discovered in association with such human diseases as Oroya fever and cat scratch disease, *Bartonella* is a somewhat uncommon bacterium that almost always requires a host animal for its existence in nature. Usually, these bacteria reside within the red blood corpuscles of their animal (or human) hosts, although they may be seen infecting other types of cells, such as skin cells. They are rather fastidious and require either blood, serum, or both for growing in artificial media. The diseases caused by these bacteria (called bartonellosis) are usually spread via insect vectors such as sandflies and cat fleas. While *Bartonella* species are known to be motile, their flagella—which occur in polar tufts—may only be seen in culture. The principal species include *B. bacilliformis*, which is only known to infect humans and causes both Oroya fever and a skin disease called verruga peruana; and *B. henselae*, implicated in cat-scratch disease. More recently, a third species named *B. quintana*—the causative agent of trench fever, formerly identified as the rickettsial genus *Rochalimaea*—was also reclassified into this genus. *See also* CAT SCRATCH DISEASE; OROYA FEVER; RICKETTSIAE; VERRUGA PERUANA.

Bassi, Agostino (1773–1856)

An Italian lawyer and naturalist generally credited as the founder of the germ theory of disease, Bassi was the first person to publish an account of an evident link between microbes and infectious disease in animals. The monograph, which he published in 1835, was based on his experimental work on the muscardine disease of silkworms that was wreaking havoc on the silk industry in Italy and France. Through careful experimentation, he was able to trace the culprit, which turned out to be a microscopic fungus belonging to the genus we now call *Botrytis*. *See also* GERM THEORY OF DISEASE.

beer

Fermented alcoholic beverage produced by the action of yeasts—*Sacchararomyces cerevisiae* and *S. carlsbergensis*—on grains such as barley and wheat. This drink has a long history. References to beer-like drinks have been found in ancient Mesopotamian tablets dating back to about 6000 BC, and it is known that the ancient Britons brewed beer from malted wheat even before the Romans introduced them to barley.

Beer-making is a complex process involving many steps to extract and then ferment the sugars and starches present in the grain. The first step, which distinguishes this beverage from other grain-based alcohol, is malting—the process of steeping the grains in water for a few days to allow the seeds to germinate. This process releases enzymes that convert the stored starches of the grain to sugars. The sprouted grains are gently heated to halt the growth process and soaked in more water so as to extract various materials such as sugars, enzymes, and amino acids into the liquid. This extract, called malt wort, is then boiled along with hops, which give a beer its characteristic bitter flavor and retard the growth of bacteria. After cooling, this mixture is seeded with yeast and left to lie undisturbed for a week or so under conditions of low aeration to allow fermentation. Since the yeasts are subject to a largely anaerobic environment, they ferment the sugars present in the malt wort to alcohol and carbon dioxide (which makes beer effervescent). Fermentation ceases spontaneously when the accumulated alcohol rises to a level sufficient to inhibit yeast growth and metabolism. At this juncture, yeast cells may be removed from the liquid by sedimentation or filtration, and the beer is ready to be consumed or bottled. *See also* ALCOHOLIC FERMENTATION; FERMENTATION; YEAST.

Behring, Emil A. von (1854–1917)

The winner of the first Nobel Prize for Physiology and Medicine in 1901, von Behring made important contributions to the development of antitoxin therapies for diphtheria. He was the first person to show that animals inoculated with the causative organism *Corynebacterium diphtheriae* produced anti-

bodies against the toxins (called antitoxins), and that serum from these immunized animals, when injected into new animals, could successfully protect them against the bacterial toxins. *See also* ANTITOXIN; *CORYNEBACTERIUM DIPHTHERIAE*; DIPHTHERIA.

bejel

A highly infectious skin disease caused by *Treponema pallidum*, the organism that also causes syphilis. Bejel predominantly affects children in Africa, the Middle East, and Southeast Asia. The primary mode of transmission of this disease is by common household contact rather than by sex; it is therefore also called non-venereal syphilis. *See also* SYPHILIS; *TREPONEMA PALLIDUM*.

Bergey's Manual of Systematic Bacteriology

A standard reference in any microbiologist's laboratory or library, this book is one of the most comprehensive manuals of bacterial classification and identification. The first edition of this compendium was published in 1923 through the efforts of David H. Bergey and R. E. Buchanan as a single volume entitled *Bergey's Manual of Determinative Bacteriology*. Since then, it has been regularly updated and has expanded considerably. The title has also been modified to reflect the changing face of the discipline. The current edition consists of four volumes. Volume 1 deals with the gram-negative bacteria, Volume 2 covers the gram-positive organisms, Volume 3 contains the most diverse group, including the photosynthesizers (cyanobacteria) and organisms we now classify as the Archaea, and Volume 4 deals with the actinomycetes. The discussions in this volume rely on the classification scheme of *Bergey's Manual*.

binary fission

Method of reproduction or replication in which a cell duplicates its DNA and doubles the rest of its contents before dividing itself into two roughly identical progeny cells by the formation of a septum or wall between the two halves. This is the simplest and most common means of replication of most single-celled prokaryotes including bacteria and mycoplasmas. The mitosis that takes place in eukaryotic cells is also a form of symmetrical binary fission. In certain rare instances, e.g., *Caulobacter* species, binary fission is asymmetrical, resulting in the formation of two progeny of which one looks and behaves considerably differently from the parent cell. Certain cyanobacteria undergo binary fission repeatedly under a common sheath or capsule so that the progeny do not separate but resemble a filament. This process is known as multiple fission. *See also* CELL DIVISION.

binomial nomenclature

System of naming living organisms developed by Carl Linnaeus in the eighteenth century and used today in naming virtually all cellular organisms. By this scheme, every organism has a two-part Latin name, in which the first part denotes the genus to which it belongs and the second part acts as a specific label for a particular species. By convention, the name of the organism is always written in italics with the generic name capitalized and the specific name beginning in lower case. After a first mention, the name of an organism may be abbreviated by reducing the genus name to its first letter. Thus, for example, *Escherichia* (genus) *coli* (species) is represented as *E. coli*. *See also* LINNAEUS, CARL; TAXONOMY.

biofilm

Found in a variety of natural environments as diverse as water pipes, the hulls of ships, the roots of plants, and the teeth and intestinal tracts of several animals, a biofilm is a community of several different microbes found living together in a physical medium consisting of a gluey substance produced by one or more of the organisms. These communities are present in the form of a film, usually attached to a solid surface. In recent times, biofilms have attracted attention from the scientific community for their potential uses in different industries, but they are not a new phenomenon. The formation of a biofilm is a common mode of existence for many microbes in nature because pure cultures are not really possible anywhere except in the controlled environs of a laboratory.

Commercially, biofilms are used in the sewage and wastewater management industries and in industrial plants for the disposal of pollutants. Understanding the dynamics of such microbial communities is advantageous to scientists because of the problems they pose in various disease situations. Examples of disease conditions in which biofilms are implicated are dental plaque, which often leads to dental caries and tooth decay, and infections associated with the use of prosthetic devices (e.g., artificial limbs).

Basically, a biofilm may form on any solid surface where there is a steady supply of moisture and nutrients. Typically it is initiated by a slime-producing organism, called the primary colonizer, which produces copious amounts of a sticky polysaccharide substance (of variable composition) and adheres to a solid surface. Following adhesion, the organism begins to multiply at the site of attachment and produces more slime, which attracts and traps other organisms (secondary colonizers). These secondary organisms, in turn, multiply, resulting in the formation of a mixed microbial community. The formation of biofilms offers many advantages to the participating organisms, such as the ability to concentrate nutrients and share various energy and carbon sources within the slime-microbe matrix, establish mutually beneficial interrelationships (whereby the metabolites of one organism provide nutrients for another), and protect themselves from various external threats. For instance, most biofilms are too large to be susceptible to phagocytosis by immune cells. The symbiotic relationships between organisms in a community, as well as different types of metabolites that are released into the film, also render biofilms more resistant to antibiotics and other antimicrobial substances. *See also* BIOLOGICAL AERATED FILTERS; DENTAL CARIES; DENTAL PLAQUE; PHAGOCYTOSIS.

biological aerated filter

A device used in sewage disposal plants for the treatment of water that has been separated from sewage solids by a prior process of simple sedimentation. The filter consists of a bed of submerged granular substance coated with mixed cultures (biofilms) of aerobic bacteria and protozoa. The treatment process involves passing the liquids through the filter from above while pumping air near the base, so that the organisms in the biofilm may efficiently metabolize and thus remove organic materials from the water. Simultaneously, the filter removes suspended solids from the sewage. Also, because the bacteria are already present as films attached to a solid substrate, their multiplication does not generate new solid waste, as is the case when activated sludge is used. The treated fluids that percolate through the filter are much clearer and have a much lower biological oxygen demand than untreated fluids.

Although used primarily in aerobic processes, biological filters may also be operated under anaerobic conditions. In this case, the biofilms contain organisms such as the denitrifying bacteria, which can degrade and remove compounds like amino acids from sewage. Thus, by operating a series of filters under aerobic and anaerobic conditions, it may be possible to remove large amounts of dissolved material from wastewater. *See also* ACTIVATED SLUDGE; BIOFILM; BIOLOGICAL OXYGEN DEMAND; SEWAGE TREATMENT.

biological oxygen demand (BOD)

Generally used as an index of pollution, biological oxygen demand (BOD) refers to the oxygen-depleting ability of a body of water such as a lake, pond, or septic tank. In general, the BOD of a sample increases in proportion to the density of its microbial population and the amount of organic material it carries. For example, sewage-polluted ponds or lakes have high BOD because they contain a high proportion of organic compounds. These compounds support the vigorous growth of various aerobic, heterotrophic bacteria (also discarded into the sewage), which is directly responsible for the depletion of oxygen from the water. The lowered oxygen tension created in the water in turn exerts several negative effects on the environment. For instance, it may mean the death of many fish and other oxygen-dependent creatures that normally reside in the water. In addition, the reduction in oxygen encourages the growth of anaerobic bacteria, which produce a number of noxious metabolic products such as methane and hydrogen sulfide, often associated with septic tanks and untreated sewage. One of the aims of sewage treatment, therefore, is to lower the BOD potential of the wastewater.

biological pest control

The use of one group of living organisms to control the growth and effects of another organism—usually insect pests such as caterpillars, aphids, flies, and mosquitoes—that poses problems for agriculture or public health. The organisms used to control the pests may be parasites, predators, or specific pathogens against the target organisms, and the net result of their use is the lowering of the population density of the pest (versus a complete eradication, which is usually not feasible). Examples of pest control agents are such species of *Bacillus israeliensis* and *B. thuringensis*, which have been used to control populations of the black fly (the vector for river blindness) and caterpillars, respectively.

biological warfare

This term refers to the use of biological materials—such as pathogenic organisms or toxic substances derived from living organisms—as weapons of war. Biological warfare typically encompasses the use of such materials to kill or cause disease not only in humans but also in animals (e.g., livestock) and plants (farm produce). Organisms that might be used as biological weapons (in "germ warfare"), include bacteria or viruses known to cause acute diseases such as anthrax, plague, or smallpox, which have an almost immediate onset with severe, incapacitating effects and a high mortality rate. Releasing the agent of a chronic disease such as tuberculosis or cancer—regardless of its devastating effects in the long term—would not serve the attacker's purpose of either killing enemies quickly or forcing them toward a speedy surrender. Among the biological warfare agents derived from microbes, the botulism and tetanus toxins are good candidates, again because of their swift and almost universal action. The rapid advances in fields such as genetic engineering and biotechnology over the past few decades make the issue of biological warfare ever more serious and frightening because the potential for developing new and stronger biological weapons has increased.

Biological warfare has a long and inglorious history. There are accounts, for instance, of some early white settlers in the Americas attempting to decimate native populations by distributing blankets infected with the material from smallpox lesions, and of riots that broke out in Paris in 1832 following the outbreak of cholera during the great pandemic because of a belief among the poor that the rich nobility were trying to poison them via drinking water. Among the earliest examples of a concerted research effort in germ warfare in the twentieth century is that of Japan in the 1930s and 1940s. Working in Ping Fan, an occupied Chinese city, Japanese scientists experimented with the mass production and delivery of pathogenic organisms, and they appear to have been successful in initiating two major outbreaks of bubonic plague by dropping "bombs" containing infected grain and fleas. During World War II, the British experimented with the potential use of *Bacillus anthracis* spores for the same purpose. In tests, they killed massive populations of sheep, but they never used *Bacillus anthracis* against human populations. Iraq acknowledged having developed biological weaponry during the 1990s, but whether they used it during the Gulf War is not known. The potential misuse of smallpox viruses as biological weapons has been one of the strongest arguments for destroying the stockpiles of these viruses held in the vaults at the U.S. Centers for Disease Control (the CDC) and its Russian counterpart. *See also* BACILLUS ANTHRACIS; VARIOLA (SMALLPOX) VIRUS.

biological washing powders

Domestic cleaning powders, particularly for clothes, in which the active ingredient is an enzyme rather than a synthetic chemical. The most commonly used enzymes in these products are proteins called subtilisins, produced by species of *Bacillus*, e.g., *B. licheniformis*. The cleaning action of these enzymes is based on their ability to break peptide bonds in proteins and ester bonds in lipids, which enables them to decompose many sources of stains and soiling. *See also* ENZYME.

bioluminescence

The property of certain living organisms to emit light without necessarily generating a corresponding amount of heat. Perhaps the most familiar examples of luminescent organisms are the fireflies and glow-worms that we see flitting about at night, but other organisms also exhibit the same property. A notable example in the microbial kingdom is the bacterial genus *Photobacterium*, species of which are symbiotic with certain deep-sea animals. These animals use the flashes of light emitted by the bacteria for a variety of such purposes as attracting prey (food) or sexual partners, or for discouraging predators. The biochemical entity actually responsible for luminescence is a long fatty molecule called luciferin. The modification of this molecule by an enzyme called luciferase (produced by the bacteria) results in the emission of light. The reaction also requires the expenditure of energy or ATP molecules. *See also* LUX GENES.

biosensor

Highly sensitive and selective electronic devices that can monitor the concentration of specific substances—such as sugars, amino acids, alcohols, and acids—in very low-volume samples. These sensors are often used to monitor levels of specific compounds in samples such as blood or serum. A biosensor is made up of two parts: a tiny electrode, or voltaic cell, which records changes in concentration of molecules in terms of electrical output and a biological entity such as an enzyme or microorganism, which imparts specificity to the device. For example, one might fashion a biosensor for glucose—useful for monitoring blood or serum samples of diabetic patients—by coupling an oxygen electrode with the enzyme glucose oxidase. Because the enzyme uses oxygen exclusively in the presence of glucose, any changes the sensor might detect in oxygen concentration would be the direct result of changing amounts of glucose in the blood sample.

biosphere

Term used by biologists to represent the zone in the earth's crust that is inhabitable by living organisms. Basically this is a region of a maximum thickness of about 40 km on either side of the surface where the earth's crust (lithosphere) meets the lower atmosphere. The density of life is highest near the surface—whether terrestrial or aquatic—and gradually decreases in both directions. The so called "higher animals" occupy the narrowest zone, with the outer edges of the biosphere occupied by special groups of microbes particularly adapted for the extreme conditions of those areas. Bacterial and fungal spores have been detected at heights of up to 32 km in the atmosphere, while in the oceans, bacteria are known to flourish as deep as 11 km.

biotechnology

The use of living cells—usually microbes, but also certain plants and animals—in various industrial processes. Examples of commercial products based on biotechnology include antibiotics, beer, wine, cheeses, acetone, single-cell protein, and even certain plastics. The development of cloning and genetic engineering has greatly enhanced the biotechnology industry in the past few decades.

Blastomyces dermatitidis

A fungus associated with a chronic, granulomatous infection of the lung called blastomycosis, also known as Gilchrist disease in southern Canada and parts of the United States, where it is endemic. In other parts of the world, blastomycosis is relatively rare, although it may occur in sporadic outbreaks just about anywhere.

B. dermatitidis is a dimorphic fungus that appears in the form of unicellular yeasts in tissue or when grown in enriched media at 37°C, but grows as a multicellular mold at room temperature (25°C). The mold form is frequently found in moist soils along waterways or in damp, undisturbed environments. In this form, the fungi propagate by forming spores, which are released into the air and may be easily spread by this agency. Primary infection in humans occurs mainly by the inhalation of the spores, which germinate and proceed to multiply in the lungs in their yeast form. This pulmonary infection may be asymptomatic due to the successful clearance of the organisms by the immune system. Sometimes there may be an acute disease with symptoms of cough and fever that resolve spontaneously 1–3 weeks after infection. More often, however, the pulmonary symptoms come on rather gradually and evolve into a chronic disease. If untreated, the fungi may spread to other parts of the body via the bloodstream and give rise to lesions in the subcutaneous tissues. This form of the disease is called chronic cutaneous blastomycosis. Systemic infection may also lead to the involvement of other organs and bones.

Depending on the clinical presentation—i.e., pulmonary or cutaneous—appropriate samples, sputum, or skin scrapings should be examined for the presence of infecting organisms. *B. dermatitidis* may be recognized by their peculiar microscopic appearance on staining—the cytoplasm of these yeasts shrinks upon taking on the stain so that the cells appear double walled. Serological tests are also useful in confirming diagnoses. The drug of choice for treating blastomycosis is amphotericin B. Treatment may prove effective in cases of chronic pulmonary or skin infections—acute infections are typically self-limiting—but systemic blastomycoses are almost always fatal.

blood-borne pathogens

A term used by public health workers to describe such pathogenic organisms as are transmitted from the bloodstream of one organism to another through "bloody" materials such as clothing, bandages, hypodermic syringes, and other sharp objects. This mode of infection is often not the natural primary route used by the organisms, and thus the infection in the new host may not follow its typical course.

bloom

A sudden, temporary, superabundance in the growth and reproduction of aquatic microbes—primarily cyanobacteria and microscopic algae—usually in response to a sustained supply of nutrient materials such as nitrates and phosphates (i.e., eutrophication) in bodies of water such as lakes, rivers, or oceans. The word bloom is very descriptive of the phenomenon, which is characterized by the formation of densely colored patches in water due to the release of large quantities of pigments by the organisms in question. A commonly used term for this phenomenon is red tide, which derives its name from the blooms of dinoflagellate algae, which tinge the waters with characteristic ruddy hues.

With the exception of situations where the concentrated growth of an algal species is desirable, e.g., in the cultivation of a particular species, blooms are associated with negative impacts on the environment and public health. As organisms in a bloom multiply rapidly, they use and deplete materials such as dissolved oxygen, which is detrimental for fish and other oxygen-consuming creatures that also live in the water. This is particularly problematic in aquaculture tanks where the fish or shellfish are held in contained tanks, and are thus unable to escape to more oxygenated environs. Another problem that arises is

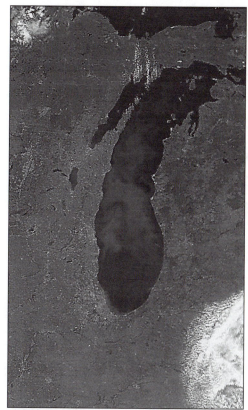

Algal bloom in Lake Michigan, September 1999. Image of Lake Michigan courtesty of ORBIMAGE. © (1999) *Oribtal Imaging Corporation and processing by NASA Goddard Space Flight Center.*

that as the bloom continues to grow, older organisms die and decay, fouling the water and making it unfit for consumption.

Perhaps the most devastating consequence of blooms is the production of toxins, which are responsible for a number of potentially lethal poisoning syndromes. Examples include paralytic shellfish poisoning (PSP), diarrhetic shellfish poisoning (DSP), neurotoxic shellfish poisoning (NSP), domoic acid poisoning (DAP, sometimes known as amnesic shellfish poisoning or ASP), and ciguatera fish poisoning (tropical fish poisoning). The link between humans and these diseases is provided by different types of shellfish—for example, mussels, clams, oysters, and scallops—which feed mostly on phytoplankton. While these animals are not sensitive to the toxins, they lack the ability to metabolize them. Thus, over time, the toxins accumulate in their flesh and cause a lethal effect when consumed by humans or other animals. Toxic blooms are also responsible for fish kills, characterized by the death of large numbers of fish, either due to algal toxins or directly by the algae. In addition to the health hazards, the economic impact of the toxic blooms is enormous, especially for fisheries and the aquaculture industry. In 1997, for instance, a toxic bloom of *Alexandrium catenella*—which produces the toxin responsible for PSP—broke out in southern Puget Sound in Washington, shutting down all the local oyster farms, one of the major sources of livelihood in that region. *See also* ALGAE; EUTROPHICATION; RED TIDE.

blue-green algae

Older term, still often used to represent the cyanobacteria. These organisms cannot be considered true algae because of their prokaryotic nature; they are therefore better classified with the prokaryotic bacteria. However, the fact that they are photosynthetic due to the presence of chlorophyll-like pigments indicates that they might be the precursors to the photosynthetic eukaryotes—i.e., algae and higher plants. *See also* ALGAE; PHOTOSYNTHESIS; PROKARYOTE.

Blumberg, Baruch S. (1925–)

Scientist who discovered the Australia antigen and established the connection between the presence of this protein and hepatitis B or serum hepatitis, as distinct from infectious hepatitis (A). Blumberg's contributions to the understanding of the pathogenicity and transmission of this disease garnered him a share in the 1976 Nobel Prize in Medicine. *See also* AUSTRALIA ANTIGEN; HEPATITIS B.

Bordet, Jules (1870–1961)

The name of the Belgian scientist Jules Bordet is probably most familiar in a slightly modified form—as *Bordetella pertussis*, the bacterial species that causes whooping cough. And while it is true that Bordet was one of

the two key scientists to demonstrate how this organism causes the disease, it was by no means his only contribution to medical science. One of his earliest scientific accomplishments was to show how antibodies function in clearing bacteria from the blood during an infection. He demonstrated that normal, non-immune serum contains a substance that is necessary for the complete action of antibodies. This was the component that we now know as complement. He also developed methods for identifying different strains of microbes on the basis of their reactivity to immune sera—i.e., the technique of serotyping. Bordet also contributed to the understanding of effector mechanisms of the immune system besides complement, such as the anaphylaxis-inducing poisons in blood. He received the Nobel Prize in Medicine in 1919. *See also* BORDETELLA; COMPLEMENT.

Bordetella

This bacterial genus is best known as the causative agent for whooping cough. The species responsible for this disease is called *B. pertussis* and was first isolated from patients in 1906 by two Belgian scientists, Jules Bordet and Octave Gengou. *B. parapertussis* is a closely related but less virulent species. *B. bronciseptica* is commonly found in the respiratory diseases of pets such as dogs and rabbits. It is markedly different from other species in that it is actively motile, due to the presence of flagella.

The different species of this genus are gram-negative, ovoid rods that are strictly aerobic. Consequently, *Bordetella* tends to cause respiratory infections and is seldom, if ever, implicated in deeper systemic infections. Although currently classified under the family Pseudomonaceae, according to *Bergey's Manual*, the taxonomic position of *Bordetella* is not completely clear. Primary isolates reveal the presence of capsules that disappear in serial cultures. These organisms are fastidious and require complex media, enriched with blood or serum, for growth in the lab. They may be selected for using penicillin, to which they are resistant. *B. pertussis* grows slowly—colonies typically appear after 48–72 hours of incubation—and form small, cohesive colonies with a metallic sheen resembling mercury. *B. parapertussis* colonies grow up to 3 mm in diameter and produce a brown pigment. *See also* WHOOPING COUGH.

Borrelia

Genus of slender, helical bacteria (spirochetes) that are responsible for arthropod-borne infections in humans. There are two main species of importance to humans: *B. burgdorferi*, transmitted by ticks of the genus *Ixodes* and believed to cause Lyme disease, and *B. recurrentis*, which causes an epidemic type of relapsing fever spread by lice. Variants of the latter species that are transmitted by ticks are often known by other specific names such as *B. hermsii* and *B. parkerii*.

Like the other spirochetes such as *Treponema* and *Leptospira*, *Borrelia* does not respond to the Gram stain and may be visualized under the microscope using Giemsa or Wright stains. These bacteria display a very characteristic flexous type of motion due to the presence of 7–30 flagella originating at one end of the cell and positioned with their free ends toward the opposite end. The localization of these flagella within the periplasmic space of the organisms, between the cell and its outer membranes, is responsible for the twisting or flexous type of motility. For growth in the laboratory, these organisms require complex media containing protein sources such as blood or serum. *See also* LYME DISEASE; RELAPSING FEVER.

botulism

Acute paralytic disease—often fatal—caused by *Clostridium botulinum*. Botulism is not an infection, rather, it is an intoxication, caused by a powerful neurotoxin produced by the organism. In fact, the physical presence of viable clostridia is not essential for the disease to occur. Three main clinical forms of botulism are distinguished according to mode of onset. Perhaps the most common disease is food-borne, caused by the consumption of foods containing the toxin. Frequent sources of disease are improperly canned foods, especially those prepared at home; unrefrigerated, oil-based foods (e.g., potatoes or garlic stored in oil); and uncooked homemade meat products such as sausage, which provide the anaerobic environment conducive for the germination and growth of clostridia. A second form of this disease, called wound botulism, is caused by the infection of cuts, sores, or other open wounds with spores of *C. botulinum*. Finally, there is infant botulism, which is seen mainly in infants up to one year of age. Contaminated foods, especially wild honey, have been implicated, but thus far, specific sources have not been confirmed.

There is an initial phase of nausea, vomiting, and diarrhea associated with food-borne botulism, which is typically absent from the wound infections. Wound botulism also has a longer incubation period—4–14 days as opposed to 8–36 hours—but the major clinical symptoms of all forms of botulism are similar and correspond to the absorption of the toxin into the bloodstream. The neurotoxin causes a descending paralysis beginning with a distortion of vision due to interference with the ocular muscles, progressing to the gut and throat and further below and outward to the extremities. Death is usually the result of the paralysis of the respiratory muscles.

Botulism therapy is aimed at neutralizing circulating toxin with antitoxin, and supportive therapy to help a patient with breathing and other functions. Antimicrobial therapy may be necessary, especially in the case of wound botulism. Although there is no antidote for wound toxin, it is possible to restore muscular functions and nerve-muscle communication by regenerating toxin-free receptor molecules in the nerve endings. Precautionary measures against botulism food poisoning include careful handling of foods, such as proper heating of canned food, which inactivates toxin, and prompt refrigeration of unused food, which discourages the germination of spores. *See also* CLOSTRIDIUM BOTULINUM.

bread

This basic staple in most of our diets is a product of microbial metabolism—most commonly due to the action of the yeast *Saccharomyces cerevisiae* on the starches present in wheat flour. In contrast to the type of fermentation that occurs in beverages such as beer and wine, the starches and sugars in this case are broken down completely so as to form carbon dioxide, which causes the leavened dough to rise. During the baking process, the gases expand and escape from the dough, leaving behind the empty spaces that give baked bread its characteristic spongy appearance. In addition to the yeasts, certain cultures (particularly in sourdough bread) also contain bacteria—e.g., species of *Lactobacillus*—which perform acid fermentation and impart a characteristic sour taste to the bread. *See also* FERMENTATION; LACTIC ACID FERMENTATION.

Brucella

Commonly found in nature as intracellular parasites of various farm animals, *Brucella* has had grave economic consequences for farmers and ranchers. While they often live in their animal hosts without any apparent effects, different species of *Brucella* are known to cause a number of serious problems including abortions, mastitis, lameness, and skin abscesses in their hosts. The occurrence and severity of disease is especially marked in cattle. As a rule, *Brucella* does not parasitize humans, but some species have been known to cause zoonotic infections known as brucelloses. Human infection occurs through the consumption of milk from infected cattle or through the handling of infected carcasses. Disease symptoms are fairly nonspecific and include recurrent fevers, shaking chills, weakness, malaise, and weight loss. Onset may be either acute or gradual, and initial diagnosis is largely dependent on clinical suspicion and recent history of exposure to animals. During the course of the infection, organisms may be found circulating in the blood of infected patients. The best available treatment is antibiotic therapy with broad-spectrum drugs such as gentamycin or tetracycline.

The brucellae are tiny rods that have the cell wall composition and biochemical characteristics of gram-negative bacteria. However these bacteria are so small that they often fail to retain the safranin counter-stain during the gram-staining process and are better visualized with a vital dye such as carbolfuchsin. They are fastidious and require complex media such as blood agar for growing in the laboratory. By and large, these bacteria are aerobic, non-motile, and do not form spores. This last property renders them susceptible to pasteurization, which is one of the most effective means of preventing the occurrence and spread of milk-borne brucellosis.

At least six different species of *Brucella* have been identified thus far. These may be distinguished on the basis of their primary animal hosts, although some overlap is known to occur. The species include *B. abortus* from cattle, *B. melitensis* from sheep and goats, *B. ovis* exclusively from sheep, *B. suis* from swine, *B. canis* from dogs, and *B. neotomae*, which is found residing in the desert wood rat. *B. melitensis* and *B. abortus* are the species most often found in human infections, although the canine and porcine (pig) species have also been isolated from human brucelloses on rare occasions. *See also* AGAR; PASTEURIZATION; ZOONOSIS.

budding

Form of cellular reproduction in which a daughter cell develops from the parent as a small bud or localized outgrowth and eventually detaches to form an independent new organism. This type of reproduction is seen mostly in yeasts like *Saccharomyces*, but also in some bacteria such as *Nitrobacter* species. *See also* CELL DIVISION; FUNGI; YEAST.

bunyavirus

This label may be applied to either the family of over 300 different enveloped RNA viruses responsible for a broad range of human diseases or to a single type of virus within the family. Among the diseases that bunyaviruses produce are various hemorrhagic fevers, (e.g., Rift Valley fever, Sandfly fever, and California encephalitis) and other febrile diseases involving joint and muscle pains, (e.g., oropouche fever). Members of this family are classified into five genera on the basis of host preference and pathogenesis, and with the exception of the rodent-borne hantavirus, all are arboviruses, i.e., they are found in various arthropods such as mosquitoes, ticks, and sandflies. All bunyaviruses have a distinctive genome structure consisting of three unequal segments, with the smallest piece encoding the nucleocapsid protein, the medium piece encoding the envelope glycoproteins, and the largest segment producing the viral enzyme for replication. The genus bunyavirus consists of over 150 individuals that are typically found in mosquitoes or gnats. *See also* ARBO-VIRUS; ENCEPHALITIS; HEMORRHAGIC FEVERS.

Burkitt lymphoma

A tumor of B-lymphocytes, first described in 1958, frequently associated with symptoms of jaw involvement and usually seen among African children. Burkitt lymphoma has been found linked to Epstein-Barr virus (EBV) infections in a significant majority of the cases, but the link is not as strong as between certain other viruses and diseases, such as the Varicella-Zoster virus with chicken pox or the poliovirus with poliomyelitis. This is because the tumor is not a direct result of viral metabolism or the interference of the virus with cellular activity, but rather due to a specific gene translocation that it induces. Other agents may also induce this same translocation, which eventually results in the activation of a cancer-causing gene called c-*myc*. Both EBV and non-E-associated lymphomas are seen to occur with a greater frequency in immuno-suppressed or compromised individuals and are hyperendemic in regions with a high rate of malaria, such as tropical Africa and Papua New Guinea. Burkitt lymphoma is also seen in many AIDS patients, with about a 30 percent association with the Epstein-Barr virus.

See also CANCER; EPSTEIN-BARR VIRUS.

butanediol fermentation

Fermentation reactions carried out by bacteria such as *Klebsiella, Erwinia,* and *Enterobacter,* in which a proportion of the pyruvic acid from glycolysis is converted by enzymatic means to butanediol (chemical formula—CH_3–$[CHOH]_2$–CH_3) via such intermediates as acetolactic acid and acetoin, which may also accumulate in certain cases. These by-products are desirable in a variety of dairy products, including butter, cheeses, and buttermilk. The relative quantities of various end products and intermediates depend on the growth conditions. *See also* FERMEN-TATION; GLYCOLYSIS.

butter

Dairy product consisting mainly of milk fat separated from other milk components. Commercially, butter is prepared by inoculating raw or pasteurized cream with starter cultures consisting of *Streptococcus cremoris* or *S. lactis,* along with heterolactic fermenters such as *S. diacetilactis* and *Leuconostoc cremoris.* During fermentation, the lactococci produce lactic acid, while the hetero-fermenters produce diacetyl, which is responsible for the buttery flavor and aroma. When the cream is sufficiently acidified, butter is separated from the rest of the milk by a process called churning. Butter made without using cultures is called sweet cream butter and typically does not last as long, at least in homemade preparations. *See also* LACTIC ACID FERMENTATION; *Lactococcus.*

buttermilk

Dairy product made from soured skimmed milk. Perhaps the simplest form of buttermilk is the watery liquid that remains after churning butter away from soured cream. For commercial purposes, however, buttermilk is prepared by inoculating skimmed milk with a starter culture of *Lactobacillus bulgaricus* and waiting until it begins to turn sour and curdle. At that point, the curd is broken up into fine particles by agitation to form buttermilk. A similar product called acidophilus milk is made using starter cultures of *L. acidophilus*. Acidophilus milk is especially beneficial for health because it acts as a probiotic in the intestines and exerts antibacterial effects against enteric pathogens such as *E. coli* and *Enterobacter aerogenes*. *See also* LACTIC ACID FERMENTATION; *LACTOCOCCUS;* PROBIOTIC.

C

calicivirus

This family of RNA viruses includes such human pathogens as the hepatitis E virus and the Norwalk virus, as well as the etiologic agents of vesicular exanthema of swine and other veterinary diseases. The human caliciviruses appear to have some affinity for the gastrointestinal tract, as indicated by the nature of the viral diseases. The name for this virus is derived from the presence of cuplike (calyx-like) depressions on the surface of the virus particles. Each particle is composed of a naked (without envelope), regularly shaped capsid made of multiple units of a single species of polypeptide. The viral genome is a single-stranded, positive sense RNA molecule of about 7.5 kilobase length. Calicivirus replication takes place entirely in the cytoplasm of the host cell, where the naked genome serves as the mRNA to synthesize first the viral proteins, and then new genomes via a negative sense template. The assembly of the virion particles may be recognized microscopically by the formation of crystalline or part-crystalline arrays of capsid proteins associated with the cytoskeleton of the host cells. *See also* HEPATITIS E VIRUS; NORWALK VIRUS.

Campylobacter

According to both the Centers for Disease Control (CDC) and the Food and Drug Administration (FDA), this organism, which is found as a resident of the intestinal tracts of both symptomatic and healthy human beings and animals, is the leading cause of bacterial diarrhea in the United States. Virtually all human diseases—typically enteritis or gastroenteritis—are caused by a single species of this genus called *C. jejuni*. The bacteria gain entry into the body through consumption of contaminated water or food and virtually all symptoms are related to the intestinal tract, although it is not completely clear as to whether the symptoms are due to an enterotoxin or the proliferation of the bacteria in the intestine. Onset is typically 2–5 days after ingestion and is characterized by watery or sticky diarrhea, which may or may not contain RBCs and leukocytes. Other symptoms may include fever, abdominal pain, nausea, headaches, and muscular pain. Infections are for the most part self-limiting and do not require antibiotics, although erythromycin appears to decrease the infective period.

Campylobacter species are slender, fragile, spiral rods that are gram-negative in character. They are microaerophile, require about 3–5 percent of CO_2 for growth, and in fact may be killed by high concentrations of oxygen. They are also immensely susceptible to drying. Unlike most other bacteria, *Campylobacter* species do not use sugars but derive their energy and carbon from either amino acids or Kreb-cycle intermediates. They are motile bacteria with single flagella present at either one or both poles. The isolation of these organisms in the laboratory requires microaerophilic conditions as well as special antibiotic-containing media to inhibit the growth of other, more robust bacteria. *See also* DIARRHEA; FOOD-BORNE DISEASES; KREB CYCLE; WATER-BORNE INFECTION.

cancer

Disease—or rather, group of diseases—whose common characteristic is the uncontrolled growth and multiplication of cells in the body.

The main reason for including a discussion of cancer in a book about microorganisms is the connection between cancer and viruses. While cancers are not what we typically think of as infectious diseases—and are certainly not contagious like most viral diseases—there are certain viruses that are capable of entering the cells of multicellular organisms and inducing changes that ultimately result in cancer. As far as we know, viruses are the only living agents that have been positively linked to cancer in humans and animals. In addition, most of our current knowledge about the cellular mechanisms underlying the onset and progression of this disease stems from the advances in virology over the past half-century or so. Regardless of this connection between viruses and cancer, it is not possible to discuss the latter without information about its clinical presentation, diagnosis, and treatment, and thus, such information is also included.

Cancer has been aptly described as simultaneously a "single disease and a hundred diseases." As stated earlier, the single common feature is uncontrolled cell growth and multiplication. Normally a cell in any part of the body "knows" when to stop growing because of the other cells around it. This property is known as contact inhibition and is important for maintaining the size and shape of cells as they form different tissues and organs. Cancer cells lose control over contact inhibition, as well as other aspects of their normal development, leading to a rearrangement or disorganization of the cells and, even-

tually, a loss of function in the tissues or organs.

What causes a cell to deviate from its normal behavior? There is no single answer to this question, which is one of the reasons for the many faces of cancer. There are numerous possible agents that may trigger the cell to go wild—viruses, chemicals, and radiation, to name some of the most obvious. In addition to the multiplicity of causes, different cancers appear and behave differently according to their initial location in the body. Although a cancer cell looks and behaves differently from the tissue of origin, different cell types give rise to distinctly different cancers. For example, breast, liver, and skin cancers usually originate in the epithelial tissue and are called carcinomas. The appearance and characteristics of the cancer cells may be different depending on whether the tissue was secretory as in the case of the epithelium in breast and liver tissues, or merely protective (as in the case of skin). Other types of cancers include the sarcomas, which arise from connective tissue; blastomas in nervous tissue; and cancers of the blood-forming cells such as leukemias, lymphomas, and myelomas. The term "tumor" is used to describe cancers of solid tissues such carcinomas, sarcomas, and certain blastomas. At the earliest stages of a cancer, only a few cells are affected, which is often difficult to detect, but as the disease progresses, the differences between normal cells and cancer cells become clearer and are easily detectable, even under an ordinary light microscope.

The formation of a cancer cell from a normal one results in a mass of undifferentiated cells in the spot where the cell occurs. This process is called neoplasia (for "new growth"), and the resultant mass of tissue is known as a neoplasm or tumor. The simplest known example of neoplasia is probably the formation of warts or polyps. These are self-limiting tumors—triggered by viruses—that cause minor irritation but do not, in general, pose great health risks, except perhaps as sites for the development of more serious neoplasias. Virtually all solid-tissue cancers originate in localized pockets of neoplasia, which may be either self-limiting, forming what is called benign growths, or may become malignant and continue to grow unchecked. Malignant neoplasia causes the destruction of the surrounding normal tissue, which in turn proceeds to shut down proper functioning of the organ where this occurs. In addition, cells from the original site of cancerous growth may break off and spread to other parts of the body, where they implant themselves and cause further damage. The spreading of cancer is termed metastasis.

There are a number of avenues for metastatic cells to travel in the body, the most obvious being the blood and the lymphatic system. Some tumors may spread simply by physical movement from one site to another, e.g., the movement of neoplastic cells from one site in the gastrointestinal tract to another. In addition, there have been instances of cancer cells being transferred from one site to another in the body via a surgeon's knife.

Pathologists can detect the differences between benign and malignant tissue via microscopic examination. Malignant tumors tend to contain a large number of poorly differentiated or undifferentiated cells, which means that the cells do not particularly resemble the normal tissues they came from. Benign cancers and those in the early stages (which may develop into malignant tumors later) often contain well-differentiated cells whose origins are easy to discern. Usually malignant tumors are more aggressive than benign growths; they grow faster and spread more rapidly.

For clinical purposes, physicians have developed a standard staging system to describe a cancerous lesion. Staging is a useful index for identifying the extent of a tumor, estimating the prognosis for a condition, and making decisions with respect to appropriate treatments. In addition, it provides a uniform system by which doctors all over the world can compare diagnoses and treatments for a specific stage of a cancer. The standardized staging system, called the TNM system, is based on three main criteria—the size of the tumor (T); the degree of spread to the lymph nodes (N); and the presence of metastasis (M). Each of these criteria is followed by a number to indicate degree of size or spread. Thus a breast cancer lesion classified as a T2 N1 M0 tumor indicates a large growth (1 inch diameter) in the breast with some involvement of the nearby lymph nodes (in the armpit) but with no observable metastasis. For a more extensive description of staging, readers are referred to publications of the National Cancer Institute and the American Cancer Society on specific types of cancers and their treatment and prognosis.

By definition, cancer is a disease of multicellular organisms. The basic cause is a disruption of the mechanisms that control the cell cycle—i.e., the growth, development, and replication of a cell. The cell cycle is governed by a complex process involving several factors both outside and inside the cell that enable the cell to sense changes in its environment and behave accordingly. Physical contact with a neighboring cell is one example of an external factor or signal for a cell to stop growing. Other stimuli, such as hormones, may serve as signals for the cell to start growth. Since the actual site of growth in a cell is the nucleus, where protein synthesis is initiated, the information about the external contact must somehow be conveyed to the appropriate genes.

The transmission of such messages occurs via an extensive communications network that relays the message across the cell membrane and through the cytoplasm and nuclear membrane. These networks, also known as the signal transduction pathways, involve multiple players such as proteins and hormones and must be extremely well regulated to ensure the proper maintenance of the cell cycle. Most of these components have active and inactive forms—like a switch that goes on or off—which enables them to start or stop their activities. A mutation in any of these signal-transduction components has the potential to lead to cancer. The cancer-causing mutations may be induced by a number of different agents or mutagens. Chemicals and X-rays are examples of such agents, as are some viruses. Viruses capable of inducing cancer are said to be oncogenic.

In addition to losing control over their growth cycle and their ability to differentiate into mature cells, cancer cells also change with respect to how they are seen or perceived by other cells in the body. This is particularly important with respect to the immune system, which is the body's way of defending itself from harmful organisms and threats from the environment. The mutations in cancer cells cause them to produce certain proteins or antigens that are not recognized by the immune system, which then tries to mount an attack against the cancer cells. In fact, this immune mechanism may serve as a built-in system to eliminate many potential neoplasms in their early stages, but when the growth is too fast, or when the cells develop some system to evade the immune system, then the cancer takes a hold in the body.

Like most illnesses, cancer is best treated at its earliest stages. Unfortunately, however, a neoplasm is often unnoticed for several months because the growth is too small to be detected or in a place that is inaccessible in routine checkups. New technologies are allowing for earlier detection, particularly in people who are identified as part of a high-risk group for a particular cancer—e.g., people with unusually large numbers of moles and a tendency to sunburn easily are considered to be at a higher risk for melanoma or skin cancer, and people who smoke "regularly" are more likely to develop lung cancer. Once cancer is suspected—regardless of the trigger—the investigation follows a more-or-less set pattern, and the patient is run through a panel of tests, including physical examinations, biochemical and other blood tests, imaging of suspect organs, and a histological examination of tissues.

A routine physical may serve as the first tip-off of the development of a cancer. Physicians familiar with their patient's medical history are likely to notice deviations from the normal pattern. Nowadays, yearly examinations include certain types of screenings—common examples are breast examination in women; the annual Pap smear to detect changes in the cervix; changes in the size of the prostate gland in men; and, as mentioned earlier, obvious changes in the appearance or size of moles, warts, and other skin lesions. Certain nonspecific blood tests performed in the course of a routine checkup may also be indicators of underlying problems—for instance, abnormalities in the blood count or the functioning of different glands such as the liver or spleen, as indicated by biochemical tests. When accompanied by physical symptoms such as bleeding, cramps, or pain in specific organs, these tests may serve as clues as to what may be wrong. In addition to these tests, there are also blood tests for detecting tumors on the basis of specific properties such as antigenicity. Besides blood tests, urine and stool examinations may also provide clues about developing tumors.

A neoplasm in any organ or tissue that has developed beyond the initial stages may be seen by various imaging techniques including X-rays, ultrasound (in the area of the abdomen), and nuclear imaging processes. These different methods are used to visualize a tumor and help doctors assess its size and the degree to which it has spread. Mammography is the specific term for imaging techniques applied to the breast tissue and is recommended as part of the examination process in screening for breast cancer. Special devices such as endoscopes and bronchoscopes enable doctors to observe the inside of an organ such as the intestine or lung without having to resort to major surgery.

The final and perhaps the most definitive diagnostic method for cancer is the microscopic examination of a piece of the suspect tissue (called a biopsy) to directly observe the changes that have taken place. There are two main types of biopsies possible—incisional biopsies, which involve taking only a small piece of the growth and stitching the area closed; and excisional biopsies, which entail removing an entire growth, as is commonly done in the case of moles. In some cases, physicians use a fine needle to aspirate or suck out a few cells from the suspect tissue. The advantage of this technique is that it is minimally invasive. However, the major drawback is that the cells cannot be viewed in the context of the surrounding tissue.

The traditional treatment of cancer includes three main approaches—surgery, chemotherapy, and radiation—the specific therapy or combination of therapies depending on the type of cancer and extent of its spread. In the past few decades, advances made in the general understanding of cancer have added another player—biological therapy—to the arsenal, greatly improving the survival, longevity, and quality of life of cancer patients. Nevertheless, the war on cancer is far from over, and to date there is no sure-fire scheme to eliminate the disease.

Perhaps the oldest of the treatments, surgery, may still be the most successful method to eliminate a tumor. This may take the form of an incisive biopsy that removes only part of the growth, or it may entail the complete excision of the neoplasm. However, it is only possible to perform the latter if the tumor is localized and in a nonvital location—if the tumor is localized in a single kidney or a part of the lung, for instance, it may be possible to remove that section of tissue completely. However, it is only possible to perform the latter if the tumor is localized and is present in a non-vital location. If the tumor is contained within a single kidney or a part of a lung, for instance, it may be possible to remove the organ or a section of the tissue completely. An organ such as the brain or heart, on the other hand, cannot be removed. Leukemias and other blood-related cancers, as well as cancers of diffuse systems such as the islet cells (in the pancreas), cannot be treated with surgery. Yet another disadvantage is that surgery is only possible after a tumor has grown beyond a certain size, which means that it does not limit the condition when the disease is still at its most treatable stage. Often a surgeon may remove a growth only to find, after some months, that there is more growth in new locations because the initial tumor had begun to metastasize before it was removed or removed or even detected.

A second option is the use of radiation, such as gamma rays or radioactive substances, to shrink a tumor. The biological basis for this approach is that actively dividing cells, such as those in cancerous tissues, are more susceptible to the damaging effects of radiation than resting cells. The major drawback of this approach is that the cells in the rest of the body are by no means inactive, which makes them susceptible to radiation damage as well. Thus, radiation therapy often has severe side effects and usually needs to be augmented with additional treatment to reconstitute the damage done to the body. The use of drugs or chemotherapy often has the same advantages and drawbacks as radiation, since these, too, are primarily targeted at growing, rather than quiescent, cells.

Biological therapy is usually used in conjunction with radiation or chemotherapy to make up for a deficiency caused by the damage to vital functions, induced by those treatments. For example, the treatment may consist of a growth factor to stimulate the growth of normal cells that are damaged during the course of radiotherapy. Alternatively, some biological therapeutic agents may serve as enhancers of the immune system, stimulating it to mount an attack on cancer cells. The development of various biological strategies against cancer is perhaps the most active area of research in cancer therapy today.

The idea that viruses could cause cancer was first suggested in the early part of the twentieth century by a medical researcher named Peyton Rous, when a worried poultry farmer brought a hen with an unusual tumor-like growth to his attention. Rous began to investigate the nature of this tumor tissue in the laboratory. To his surprise, he found that it was possible to transmit the sarcoma—i.e., cause it in other birds of the same species—simply by transferring an extract from the tumor tissue to uninfected birds. Upon further probing, he discovered that the tumor-causing agent was filterable and remained active even when the tumor tissue was treated under conditions designed to break all intact cells. Rous first published his results in 1911, but his findings were largely ignored for some decades. The virus he discovered was later named the Rous sarcoma virus (RSV) in his honor, and eventually, in 1966, he received a Nobel Prize for his research.

Scientists started to more seriosly consider viruses as possible cancer agents after researchers began to find links between viruses and mammalian cancers. But even after they accepted that viruses could induce certain types of tumors, questions remained. How could such a small entity—containing only a few genes—wreak such havoc in the workings of a machine as complex as a cell?

The answers to this question and others like these evidently lay in the genes of the cancer-inducing viruses, since it was the viral nucleic acid that infected the cells and made them go wild. However, such questions could only be addressed after scientists developed the means to culture these viruses in the laboratory outside of living hosts, and to assay them. In 1958, researchers Howard Temin and Harry Rubin found a way to do precisely that. They found that when certain types of cell cultures were infected with tumor viruses, they underwent characteristic changes in their growth and morphology. This cellular "transformation" could then be detected under the microscope. Furthermore, the number of such regions of transformation depended on the amount of virus used. Thus they had an easily detectable assay system for the tumorigenic viruses.

Temin and Rubin's assay system enabled scientists to delve into the mechanisms that underlie cancer. The special nature of viruses as cellular "hijackers" made them more attractive for laboratory use than chemicals and radiation. The latter cause mutations by direct interaction with the cell's DNA, inducing a change in the sequence of a gene. Since the human genome is very large, the chances of a chemical or X-ray targeting the same gene, much less the same spot on a gene, with every exposure of the cells to the agent is very rare. On the other hand, because the mutation-inducing agent in viruses is a nucleic acid, the outcome of introducing it into a cell was largely predictable. For instance, some viruses have the ability to integrate their DNA into the cell's genome. Because this typically occurs in a sequence-specific manner, the virus would tend to integrate at the same location each time, thus disrupting the structure and function of the

same gene every time. Another mechanism by which viruses induce tumors is by bringing along a mutant version of a normal cell-cycle gene as part of their genome. This "stowaway" then induces cell transformation. Again, specific viruses bring along specific genes. Thus, the virus-induced mutations were more trackable.

Nearly all the oncogenic viruses discovered are known to exert their effects via the mutagenesis of a signal transduction component, or alternatively, a gene or protein that interacts with such components for regulating their expression or activity. For example, the Rous sarcoma virus induces tumors because it carries a mutated gene for one of the signal transduction proteins. As scientists discovered, the mutation had resulted in the malfunctioning of an internal switch or brake that the normal protein contained to keep its activity in control. Thus, the mutant protein in an RSV-infected cell is permanently "on," resulting in a constant signal to the cells to grow and multiply. *See also* AVIAN SARCOMA VIRUS; BURKITT LYMPHOMA; CHEMOTHERAPY; EPSTEIN-BARR VIRUS; HERPESVIRUS; HUMAN T CELL LEUKEMIA VIRUS; KAPOSI'S SARCOMA; ONCOGENIC VIRUSES; ROUS, PEYTON; TRANSFORMATION (CELLULAR); VIRUS.

Candida

This fungus (yeast) is the culprit in such common conditions as vaginal yeast infections and thrush, a superficial throat infection of children. Candida is found widely in the environment, often as part of the normal surface microflora of many animals, although not necessarily humans. Despite the fact that this genus requires only dead organic matter (i.e., it is saprophytic in nature), many species have an obligate requirement for animal hosts and cannot establish stable populations in other habitats. With regards to humans, various *Candida* species are opportunistic pathogens, capable of inducing a number of different nonspecific superficial or systemic infections (candidiasis). Infection is presumably by simple contact or inhalation of fungal cells in the air, since *Candida* does not form spores. Whether or not a disease takes hold depends on the initial dose of the yeast cells as well as the general immune status of the exposed individual. In certain rare instances, children may develop a delayed-type hypersensitivity response to the yeast antigens and develop a condition known as candidal granuloma, characterized by the appearance of crusted papules of up to 2 cm diameter, which may also form horn-like protrusions. As this is an immune reaction, these lesions are not infective and they cannot be treated with the topical medications effective against other *Candida* infections, such as vaginal yeast infections.

Although it does not form spores, *Candida* is dimorphic in nature and may exist either as single cells or as mycelia with flat, horizontal hyphae. Both forms reproduce by budding and there are no sexual forms in the life cycle. The most common species is *C. albicans*. Other species commonly encountered in clinical situations include *C. tropicalis, C. krusei, C. parapsilosis,* and *C. glabarata* (which has also been named as a separate genus, *Torulopsis glabarata*). *See also* BUDDING; HYPERSENSITIVITY; SAPROPHYTE; THRUSH; VAGINAL YEAST INFECTION; YEAST.

canning

Method of food preservation, used primarily for fruits and vegetables but also for meat, in which the food is first processed by blanching, cooking, or pickling in brine, vinegar, oil, or (in the case of fruit) in sugar syrup, and subsequently sealed hermetically within sterilized containers. Canning increases the shelf life of many foods by months or even years because in addition to maintaining the food within a sterile environment, it also prevents any significant changes in the physical, chemical, and organoleptic qualities of the food product after it is canned—until it is reopened. *See also* FOOD PRESERVATION; FOOD SPOILAGE.

capsid

The protein coat of a virus particle which, together with the nucleic acid, forms the viral core. *See also* VIRION; VIRUS.

capsule

Outer sheath of a cell, situated on the outer surface of the cell wall, found in various species of bacteria. The main function of this structure is to protect the organism from various threats in its environment such as desiccation, toxins, bacteriophages, and phagocytes. In some species, the capsule serves as a nutrient reserve, while in others it facilitates functions such as adhesion to surfaces. In addition, the capsule also plays a role in the virulence of pathogenic bacteria, as evidenced by the fact that capsule-less strains (called R strains) of bacterial species that normally possess capsules are avirulent.

Chemically, bacterial capsules are made up of gummy or starchy polysaccharide complexes. (*Bacillus anthracis*, which has a protein capsule, is an exception). Thus, capsules do not bind to standard bacterial stains such as crystal violet, which are proteins. To visualize a capsule, the cells are treated with a protein mordant, such as milk, which coats the cells and demarcates their outline, prior to staining in a normal fashion. The capsule itself may be visualized by staining with a solution of copper sulfate.

In addition to the myriad uses to the organism itself, the chemical makeup of the capsule can be used to differentiate between closely related species or among strains of the same species. Differences arise from variations in the sugar composition of the capsule, which results in changes in the antigenic specificity. Bacteria that are identical in all other respects may have different capsular antigens, giving rise to multiple serotypes within a species. An interesting aside about capsules is that the discovery of DNA came about from a study of capsular antigens of the organism *Streptococcus pneumonii*, the cause of serious pneumonia infections during World War II. *See also* BACTERIA; NEGATIVE STAINING; TRANSFORMATION (BACTERIAL); STAINING.

carbon cycle

The process, or series of processes, through which carbon atoms are cycled through living and nonliving matter on earth. Although invisible to our eyes, microbes form one of the most important players in this cycle, participating in nearly all the different types of conversions from inorganic to organic carbon and back again. The main source from which inorganic carbon is incorporated into the biomass (living things) is the atmosphere. Various photosynthetic beings, including the cyanobacteria, algae, and green plants, assimilate carbon dioxide and convert it into different organic molecules, for example, into sugars, proteins, and nucleic acids. While most animals cannot use carbon dioxide directly, they use molecules produced by plants or microbes, in particular the sugars, to derive both energy and carbon for their own macromolecules. The organic carbon in these life forms is returned to the atmosphere, either directly by activities such as respiration or via a series of metabolic reactions on the part of different heterotrophic microbes. Various human activities, such as the burning of coal and fossil fuels (which are themselves the products of accumulations of dead organic material over centuries), also play a significant and ever-increasing role in furnishing carbon to the atmosphere. *See also* ASSIMILATION; PHOTOSYNTHESIS; RESPIRATION.

cardiovirus

Group or genus of picornaviruses that are similar in most respects to the enteroviruses, except in their affinity for cardiac or heart muscle tissue. *See also* PICORNAVIRUS.

carrier

Person who is infected with a pathogenic organism—bacterium, virus, or protozoan—without manifesting any symptoms of the disease, but who is capable of transmitting the organisms and disease to another susceptible host. Carriers who are asymptomatic for the duration of the infection are called healthy or asymptomatic carriers. People in the asymptomatic, incubation, or convalescent phases of a disease may also serve as carriers. The carrier state might be temporary—as is usually the case during the incu-

bation period of a disease—or extend for long periods of time. A famous story concerning a healthy carrier is that of "Typhoid Mary," a woman named Mary Mallon (1869–1938) who worked as a cook at a New York City hospital and transmitted the *Salmonella* organisms to patients and coworkers while never exhibiting any signs of typhoid fever herself. *See also* TYPHOID MARY.

case fatality rate

An epidemiological measure of the number of deaths that occurred due to a specific disease, in a given population, during the course of an outbreak of that disease. Mathematically, case fatality rate is expressed as the percentage of people diagnosed with the disease in question who died within the period of the outbreak. Data on case fatality rates may be used to estimate mortality rates for a given disease or population, but the two concepts should not be confused, because the former is specific to a given incident. *See also* EPIDEMIOLOGY.

cat scratch disease

This bacterial infection, caused by *Bartonella henselae*, was first recognized in the 1930s in both the United States and France as a disease associated with prolonged exposure to cats. The typical course of the disease was the development of a hard, red papule (nonpus forming lesion) at the site of a cat scratch or bite, followed in 1–7 weeks by systemic symptoms of fever, malaise, and lymphatic tenderness. Although infected cats show no outward signs of disease, the infecting bacteria can be found circulating in their bloodstream and may be transferred to humans via scratches or bites. Cat fleas have also been implicated in transferring infections from the felines to humans. Cat scratch disease is typically self-limiting and does not require treatment. However, diagnosis is necessary because its symptoms are similar to other bacterial infections such as brucellosis and tularemia that do require treatment. Serological tests to detect antibodies to the organism are the most reliable indicators of infection. *See also* BARTONELLA.

catalase

Enzyme that catalyzes the breakdown of hydrogen peroxide (H_2O_2) to water and gaseous oxygen. It is produced by many aerobic and aerotolerant bacteria to increase the amount of oxygen in their immediate surroundings. The presence or absence of this enzyme is often used as a parameter for bacterial classification.

Caulobacter

This organism is unique among the bacteria both in cellular structure and in mode of cell division/replication. Unlike most other bacteria, which are free to move about either actively (using flagella or cilia) or passively (due to Brownian motion), *Caulobacter* is a stationary organism, consisting of a non-motile cell that is anchored to substrate by a short, stalk-like outgrowth. The free end is sometimes sparsely flagellated. Like other bacteria, the basic mechanism of cell division in this organism is binary fission (namely, the splitting of a single parent into two progeny cells), but unlike the fission of other bacteria, *Caulobacter* cell division is asymmetrical. Shortly before the division actually takes place, the bacteria distributes its cellular components in an unequal manner, and, after DNA replication (which is normal and symmetrical), the bacteria splits to give rise to two distinct types of progeny cells. One of these cells remains attached to the stalk and is a replica of the immobile (sessile) parent cell. The other cell, called a swarm cell, is a heavily flagellated motile entity with no stalk. Although it contains the full complement of *Caulobacter* DNA, the swarm cell is incapable of cell division. It swims around for awhile and eventually settles down on a solid substrate, loses most of its flagella, and forms a stalk near one of its poles. After this step in its differentiation, the organism is capable of a new round of fission. *Caulobacter* is freely distributed in nature in soil and water, but it has never been implicated in any human con-

ditions. It is gram-negative, strictly aerobic, and has a heterotrophic mode of nutrition. *See also* BINARY FISSION; CELL DIVISION.

CDC. *See* CENTERS FOR DISEASE CONTROL AND PREVENTION

cell

The smallest individual entity capable of sustaining life by itself. In this context, life may be defined in terms of two components: first, a capacity to exist independently and sustain oneself using energy from one's surroundings; and second, the ability to create another entity, identical or nearly identical to oneself, which in turn is capable of sustaining itself. Thus, a cell may be considered as the structural and functional unit of life.

All living things are made up of one or several cells. The only exceptions are the viruses and prions, which are not considered as true living beings. Many different types of cells vary widely with respect to their size, shape, and function. Basically, a cell contains a set of instructions for living (in the form of DNA), along with the molecules (usually proteins) that perform these functions, encased within a finite boundary provided by a membrane. Different structural components are made up of various molecules, including proteins, carbohydrates, lipids, or complexes of more than one of these molecules. The structure and organization of various cellular components is one of the most important criteria for classifying living organisms. *See also* CELL MEMBRANE; CYTOPLASM; NUCLEUS.

cell division

The process by which a single cell (called the parent) replicates its genetic material and then physically splits into progeny cells containing all the genetic information of the parent. In most cases, a single cycle of cell division gives rise to two progeny or daughter cells. In the case of single-celled organisms, the progeny are identical or nearly identical to the parent, with differences (if any) arising through mutations in the genome, which have occurred during replication. In multicellular organisms, dividing cells do not always produce identical progeny, especially during the developmental stages of the organism. The difference in appearance of parent and progeny cells is not a function of their genetic makeup (which remains unchanged) but, rather, due to differences in gene expression in the different cells.

There are two modes of cell division. The simple duplication of DNA and its equal distribution to the progeny is termed mitosis. The second mechanism of division, called meiosis, takes place only when a cell undergoes a transition from the diploid to the haploid state; this transition typically occurs just before an organism is about to enter the sexual phase of its life cycle. Meiotic cell division involves the separation of chromosome pairs and reduction of the total number of chromosomes in a cell by half. Each progeny may receive chromosomes from either parent, whose nuclei would have fused when the cell went from a haploid to a diploid phase. *See also* BINARY FISSION; BUDDING; DIPLOID; DNA; HAPLOID; SCHIZOGONY.

cell membrane

Structure that provides the physical external boundary of a cell. The main function of the cell membrane is to preserve the integrity and individuality of the cell and at the same time enable communication between the interior of the cell and its external surroundings. Thus, it needs to be flexible and semipermeable—which means that it allows only certain materials to enter and leave the cell. Chemically, the membrane is made up of a double layer of phospholipid molecules, in which different proteins are embedded. The lipid bilayer, which is hydrophobic in nature, provides a fluid but largely impenetrable boundary between the cell and the environment, while the membrane proteins perform functions such as permeability and communication. *See also* CELL.

cell wall

Outer covering of a cell, situated outside the cell membrane. This structure is present in bacteria as well as in eukaryotic organisms such as fungi, algae, and plants, but not in the smaller prokaryotes like the mycoplasmas or in protozoa or higher animals. The cell wall gives an organism its characteristic shape and serves as protection against the external environment. It is especially important in single-celled organisms because it protects a cell from swelling and ultimately bursting due to excessive osmotic pressure built up by the diffusion of fluid into the cell.

The bacterial cell wall warrants special mention because its structure is the basis of the Gram stain, which is one of the most important criteria in the taxonomy of these organisms. The primary component of the bacterial cell wall is a polymer called peptidoglycan, although some species may also contain other chemicals that endow the cell with additional physical and chemical properties, e.g., mycolic acids in mycobacteria. *See also* CELL; GRAM STAIN.

cellulase

Enzyme, or rather a class of enzymes, capable of degrading cellulose, the polysaccharide that makes up most of the structural matter of higher plants. This enzyme is produced by a number of microbes in nature, predominantly in the soil and in grazing animals such as cows and sheep. Examples include species of *Bacillus* such as *B. subtilis* and *B. licheniformis*, which are often found degrading cellulose in compost heaps, and species of *Ruminococcus,* which are found in the rumen of cattle.

Cellulomonas

Gram-positive, non-spore-forming bacteria classified in *Bergey's Manual* under the category of the irregular gram-positives. A special feature is their ability to degrade cellulose by the production of extracellular cellulases, which break down cellulose to smaller monosaccharide and disaccharide units.

These bacteria also degrade starch. *Cellulomonas* species are found in soil and occasionally in the rumen of certain animals. They exhibit a wide range of morphologies—rods, cocci, or filaments—may be either aerobic or facultatively anaerobic, and capable of both respiration and fermentation. The type species is *C. flavigena*.

Centers for Disease Control and Prevention (CDC)

Better known as the CDC, this federal agency is the branch of the U.S. Department of Health and Human Services responsible for monitoring and administrating public health in the nation. The main mission of this agency is "to promote health and quality of life by preventing and controlling disease, injury, and disability." To this end, the CDC maintains 11 separate institutes and offices and employs nearly 8,000 people in various capacities. Different people conduct both laboratory and field research on different diseases and public health issues, analyze the data obtained, and disseminate it to the public. The CDC works closely with state departments of health and other governmental and non-governmental agencies to conduct its duties in the most efficient manner possible. Its headquarters are in Atlanta, Georgia. For more information about specific projects, publications, and other aspects of the CDC, visit the CDC Web site at <http://www.cdc.gov>.

centrifugation

Technique of spinning fluids at high velocities (i.e., subjecting them to high centrifugal force), usually under a vacuum, so as to separate various dissolved and particulate components by their size and density. The separation occurs because particles of different sizes and densities settle down at different rates when subjected to centrifugal force. Bacteria, for instance, will settle at the bottom of a tube if centrifuged for 10 minutes at 300 revolutions per minute. Centrifugation can be adapted for various purposes, e.g., ultra-centrifugation and density gradient centrifugation. *See also* DENSITY GRADIENT CENTRIFUGATION, ULTRA-CENTRIFUGATION.

cephalosporin

Broad-spectrum antibiotic containing β-lactam groups, first isolated from a species of Mediterranean fungus called *Cephalosporium acremonium*. *See also* ANTOBIOTIC; BETA LACTAM ANTIBIOTICS.

Chagas disease

Zoonotic disease caused by the protozoan *Trypanosoma cruzi*, also known as American trypanosomiasis. Chagas disease is seen only in the New World, mainly in Latin America, where an estimated 100 million people are at risk of contracting infections. The number of actual cases is in the vicinity of 16–18 million. In North America, Chagas disease is seen primarily in Texas and California. The disease is most severe in children under the age of 5 and tends to occur in a milder, chronic form in older children and adults.

Chagas disease is transmitted to humans when the feces of the insect host—kissing bugs or *Triatoma*—is rubbed into bite wounds or exposed mucus membranes. Animal hosts that serve as primary sources of the trypanosomes include household pets such as cats and dogs. In addition to insects, infections may occur through contaminated blood transfusions, via the placenta, or due to the accidental ingestion of infected bugs. A distinctive feature of these infections, in contrast to those of *T. brucei*, is that the multiplication of the non-motile amastigote form of the parasite, rather than the motile hemoflagellate form, is responsible for the principal disease symptoms. Soon after entering the wound of the victim—usually on the face but also on arms, legs, or other exposed surfaces—the trypomastigotes enter the local histiocytes either actively or by phagocytosis and proceed to invade the surrounding fat and muscle tissue wherein they form amastigotes. They also spread to the local lymph nodes, which become enlarged, hard, and tender. An inflammatory reaction at the site of initial infection may result in the development of a painful, subcutaneous nodule called a chagoma, which usually takes 2–3 months to subside. Often this original chagoma occurs in or near the eyes, resulting in conjunctivitis and edema of the eyelids, a symptom called Romaña's sign.

Meanwhile, the amastigotes multiply within the cells in and around the chagoma and rupture these cells to release new trypomastigotes and amastigotes, which travel to virtually all parts of the body via the bloodstream. Parasites show a preference for colonizing cells of the blood vessels, the muscles, (cardiac, skeletal, and smooth muscles), and the neuroglia, where they form lesions similar to the initial chagoma, with devastating consequences. The proliferation of amastigotes in the cardiac muscles, for instance, induces defects in the heart's ability to contract, which can lead to a fatal cardiac arrest. Circulating trypanosomes are not capable of dividing and hence diminish to smaller numbers later in the infection. Insects become infected by these forms.

In the acute form of Chagas disease, the trypanosomes spread rapidly in the body, leading to manifestations of cardiac arrest or other potentially fatal outcomes within a matter of a few weeks. However, individuals whose immune systems are slightly better equipped to deal with infections may recover to develop a chronic form of the disease, which is diagnosed more commonly than the acute type.

Early diagnosis is important for preventing fatalities in case of the acute Chagas disease and forestalling the establishment of chronic conditions. Medical histories play an important role in making definitive diagnoses. A history of consistent exposure to the bugs—due to residential conditions, travel, blood transfusions, or some other cause—is a good indicator of exposure to the parasite. Because the chagomas may often resemble other insect bites, they are not always quickly detected. Medical experts advise a differential diagnosis of this disease against a backdrop of brucellosis and toxoplasma infections, connective tissue diseases, tuberculosis, and leukemia. Only trained medical personnel should handle infectious blood and tissues because of the high infectivity of the parasites. In addition to blood examinations, serological tests are important because the low number of circulating parasites can give false negative reactions. As yet, there are no standardized drug therapies for Chagas disease, although a drug called Nifurtimox has yielded encouraging results in preliminary clinical tests, especially against the acute disease. No vaccines are available against this organism either. Control measures are largely aimed at reducing the insect populations so as to minimize opportunities for exposure. *See also* TRYPANSOMA CRUZI.

Chain, Ernst Boris (1906–1979)

Chemist who collaborated with Lord Howard Florey to purify, test, and standardize the efficacy of penicillin in animals and humans. He shared the 1945 Nobel Prize with Florey and Sir Alexander Fleming, who is credited with the discovery of the antibiotic as well as with the early testing of its potential use in animals. *See also* ANTIBIOTIC; PENICILLIN.

cheese

Dairy product made by the action of various bacteria or fungi on coagulated (curdled) milk. Over 400 different varieties of cheese are produced in different parts of the world. The varieties differ from one another in several respects, including the manufacturing process, the source of milk (from cows, sheep, goats, or other animals), the type of milk used (skimmed or whole), and the organisms used for fermentation. The end products, which have diverse tastes and textures, form the basis of a flourishing international industry.

The basic process of cheese-making involves two steps. First, milk is inoculated with a starter consisting of one or more lactic acid bacteria that, along with the enzyme renin (extracted from the stomach lining of cows), causes the milk to turn sour and curdle. A number of different bacteria may be used in this initial step. A common choice is *Streptococcus thermophilus* because of its ability to

Assorted cheeses. *Courtesy of the American Dairy Association.*

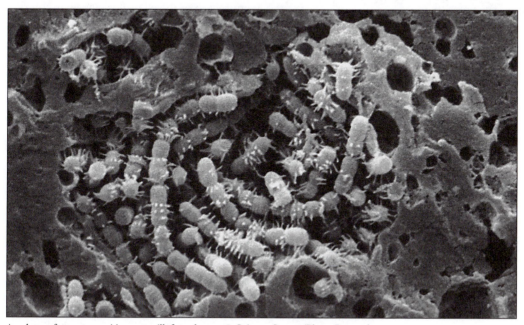

A colony of streptococci in goat milk feta cheese. © *Science Source/Photo Researchers.*

withstand heat and therefore stimulate curdling in cooked curd. The second step involves the processing or ripening of the curdled milk. Cream cheese and cottage cheese undergo the least amount of processing and typically do not involve any further microbial action. These are known as the unripened cheeses and do not have a long shelf life. During the ripening process, the proteins and carbohydrates of the curd (cheese) are further metabolized by various organisms present in the mixture. While the predominant actors are the lactic acid bacteria, a number of well known cheeses owe their peculiar nature—their taste and texture—to other specific organisms that grow during the ripening stage. The characteristic holey appearance and nutty flavor of Swiss cheese, for example, are due to gas and propionic acid produced during fermentation by *Propionibacterium shermanii*; the blue veins seen in Roquefort cheese are due to spores of the mold *Penicillium roqueforti*; and the soft Camembert cheese is made by inoculating the curd with spores of *Penicillium camemberti*.

A practical way of classifying ripened cheeses is by their moisture content and texture. Hard cheeses such as cheddar, romano, and edam, for example, have very low moisture content and are all made by ripening with bacterial cultures over a period of 2–16 months. Softer varieties, such as gouda and muenster, called semi-hard cheeses, require somewhat shorter (1 to 8 months) ripening periods. Soft cheeses, which have a high fat and moisture content, include varieties ripened by both mold (e.g., camembert and brie) and bacteria (e.g., limburger). *See also* FERMENTATION; LACTIC ACID FERMENTATION.

chemoautotroph

Free-living organism that derives energy by processing chemical sources in the environment. Most bacteria in this category are lithotrophic, and they use inorganic chemicals such as sulfur, ammonia, and hydrogen as their primary energy sources. Examples of such organisms, also known as chemolithoautotrophs, include nitrifying bacteria as well as archaebacteria such as *Thiobacillus* species and methanogens. *See also* AUTOTROPH.

chemoorganotroph

Chemotrophic organism that derives energy, carbon, or both from organic material in the surrounding environment. The chemo–organoheterotrophic bacteria are among the most abundant and widespread organisms on earth. They use a number of substrates—such as sugars or organic acids—in processes such as fermentation and respiration. *See also* AUTOTROPH.

chemotaxis

Specifically directed movement of a cell in response to a chemical stimulus. Microbes typically use chemotactic movements to move closer to nutrient sources or away from harmful chemicals. Chemotaxis may also be observed in higher organisms; for instance, the movement of tissue macrophages toward the site of infection is a direct response to specific chemical substances such as bacterial endotoxin or to cell wall or capsular components released at the site. In these organisms, chemotaxis functions are an important defense mechanism against infectious agents. *See also* PHAGOCYTOSIS.

Chemotherapy

The use of chemical substances, such as antibiotics, antidepressants, stimulants, and various other drugs, to control or treat disease. Technically, chemotherapy applies to all types of disorders, including infections, tumors, and adverse symptoms of nonspecific or unknown causes (headaches and fever, for instance). However, the term is also often used to represent chemical-based therapies for cancer, especially outside the medical profession. A related concept is chemoprophylaxis, which involves the administration or use of chemicals as a means of *preventing* the occurrence or transmission of a disease, rather than treating it after it has taken hold. *See also* CANCER.

chemotroph

Living organisms that derive their energy from chemical sources in the environment rather than from the sun's radiant energy. Except for the cyanobacteria and the purple photosynthetic bacteria, most prokaryotes, whether autotrophs or heterotrophs, are chemo–trophic. Among the eukaryotes, the fungi, most protozoa, and the entire animal kingdom fall in this category. *See also* PHOTOTROPH.

chickenpox

Highly contagious disease characterized by the formation of vesicular lesions on the skin as a result of a primary infection by the Varicella-Zoster virus (VZV). Although it is likely to have existed since antiquity, historical records on chickenpox are somewhat doubtful, because it was not differentiated from smallpox until nearly the twentieth century.

Chickenpox has a high rate of incidence, with over 95 percent of Americans exposed to infection before they reach adulthood. Chickenpox often occurs in epidemic-like outbreaks during the late winter and spring. It occurs predominantly in children under 10 years of age, in whom it causes a mild disease, while adults usually suffer from more severe symptoms and sequelae, even death. Infection is by means of inhalation. After an initial phase of replication in the mucus membranes of the mouth and throat, there is a transient viremia, followed by the dispersal of the virus to the skin. This is marked by the eruption of the characteristic itchy rash, beginning on the trunk and scalp and spreading outward to the extremities. The rash progresses to form vesicles, which become encrusted scabs over a period of days. A hallmark of this disease is the appearance of the rash in crops, so that multiple stages may be present at the same time. This is markedly different from smallpox, in which all lesions occurring at one time are in the same stage of advancement. Chickenpox patients are most infective in the period beginning 1–2 days before the appearance of the rash until some 5–6 days after the final crop. Such immunity that does develop appears to be for life, although chickenpox does not prevent shingles.

Treatment is seldom necessary for chickenpox, other than offering temporary relief for the itchiness. The only exceptions are immuno-suppressed individuals and newborns exposed *in utero* or at birth. The drug acyclovir is used to prevent dissemination of the virus, while some immunoglobulin therapy is available for infants of mothers who acquired the infection a week or so before or after delivery. A live-attenuated vaccine against VZV was approved in 1995 and is recommended for all children at 12–18 months of age, all susceptible children until the age of 13, and for adults who have not had chickenpox but are considered at risk. *See also* SHINGLES; VARICELLA-ZOSTER VIRUS.

Chlamydiae

Group of infectious organisms that, along with the rickettsiae, are the smallest entities that are still considered by scientists to be bonafide living creatures. Only the viruses (which are considered to exist at the threshold of life) and prions (whose existence is still disputed in some circles) are smaller. The chlamydiae are obligate intracellular parasites that are found throughout the animal kingdom. All organisms of this group belong to a single genus, *Chlamydia*, many species of which are important animal pathogens. Three species, *C. trachomatis*, *C. pneumoniae*, and *C. psittaci*, have been implicated in human disease and are discussed in individual entries.

Structurally, the chlamydiae are discrete cells, about 0.1–0.2 μm in diameter, with walls that lack peptidoglycan but have outer membranes akin to those found in gram-negative bacteria. Upon Gram staining, the chlamydia yield a negative reaction. They contain the cellular machinery for making their own macromolecules, namely DNA, RNA, and proteins, but they lack the ability to make ATP or other forms of energy. Thus, they depend on the host cell as the source for this commodity. These organisms have an unusual developmental

Table 2.
A Comparison of the Properties of Chlamydiae, Rickettsiae, and Viruses

Viruses	Chlamydiae	Rickettsiae
Obligate intracellular parasites.[1]		
Uncertain status in living kingdom. Considered to exist at the "threshold" of life.	True living organisms, classified with the bacteria.	
Particles consist of only one type of nucleic acid.	Cells contain both types of nucleic acids, proteins and have a distinct membrane.	
External envelope derived from host's cell membrane; Consequently do not display any Gram staining reaction and are resistant to most antibiotics.	Possess distinct cell membrane and cell wall; Exhibit a Gram negative staining reaction.	
	Lack peptidoglycan	Possess peptidoglycan.
	Sensitive to penicillin as well as tetracycline.	Resistant to penicillin but sensitive to tetracycline.
Typically enter host cell as naked nucleic acid and thus evade the host's protein-degrading enzymes; Some viruses integrate their DNA with the host's genome.	Escape host cell enzymes by remaining sequestered within vacuole (phagosome) and preventing fusion of these with lysosomes.	Survive in host cell by breaking down the membrane vacuole and existing free in the cytoplasm.[2]
Replication controlled by enzymes of the host cell.	Replication is, for most part, autonomous and organisms produce their own replicative enzymes.	
	Distinct life cycle with alternating infectious (elementary bodies) and non infective forms (reticulate bodies).	Most have simple replicative cycles like other bacteria.[3]
	Require ATP (energy) from host cells.	Capable of producing own energy, although can and do use host's ATP.
	Humans and other vertebrates are primary hosts.	Mainly arthropod-borne organisms, for which humans are only accidental hosts.
	Never transmitted to humans via insects.	Transmitted to humans via insect bites.[4]
	Show a preference for moist, membranous tissues; Thus, chlamydial diseases typically occur in the region of eyes, genital tract and respiratory tract.	Most organisms show an affinity for vascular tissue and cause hemorrhagic fevers and rash.[5]

Notes:
[1] The rickettsial genus *Rochalimaea* is an exception to this rule and can live extracellularly.
[2] The characteristics described apply to the genus Rickettsia. Ehrlichia behaves more like the chlamydiae while Coxiella is naturally resistant to lysosomal enzymes.
[3] An exception is *Coxiella burnetii* , which has a life cycle with distinct endospore and vegetative states.
[4] *C. burnetii* is transmitted to humans via aerosols rather than insect vectors.
[5] *C. burnetii* is an exception and causes a lung infection without any rash.

cycle, with two distinct phases. Inside the host cells, the chlamydiae are present in the form of metabolically active reticulate bodies, present within protected membrane-bound compartments called phagosomes. This is the form of the organism that undergoes active multiplication by simple binary fission. The reticulate bodies are noninfective. After a certain number of rounds of multiplication, the reticulate bodies undergo a reorganization of their cells to form structures called elementary bodies, which make up the infective phase of the chlamydial life cycle. The formation of adequate numbers of these metabolically inert elementary bodies induces cell lysis, and the particles that are released can then infect new host cells. Infection occurs via a process of modified phagocytosis, in which the organisms are, as usual, internalized into the host cell but remain protected from the host enzymes within the phagosomes.

Chlamydia pneumoniae

First isolated in 1965, *C. pneumoniae* is a widely prevalent chlamydial species, responsible for causing a variety of acute respiratory-tract diseases such as bronchitis, pharyngitis, and sinusitis, especially in young adults. Infections are spread from person to person via close contact, aerosols, and fomites, although details about features such as their pathogenesis, onset, and incubation period are unknown. About 40–60 percent of the world's adult population contains antibodies against this organism, thus showing evidence for past or current infection. *C. pneumoniae* follows the basic life cycle of alternating elementary and reticulate bodies, as described in the Chlamydiae entry but it appears to differ from other members of the genus in the mechanisms by which it attaches to and enters host cells, although these mechanisms are not yet fully understood. *See also* CHLAMYDIAE.

Chlamydia psittaci

An avian species found most commonly in parakeets, parrots, and lovebirds, and, to a lesser degree in pigeons, poultry, and seabirds, *C. psittaci* emerged as a serious human pathogen after the appearance of AIDS. Prior to this, cases of infection with this organism, called psittacosis, were known to occur only sporadically among humans beings. Psittacosis is an acute condition typically localized in the respiratory tract but with generalized symptoms of infection such as fever, chills, headaches, rash, and muscle aches. The respiratory symptoms appear relatively mild, although there is extensive pneumonia, which may be revealed by chest X-rays. Primary infection occurs by the inhalation of the chlamydiae, which are released into the air via desiccated bird droppings, pieces of feathers, and various secretions. The infection typically manifests itself after an incubation period of 1–4 weeks. Psittacosis is diagnosed by the isolation and identification of chlamydiae in sputum or blood. Various broad-spectrum antibiotics may be used to treat these infections, although this is, at best, a temporary measure in dealing with psittacosis in AIDS cases.

Chlamydia trachomatis

This organism is perhaps the most important and widespread human pathogen among the chlamydiae. Depending on the site of infection, it has been identified as the cause of eye infections (trachoma and conjunctivitis), genital diseases (*C. trachomatis* is the leading cause of sexually transmitted diseases in the United States), and pneumonia. The underlying pathogenic principle in all these diseases appears to be the infection and irritation of the mucus membranes. Trachoma is the most severe form of chlamydial eye infection and often develops into a chronic state, eventually leading to blindness. Inclusion conjunctivitis is the term given to *C. trachomatis* infections that are milder and do not result in blindness. This form of the disease may often be seen in infants, who most likely acquire the organisms during birth from the genital tract of infected mothers. Such children may also develop chlamydial pneumonia.

C. trachomatis is the primary pathogen in what is called nongonococcal urethritis (NGU) in men, which may also be caused by certain mycoplasmas, bacteria, and fungi. Identification of the pathogen becomes important when trying to decide on a suitable course of antibiotics and other therapies. The disease is spread by sexual contact and characterized by a white urethral discharge some 2–3 weeks after infection. Fortunately, *C. trachomatis*-induced NGU is self-limiting; infections cannot be treated with antibiotics such as penicillin that are effective against other genital pathogens such as the gonococci. Genital disease in women most often takes the form of cervical infections. While many of these resolve themselves, a small but significant fraction—2–20 percent—develop more serious consequences such as pelvic inflammatory disease (PID), which is characterized by inflammation of the area followed by destruction of tissues and blockage of the fallopian tubes, leading to infertility. A number of different organisms may cause PID, although *C. trachomatis* appears to be the leading cause, at least in the United States. In addition to these nonspecific infections, this organism also causes another genital disease called lymphogranuloma venereum. *See also* LYMPHOGRANULOMA VENEREUM; TRACHOMA.

chloramphenicol

A broad-spectrum, bacteriostatic antibiotic that binds to the ribosome and inhibits protein synthesis specifically by inhibiting the addition of amino acids to a growing peptide chain. *See also* ANTIBIOTIC.

chlorophyll

Light-trapping pigments present in photosynthetic organisms. The most abundant type of chlorophyll (chlorophyll a) is green in color and present in higher plants, algae, and the cyanobacteria. Certain purple bacteria contain different types of chlorophyll that absorb light rays of a different wavelength than chlorophyll a. *See also* CYANOBACTERIA; PHOTOSYNTHESIS.

chloroplast

Organelle that serves as the site of photosynthesis in eukaryotic cells, such as those of algae and green plants. It is composed of a double membrane, the inside of which is associated with chlorophyll-containing vesicles, not unlike the thylakoids of the cyanobacteria. In fact, some scientists postulate that the choloroplasts evolved from cyanobacterial endosymbionts. Further strengthening this theory is the presence of small, circular, DNA molecules, which lend the organelle some measure of autonomy, though not complete independence, from nuclear control. *See also* CELL; CHLOROPLAST; EUKARYOTE; PHOTOSYNTHESIS.

cholera

Acute, sometimes fatal, gastrointestinal disease characterized by profuse vomiting and diarrhea caused by *Vibrio cholerae*. Endemic in certain parts of India and the Middle East, cholera was unknown in the West until the nineteenth century, after which it has broken out several times, usually in epidemic or pandemic episodes. Organisms enter the body through contaminated drinking water or food. In endemic areas, the disease seems to occur with the highest frequency among the lowest socioeconomic groups, reflecting their greater exposure to unsanitary living conditions and contaminated drinking water. Incidence is considerably lower in developed countries such as the United States where outbreaks have typically been traced to the consumption of raw or undercooked seafood.

The vibrios multiply in the gut, where they produce an enterotoxin that causes the primary symptoms of the disease. The toxin acts on the cells of the small intestine, stimulating a specific enzyme pathway, that causes the cells to release copious amounts of water and electrolytes—primarily potassium and bicarbonate—into the lumen of the intestine. This gives rise to the "rice-water" stools characteristic of cholera, as well as violent, often painful vomiting, resulting in the loss of as much as 25 liters of fluids from the body in a matter of days. This dehydration is the main cause of death from cholera. Despite its early association with violent, gruesome deaths due to severe diarrhea, vomiting, and dehydration, cholera's death rate is relatively low, less than 1 percent of the total affected population. In most cases, the organism disappears from the body after the course of an infection and rarely establishes a carrier state. *V. cholerae* has not been found associated with any extraintestinal infections. Probably because of this localization, the immune response to cholera infections is relatively weak, and there is little lasting immunity against the disease. Consequently, the same person may suffer multiple attacks, especially in endemic areas.

Cholera is diagnosed by the isolation and identification of the vibrios from stools. Although the acute symptoms are distinctive, it may be necessary to differentiate *V. cholerae* from other vibrios, as well as from other enteric pathogens such as *Pseudomonas* and members of the Enterobacteriaceae family, using biochemical tests such as IMViC and salt tolerance/requirement. The treatment of cholera is twofold, simultaneously aimed at stopping the organisms with antibiotics and compensating for lost fluids and electrolytes with saline drips. There are no effective long-term vaccines against this disease, although there has been some success with two types of oral vaccines, which are now available in a few countries. By and large, however, efforts to prevent and control cholera are aimed at improving sanitation and wastewater disposal mechanisms. Cholera is a reportable disease and it is important that public health officials locate the source of the infection and institute adequate containment measures as quickly as possible. Quarantine is not recommended since ingestion, not contact, is necessary for the spread of the bacteria. However, people in close contact with affected individuals should be especially vigilant about cleanliness and about treating their drinking water.

As mentioned earlier, cholera or (Asiatic cholera, as it was called earlier), was virtually unknown in the West until the nineteenth century. When the disease did appear, it arrived with all the fury and destruction of a natural disaster, occurring on four separate occasions in the nineteenth century. No other disease, except for influenza, has covered as large a territory as rapidly. Each of the pandemics—beginning in 1826, 1840, 1863, and 1883—could be traced to outbreaks in India, where the disease is endemic. What, then, was special about the nineteenth century that spurred the disease to break out of its geographical cocoon and embark on a global rampage? Medical historians suggest the mass movements of people, especially armies, across continents, and the advances in transportation, including long-distance railroads and trans-oceanic travel. At the time, little was known about the disease, its cause, and its mode of spread. The British physician John Snow is credited with pinpointing the source of cholera to a single water pump in 1854, and Koch discovered the bacillus in 1883, but even these discoveries did not serve to

eradicate the disease entirely. Today, the disease continues to remain endemic in parts of India but occurs only sporadically in the West. *See also* ENDEMIC; FOOD-BORNE DISEASES; INFECTION/COMMUNICABLE DISEASE CONTROL; PANDEMIC; SNOW, JOHN; *VIBRIO CHOLERAE*; WATER-BORNE INFECTIONS.

chromomycosis

Chronic, spreading fungal infection (mycosis) of the skin and subcutaneous tissues caused by a number of different fungi or yeasts, such as *Phialophora verrucosa, Fonsecaea pedrosoi, F. compacta, Cladosporium carrionii,* and *Rhinocladiella aqua–spersa*. Typically, the condition is seen in feet and legs and is characterized by discolored lesions that become raised and begin to resemble cauliflowers if left untreated. Rarely, if ever, do these infections progress to systemic invasion. Organisms often gain access to the body through the soil, usually through an injury such as a minor cut or a splinter of contaminated wood. Fungi are not transmitted from one person to another by simple contact. Chromomycoses appear to predominate in tropical and subtropical regions where a higher proportion of the population goes barefoot. Surgical removal of the lesions is perhaps the most effective manner of dealing with these infections, especially during the early stages.

chromosome

A discrete piece of linear or circular DNA that contains multiple genes as well as noncoding sequences. Prokaryotic organisms typically possess a single chromosome that contains all or most of their genetic information and exists free in the cellular matrix (cytoplasm). Eukaryotes have multiple chromosomes (the number is variable), which are present within a separate membrane-bound nucleus in the cell. In addition, eukaryotic chromosomes are often found in association with special nuclear proteins. *See also* DNA; GENE; NUCLEUS; PLASMID.

chronic fatigue syndrome

Disease or group of diseases whose predominant characteristic is a persistent, unexplained, clinically evaluated, often debilitating fatigue of at least six months duration. This tiredness is often accompanied by other nonspecific symptoms, such as sore throat, unusual headaches, muscle and joint aches, and unrefreshing sleep, which persist for a similarly long period of time. The cause (or causes) of chronic fatigue syndrome is unknown. Although there are reports of some correlation of the disease with preceding bouts of acute infections, or with physical and even emotional trauma, no clear etiology has been established. Despite the vagueness of both symptoms and etiology, however, chronic fatigue syndrome is a genuine illness that results in a substantial reduction in energy level and has a serious impact on various personal, occupational, and social activities.

Cases of pathological chronic tiredness have appeared in the medical literature since the 1930s, and chronic fatigue syndrome may have an even longer history. It has been compared to fibromyalgia, a rheumatalogical disorder described in the nineteenth century, as well as to depression. The name chronic fatigue syndrome was chosen in 1988 by a group of medical experts because it was the most descriptive with respect to the major symptom. The disease has been linked with viral infections such as polio—in that case it is known specifically as myalgic encephalomyelitis—and the Epstein-Barr virus, as well as with bacterial infections such as Lyme disease. Diagnosis is based on thorough and careful examination of a patient's history and symptoms, which is particularly important for eliminating a wide spectrum of possibilities, including, but not restricted to, hyperthyroidism, unresolved hepatitis infections, early stages of lupus, multiple sclerosis, cancers, and depression, as well as other psychiatric disorders. Although debilitating, chronic fatigue syndrome is not progressive. Therapy includes both treatments for the alleviation of symptoms as well as emotional support.

cilia

Hair-like projections found on the free surface of either prokaryotic cells (e.g., *Paramecium*) or eukaryotic cells (e.g., the epithelial cells lining the nasal passages). Cilia may function as organs of motility, or, as in the nasal passages, to direct the flow of fluids or particles in a specific direction. *See also* MOTILITY.

cloning

The replication of either a gene or an entire genome (and therefore an organism) to make one or more identical copies or exact replicas of the parent. Viruses and bacteria are among the easiest to clone because of their relatively small genome size and their haploid mode of existence. The cloning of diploids is often more difficult because the genes from both parents are required. An additional complication in cloning eukaryotes is posed by organelles such as the mitochondria, which have their own DNA but are derived *in toto* from the maternal parent.

Clostridium

Bacterial genus comprising over 50 species of gram-positive, spore-forming, anaerobic, fermentative rods. While all clostridia are anaerobic, individual species differ from one another with respect to their ability to withstand oxygen—some species are aerotolerant, while others are strict anaerobes that may be killed or inactivated by oxygen. Different species also vary in their choice of substrates for fermentation and the pathways they use to metabolize carbohydrates and proteins. A small number of the total clostridia are known to be pathogenic and cause such destructive diseases as tetanus and gas gangrene. It should be noted that even these infections are opportunistic in nature, since the clostridia are not parasites and the destruction of host tissues is not necessarily beneficial to them. Most clostridia are relatively innocuous; for example, *C. ramosum*, a common resident of the human gut, is found in the feces at densities of up to 10^{11} per gram of stools.

Clostridium acetobutylicum

This organism was discovered in England during World War I by Chaim Weizmann, a chemist who was looking for a way to produce acetone, an important chemical for manufacturing explosives. Instead of looking for alternative chemical methods—the prevalent procedure in use then was the distillation of acetone from wood—Weizman followed Pasteur's example (of using yeasts to produce alcohol) and attempted to find a microbe that could synthesize acetone. In a matter of weeks, he had succeeded in isolating, from the natural microflora of the maize plant, an anaerobic bacterium "capable of transforming the starch of cereals, particularly maize, into a mixture of acetone and butyl alcohol." Named for these precious end products, *C. acetobutylicum* is still used in countries like South Africa for the commercial manufacture of these solvents from molasses.

Clostridium botulinum

Typically found in soil and aquatic environments, this organism is responsible for the paralyzing disease botulism. It produces an exotoxin, which is one of the most potent neurotoxins known to man. There are seven different strains of this species (designated A-G), corresponding to the type of toxin produced. Regardless of the type, the botulinus toxins recognize and bind to the nerve endings and prevent the nerve cells from communicating with muscles, ultimately causing muscular paralysis. Despite its lethal activity, the botulinum toxin type A is also used to treat such conditions as blepharospasm and strabismus, which are characterized by uncontrollable spasms of various facial muscles. Marketed under the brand name "Botox," the neurotoxin is injected in minute doses—under strictly controlled conditions and into specific locations—to relax the affected muscles and prevent their twitching, thus preventing blindness and speech loss. Currently, Botox is also being evaluated in clinical trials, both alone and as a supplement to cosmetic surgery, as a means of reducing facial wrinkles such as crows' feet and forehead furrows by relaxing the muscles in those areas. *See also* BOTULISM.

Clostridium difficile

A member of the normal microflora in a small fraction (< 5 percent) of human beings, this organism has been found associated with cases of diarrhea—with or without pseudomembranous colitis—resulting from the prolonged use of antimicrobial agents, particularly the antibiotic clindamycin. This disease, also known as CDAD (for *C. difficile* associated diarrhea) is one of the most common causes of hospital-acquired infections, and occurs at a rate of up to 30–40 cases per 1,000 discharges in the United States.

C. difficile produces two large toxic proteins, which are responsible for the major symptoms of the disease. One of these is an enterotoxin, which acts on the cells lining the intestinal wall and causes symptoms of diarrhea. The second protein is a cellular toxin that interferes with the cytoskeleton, resulting in the collapse of normal cell shape. Together these toxins induce a reaction in the intestinal wall, which can lead to the formation of a pseudomembrane consisting of immune cells, bacteria, mucosal cells, and clotted fibrin (a component of blood). Mortality rates in CDAD may be as high as 60 percent in untreated cases. The disease may be treated with either vancomycin or metronidazole. The limited use of clindamycin, as well as proper hygienic hospital practices and the elimination of electronic thermometers, have proved useful in keeping the rates of *C. difficile* infections under control. *See also* NOSOCOMIAL INFECTION.

Clostridium perfringens

This is one of the most abundant species of *Clostridium*, second only to *C. ramosum*. It is found at concentrations of about 10^9 bacteria per gram in about 70 percent of human or animal fecal samples. Although nonpathogenic in this environment or on unbroken skin, this organism can cause the deadly gas gangrene when it infects wounds or otherwise gains entry into the bloodstream. *C. perfringens* produces a number of different extracellular toxins (designated as alpha, beta, epsilon, and iota toxins), which have different necrotic and hemolytic activities. *See also* GAS GANGRENE.

Clostridium tetani

A normal inhabitant of soil, this organism causes the disease tetanus or lockjaw when it infects humans or animals. The main pathogenic effects of *C. tetani* are due to an exotoxin called tetanospasmin. Like the *C. botulinum* toxin, tetanospasmin is a potent neurotoxin, but it has the opposite effect, i.e., it causes the uncontrolled stimulation of the muscles rather than blocking it. This stimulation leads to the muscular spasms or twitching characteristic of tetanus. Toxin production is under the control of a plasmid. *C. tetani* also produces a second exotoxin with hemolytic activity whose exact role in pathogenesis is not clear. *See also* TETANUS.

coagulase

Enzyme produced by certain pathogenic bacteria, e.g., *Staphylococcus aureus*, which induces the clotting of fibrin in the blood of the host (usually human). This clot functions as a shield from phagocytosis and antimicrobial mechanisms by the body's immune system.

Coccidioides immitis

This fungus has become increasingly visible as a human pathogen with the outbreak of AIDS and the rise in opportunistic infections. Generally confined to the Western hemisphere, coccidiodomycosis, as the human condition is known, is a highly endemic disease associated with hot and dry climates; it is seen most often in the American Southwest—southern California, Arizona, New Mexico, and Texas—and all over South America. Normally *C. immitis* exists in soil as a free-living saprophyte with a filamentous structure, but under special conditions, it may exhibit a Jekyll-and-Hyde nature. When sub-

jected to various environmental stresses, individual cells of the fungus produce thick-walled spores, which are then released into the air when the hyphae dry up and disintegrate. Depending on where they germinate, these spores will either give rise to mycelia as before, or to single-celled life forms that are parasitic in nature. Common habitats for the parasitic form of *C. immitis* include humans, cattle, household pets, and llamas.

The epidemiology of coccidiodomycosis suggests that primary infection with the fungus most likely occurs via the inhalation of spores. Infectious outbreaks are often associated with dust storms and are seen to predominate among construction workers, laborers in mines and quarries, and archaeologists, who are especially exposed to spore-laden dust because of their occupations. The disease is not known to spread from one human or animal to another. For reasons that are not entirely clear, the disease is more severe in darker-skinned people. The initial disease may be asymptomatic or it may take the form of a nonspecific upper respiratory infection. In certain cases of symptomatic disease, chest X-rays may reveal the formation of a pulmonary nodule characterized by the presence of fungal cells surrounded by immune cells such as neutrophils and macrophages, with some degree of calcification. This nodule serves as the source of dissemination of the fungi, which leads to the systemic disease in about one in 1,000 cases. Primary symptoms of the systemic infection are manifestations of a delayed type hypersensitivity reaction involving the skin, subcutaneous tissues, bones, and various organs, which usually culminate in death if not treated properly. The disease is diagnosed by the demonstration of fungus in clinical specimens—such as sputum or skin scrapings—followed by confirmation of the infecting species with appropriate serological tests. A characteristic feature of coccidioidomycosis is a skin-sensitivity test to a fungal antigen, not unlike the tuberculin sensitivity test, which is apparent some 1–3 weeks after initial infection. To date, the most effective treatment against the systemic disease is amphotericin B administered intravenously. Cutaneous lesions may be treated with topical drugs. A new drug called fluconazole shows promise for the treatment of meningitis, a rare but fatal outcome of systemic coccidioidomycosis. *See also* FUNGUS; HYPERSENSITIVITY; YEAST.

coccus

A general term for a spherical bacterium, derived from the Greek word meaning "berry."

coenocyte

Aseptate tubular cell form assumed by certain filamentous bacteria, e.g., *Streptomyces* species, in which the cytoplasm is continuous from one nucleoid to the next.

colicin

Toxic protein or bacteriocin specifically produced by different strains of *Escherichia coli* to discourage the growth of related strains. These toxins may be under the control of a plasmid and are often used to "type" or identify and differentiate related strains of the organism. *See also* BACTERIOCIN.

coliform

A general label used for gram-negative, non-spore-forming, facultatively anaerobic bacteria that ferment lactose and are usually found residing in the intestinal tracts of humans and other animals. An example of a typical coliform is *Escherichia coli,* but most members of the family Enterobacteriaceae fit this description. The presence and numbers of coliforms are used in many countries as indices for monitoring the purity and cleanliness of domestic water supplies.

colony

Visible mass or clump of microbial cells—usually bacteria, but also, in some cases, fungi—arising from the growth and multiplication of a single organism on a solid medium. Colonies may be produced in the laboratory by diluting a small amount of cell suspension over a solid surface containing media suitable for supporting microbial growth. Standard techniques include spreading small amounts of bacterial culture on a petri dish with solidified medium either by streaking with a sterile metal loop—particularly useful for isolating colonies from mixed populations—or by spreading a measured volume of suspension with a sterile glass rod bent in the shape of a hockey stick. The latter is a more useful technique for quantifying bacteria in samples.

Under a given set of conditions, different bacterial species form colonies that may be distinguished from one another on the basis of shape, size, color, consistency, and surface. However, these properties might vary within a single species if the growth conditions are different. For instance, if the bacteria are crowded to begin with, there is greater competition among them for the nutrients in the medium and the overall rate of growth and multiplication will be slow, leading to the formation of small colonies. The same species might however produce larger colonies under less crowded conditions. Colonies of pigment-producing bacteria will take on the color of the pigments, e.g., *Serratia marsecens* forms bright crimson colonies, while nonpigmented bacteria form whitish, cream, or gray colored colonies. The consistency of a colony is similarly reflective of the extracellular substances—for instance, organisms with thick capsules, like *Klebsiella pneumoniae*, form shiny mucoid colonies. *See also* BACTERIA; PURE CULTURE.

Colorado tick fever

An acute arthropod-borne viral disease caused by a reovirus of the genus coltivirus. These viruses enter the body via the bite of a wood tick that has fed on an infected rodent such as a squirrel. After an incubation of 3–6 days, signifying the time in which the virus proliferates in various cells of the body, there is a sudden onset of fever, muscle weakness, and aches that are especially severe in the region of the back and legs, and retro-orbital pain—i.e., in the muscles at the back of the eyes. About 5 percent of the cases (predominantly children) may develop more serious complications such as meningitis or hemorrhagic fever. Treatment is mainly symptomatic and control measures are aimed at minimizing exposure to ticks and attempting to control tick populations. *See also* COLTIVIRUS.

coltivirus

A member of the reovirus family of double-stranded RNA viruses, this virus is the causative agent of what is known as Colorado tick fever among humans in the United States. Coltiviruses differ from all other reoviruses in that the genome is composed of 12 segments (and hence genes) in contrast to the 10 or 11 found in most other members of this family. The virus is normally found residing in reservoirs such as squirrels, chipmunks, and other small rodents, and is transmitted to humans via the bite of the wood tick. A remarkable feature of this virus is that it can be isolated from the red blood corpuscles of infected human, even though erythrocytes lack a nucleus and ribosomes and hence have no means of supporting viral growth. It is believed that the viruses infect the nucleated precursors of these cells in the bone marrow and continue to survive well after the cells have differentiated and matured into erythrocytes. A second species of this genus called the Eyach virus has been found in ticks in Europe but has not been conclusively linked with any human disease. *See also* ARBOVIRUS; COLARADO TICK FEVER; REOVIRUS.

commensalism

Symbiotic association between two living organisms, typically from different groups, in which one of the organisms (the commensal) gains benefit from the other (the host), which remains largely unaffected by the relationship. *See also* HOST; SYMBIOSIS.

complement

A complex system of more than 20 different interacting serum proteins whose major function is to aid the immune system in clearing bacteria and other infectious agents from the bloodstream. The complement system acts mainly in concert with antibodies, meaning it is an effector of the humoral (antibody) arm of the immune system.

The activation of complement by the immune system, called complement fixation, occurs via two major pathways. In the classical pathway, the triggering event is the formation of antigen-antibody complexes. Antibodies contain a specific complement-fixing site, which is made accessible when they bind to the antigens. A specific complement protein (designated Complement 1 or C1) binds to this site and initiates a cascade of reactions involving the serial binding and cleavage of various complement proteins (C2, C3, etc., in sequence). Each step of the cascade results in the generation of various biologically active substances, which, in turn, set off other defense mechanisms in the body. One result of the classical complement cascade is the lysis of the bacterial cell because of the damage to the cell membrane by the cleaved complement proteins. The binding of complement fragments (specifically, C3b) to bacteria also promotes the susceptibility of these organisms to phagocytosis by attracting cells such as macrophages, which have special complement-recognizing receptor molecules on their surfaces.

The second complement-activating pathway, known as the alternate pathway, is not antibody-specific in nature. This cascade may be directly activated by components of the bacteria—such as their surface polysaccharide or polysaccharide molecules—and begins with the binding of the C3 protein to the target molecule. The activation of the alternate pathway is quicker than the classical pathway and is important in conferring nonspecific resistance to the host.

The importance of complement in resistance to bacterial infections is evidenced by the fact that a deficiency in complement proteins predisposes people to various infections.

For example, deficiencies in the C1, C2, or C4 proteins have been linked to a higher frequency of infections with *Streptococcus pneumoniae* and *Hemophilus influenzae*, while individuals deficient in components C5–C9 are in a high-risk group for certain meningococcal infections. On the flip side, complement has been implicated in the development of autoimmune disorders, as well as allergic and anaphylactic reactions. *See also* IMMUNE REACTION; PROTEIN A.

Compost

Composting is the process by which nutrients present in organic material such as plants are gradually recycled back to nature by the action of microbes, other living organisms, and certain abiotic processes. Composting is a natural, ongoing process that has been taking place in the biosphere for millennia, e.g., with the decomposition of fallen leaves. However, the most common usage of this word is in its restricted sense, as a means of harnessing microbial/environmental decomposition to decay various plant-derived wastes. Depending on the source of waste, composting may vary from a small-scale project in the form of a backyard heap or may extend over several acres of land. The basic process and microbiology, however, are the same in both cases. The eventual products of the decay include water, CO_2, heat, and a relatively stable mass of organic material of variable composition called humus.

The microbial breakdown during composting proceeds through three main phases. The first phase, which lasts for a few days, takes place at moderate, temperatures (4–40°C). Different mesophilic organisms present in the soil rapidly decompose soluble and other easily decomposable compounds in plant material. As these organisms grow, they also generate heat, which builds up within the compost heap, creating the higher temperatures required for the second phase. This is a thermophilic stage which may last from a few days to several months. During this time, the temperatures in the compost will rise to levels of 55–65°C, which aids in

the breakdown of various complex molecules such as lipids, proteins, and complex carbohydrates such as lignin and cellulose. The most abundant organisms in the compost during this phase, at least until about 60°C, are species of *Bacillus*. Thermophilic actinomycetes contribute to the breakdown of cellulose. Species of the heat-loving bacteria normally found in hydrothermal vents and hot springs have also been isolated from compost heaps during this phase. As the supply of organic molecules begins to decrease, the activity of the thermophilic organisms begins to diminish, and the compost temperatures decline to atmospheric levels, thus initiating the final phase in the process. This "curing" or maturing phase typically lasts for several months. Conditions are once more conducive to mesophiles. The microflora of the compost during this final phase are derived from the germination of spores that survived the thermophilic phase and from the surrounding air and soil. Fungi, actinomycetes, and bacteria all play a role in the formation of the final humus.

Humus formed as a result of composting makes a good soil amendment or fertilizer. When produced at commercial-scale composting heaps, it is repackaged and sold as soil for potting and agriculture. Thus, composting is an environment-friendly process in more than one way—not only does it provide an efficient means of waste management, but it also produces usable end products. *See also* ACTINOMYCES; BACILLUS; THERMOPHILE.

conjugation

The process of the transfer of genetic information from one bacterium to another by direct contact. The genes—preferably on plasmids, although chromosomal DNA may also be transferred this way—are transported between the two organisms by way of a tubular passageway formed by the fusion of the sex pili of one of the bacteria (called the donor strain) with the second (recipient) organism. Conjugation is the closest that bacteria come to sexual behavior. Although there is no true fertilization, there is a transfer of genetic information between two individuals. The ability to conjugate is encoded in a special gene called the F factor (for fertility), which is usually found on a plasmid. Bacteria that possess the F factor are called F+, donor or male strains, and those that lack it are called F−, recipient or female strains. An organism can undergo a "sex change" merely by transferring its F plasmid. Strains that carry the F factor on their chromosome are called high frequency recombinants or Hfr strains. *See also* PILUS; PLASMID.

contact microradiography

A method of preparing samples for observation through an electron microscope. In this method, specimens are placed on a photoresist—a certain type of plastic material—and exposed to high-intensity X-ray beams. Photoresists have the ability to melt slightly in proportion to the amount of radiation that strikes them, and to then resolidify quickly. Thus, exposure to radiation in the presence of a solid specimen causes them to form a cast or mold that matches the topography of the specimen. These casts can then be coated with metal and observed under an electron microscope. The advantage in this method is that fixing procedures are bypassed, which means that fully hydrated specimens—which are not amenable to conventional electron microscopy—can be observed. Until now, contact microradiography has been the most useful application of X-rays in microscopy. *See also* ELECTRON MICROSCOPE.

contact tracing

An epidemiological term for identifying and locating both the primary source of an infectious disease and the means by which it is spread. Basically, it involves the identification of contacts—namely, the people or animals who had the chance to acquire an infection due to contact or exposure to an infected person, animal, or environment. Contact tracing is often conducted through interviews between public health officials and the population. This technique is often employed to monitor the spread of sexually transmitted diseases. *See also* EPIDEMIOLOGY.

contaminant

Usually used in the context of quality control in food, industrial, or public health microbiology, a contaminant is a microbe or chemical substance whose presence is undesirable. Mycoplasmas, for instance, are common contaminants in tissue culture, where their presence can interfere with an experiment, e.g., the growth of a particular virus or its cytopathic effects on the tissue culture cells. Bacterial or fungal spores in the air might play a similar contaminating role in tissue cultures and liquid media. Spores and microbial cells that enter food products or the water supply from different sources such as the air, water, and healthy carriers, and which escape destruction during various sterilization and disinfection techniques, are also regarded as contaminants. In some cases, a contaminant might also be a microbial toxin, or other potentially harmful product whose presence is unwanted. *See also* FOOD SPOILAGE.

contrast

The difference between an object and its surrounding medium, or between two objects, which enables us to tell them apart visually. In general, the greater the contrast, the better our ability to see an object. This is why, for instance, we can see stars against a dark night sky better than we can by day, although they are present in the sky all the time. On the other hand, colored objects, such as flowers and trees, are perceived much more clearly against a light background. Because individual microbes are more or less transparent, it is difficult to see them under a microscope, and we need to provide or enhance contrast between them and their surroundings for proper visualization. *See also* ELECTRON MICROSCOPE; MICROSCOPE; OPTICAL MICROSCOPE; STAINING.

coronavirus

One of the most common causes of colds in humans, the coronaviruses comprise a large group of RNA viruses, different species of which have been isolated from a variety of hosts including cattle, swine, dogs, cats, rodents, and birds. First isolated from chickens in 1937, the coronaviruses have since been found associated with serious disease in many of their natural hosts. Scientists only began to suspect a role for these viruses in human pathogenesis in the 1950s after the discovery of the rhinoviruses, which could only account for about 50 percent of cold cases. In addition to colds, the coronaviruses have also been implicated in other diseases including gastrointestinal disorders of infants and, in rare instances, certain fatal neurological complications.

Viewed through an electron microscope, the coronaviruses have a crown or corona-like appearance due to the presence of club-shaped envelope proteins. This envelope encloses an mRNA molecule that functions as the viral genome and is bound to capsid proteins in an irregular manner. Replication takes place in the cytoplasm, but details are not clear because these viruses, especially the human strains, do not grow well in tissue culture.

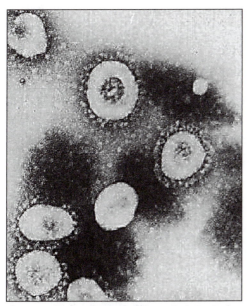

Micrograph of coronavirus OC43. *CDC/Dr. Erskine Palmer.*

Corynebacterium diphtheriae

The causative agent of diphtheria in humans, this bacterium is also called the Klebs-Loëffler (or K-L) bacillus after the scientists who first isolated and identified the bacteria in specimens from the throats of diphtheria patients. The scientific name for this genus is derived from the club-like seen of the bacteria, which is frequently seen under an ordinary light microscope.

These bacteria stain gram-positive, are non-motile, lack capsules, and do not form spores. Although largely aerobic, they can nevertheless ferment certain sugars—glucose and maltose, but not sucrose—to acid. They are fairly hardy organisms, with the ability to withstand drying and exposure to light better than most other non-spore-forming bacteria. A characteristic feature of the corynebacteria is their ability to reduce potassium tellurite. Because this compound inhibits the growth of most other organisms, it is used in selective media for *Corynebacterium* species, which form grayish to black colonies due to the reduction of tellurite. These bacteria also consist of complex, phosphate-rich molecules called volutin granules that endow them with characteristic club (and dumbbell) shapes and which are believed to be food reserves.

Pathogenic strains of *C. diphtheriae* produce an exotoxin called diphtherotoxin, which is responsible for the typical diphtheria symptoms. Strains that lack this toxin may cause sore throats and tonsillitis but seldom anything more severe. They are called diphtheroids. The ability to produce the diphtheria toxin is under the control of a gene that may be transferred from one strain to another via a bacteriophage (bacterial virus). Only strains lysogenic for this phage can produce diphtherotoxin. Studies have shown that the organisms require some amount of iron in the medium to produce the toxin, but, at iron concentrations of more than 0.5 mg/ml, toxin production is reduced to virtually undetectable amounts. In addition to the diphtheria bacilli and diphtheroids, this genus also contains another species called *C. renale*, which is implicated in kidney disease of cattle and differs from all other species by its particular affinity for urea. *See also* BACTERIOPHAGE; DIPHTHERIA; DIPHTHEROID; EXOTOXIN.

cowpox

This disease is primarily a viral disease of bovines (cows); its etiologic agent is the vaccinia virus of the poxvirus family. The most common symptom of the disease is the appearance of vesicular lesions called pocks in the region of the udder and teats. The virus does not usually infect humans but is capable of causing a mild disease (once known as milkmaid's disease) if it inadvertently gains access into the body, e.g., via a scratch or open cut on the hand of a person handling the cows. The symptoms of human disease are like a milder form of smallpox, with a rash, enlarged lymph nodes, and the development of a self-limiting pock-like lesion at the site of viral entry. The occurrence of cowpox confers immunity against the much more serious smallpox virus, a fact noticed by Edward Jenner, who took advantage of it in developing the first successful vaccine against an infectious disease. *See also* JENNER, EDWARD; POXVIRUS; SMALLPOX.

Coxiella burnetii

The causative agent of the respiratory disease called Q fever, this member of the family of rickettsiae deserves separate mention because of its unique mode of existence. Like most other genera in this family, *Coxiella* is an obligate intracellular parasite, but it displays marked differences in its life cycle, biochemistry, resistance to the environment, mode of transmission, and pathogenic capabilities. Of all the bacteria, *Coxiella* possesses a resistance to lysosomal enzymes, which are produced by phagocytic cells in certain animals to kill various infectious organisms. In contrast to other rickettsiae, which are highly sensitive and are easily killed by chemical disinfectants and changes in their surroundings, *C. burnetii* is highly resistant and has been known to survive in milk and water for up to three years. The resistance of this organism is believed to be related to its developmental

cycle, which is known to alternate between a vegetative phase and a resistant, inert, spore-like state. *See also* Q FEVER; RICKETTSIAE.

Coxsackie viruses

A large group of viruses that have been implicated in a number of different human diseases such as the hand, foot and mouth disease; a vesicular pharyngitis called herpangina; paralysis; rashes; aseptic meningitis; myocarditis; an epidemic form of acute hemorrhagic conjunctivits; and a febrile illness that may or may not involve the respiratory tract. The Coxsackie (Cox) viruses are members of the picornavirus family but genetically distinct from the polio and other enteroviruses and also different from the echoviruses, which cannot cause disease in mice. All together, there are about 30 different serotypes of these viruses, which are classified into two main subgroups on the basis of their effects when injected into mice: the Cox A virus, which produces a generalized inflammation of skeletal muscles (myositis) resulting in flaccid paralysis; and Cox B, which tends to produce foci or localized myositis and a paralysis characterized in mice by muscular spasms. In addition, the Cox B viruses grow well in cell culture while the Cox A viruses grow poorly, if at all. All the Cox viruses (except for Cox A) may be isolated from clinical samples by growth on specific cell cultures. The growth of the Cox B viruses results in the rapid development of cytopathic effects characterized by the rounding off and detachment of the tissue culture cells. Cox A infections are often overlooked because of the difficulty in obtaining a proper diagnosis by culturing. While treatments for infections are largely palliative, correct diagnosis and identification is important from an epidemiological standpoint. *See also* HAND, FOOT, MOUTH DISEASE; PICORNAVIRUS.

Creutsfeld-Jacob disease

A rare, and usually fatal, neurodegenerative disease of the spongiform encephalopathy group believed to be caused by prions. There is a progressive dementia, which may be accompanied by tremors and a wasting of muscle tissue. The spongiform brain lesions predominate in the area of the cerebrum. For the most part, Creutsfeld-Jacob disease occurs sporadically without any apparent geographical preference, with a frequency of about 1 in a million per year. It usually occurs later in life, at around the age of 60. About 10–15 percent of the cases, however, are inherited. In addition, the disease is known to be transmissible; the best documented cases being iatrogenic infections, e.g., from corneal transplant operations, during surgeries involving implantation of material into the brain, and via growth hormone injections prepared from human pituitary glands. A version of Creutsfeld-Jacob disease seen in much younger people is believed to be caused by the consumption of beef from cows exposed to mad cow disease. *See also* GAJDUSEK, D. CARLETON; KURU; PRION; PRUSINER, STANLEY; SPONGIFORM ENCEPHALOPATHY.

Crick, Francis (1916–)

British scientist, who along with James Watson, proposed the double helical structure of DNA. In their model, the scientists suggested that the opposite strands of a DNA molecule were held together by hydrogen bonds between specific pairs of bases. This type of base-pairing implicated a complementarity between the DNA strands, which has important implications for its replication. Their ideas were based on, and supported by, the X-ray crystallography work on DNA molecules conducted mainly by Maurice Wilkins. The three scientists shared the Nobel Prize in 1962. *See also* DNA; WATSON, JAMES.

crown gall disease

Bacterial disease of a wide variety of herbaceous and woody plants, including roses, maple, and willow, caused by *Agrobacterium tumefaciens*. The major disease symptoms are caused because the bacterium transfers a part of its DNA to the genome of the host plant, resulting in the transformation of the plant cells to a cancerous form. The multiplication of the transformed plant cells results in the development of large tumor-like galls on the stems and stalks. *See also* AGROBACTERIUM.

Cryptococcus

Unicellular fungus or yeast associated with a range of human infections, often of the brain and meninges, lungs, and skin, particularly in severely immuno-compromised individuals. Although the first cases of these infections (cryptococcosis) were first reported as long ago as the 1890s, they occurred rarely and only became widespread after the emergence of AIDS. Cryptococcoses are frequently associated—at least in the minds of people dealing with HIV/AIDS in urban settings—with exposure to pigeons and pigeon droppings. This is because the fungus grows well in urea, which is often provided to soils via the agency of bird droppings or guano, and in urban environments pigeons are among the most abundant bird populations. Today, *C. neoformans* is one of the most commonly isolated cause of adult meningo-encephalitis in hospitals among AIDS patients, transplant recipients, and people undergoing heavy chemotherapy.

Cryptococcus was originally identified as a species of the common yeast *Saccharomyces*, but later was renamed and reclassified on the basis of its inability to form asexual spores (the ascospores). The organism is a unicellular fungus that is seen to have both asexual and sexual phases in its life cycle, although only the haploid, vegetative forms have been isolated from nature. The vegetative cells are capable of both saprophytic and parasitic modes of existence. A characteristic feature of *Cryptococcus* species is the presence of a polysaccharide capsule, which plays a major role in the virulence of the organism by inhibiting phagocytosis by the immune cells of the body. It produces the enzyme urease. Over 37 species of this organism have been identified in various parts of this world, but only *C. neoformans* is known to be associated with human infections. There are two main distinct antigenic varieties of this species—*C. neoformans* var. *neoformans*, which is found in the eastern United States and Europe and appears to confine itself to immune-suppressed individuals; and *C. neoformans* var. *gatti*, which is more widespread in the tropical and subtropical regions of the world and is more frequently associated with the infection of healthy people.

The most likely route of infection of this fungus is via the inhalation of yeast cells. The subsequent course of infection is different, depending upon the immune status of the individual. Because the fungus is widespread and the relative number of infections is low, it may be assumed that a large number of primary infections of healthy people are asymptomatic. In susceptible individuals, there is an initial episode of pulmonary infection, followed by systemic involvement. The fungi appear to have some preference for nervous tissue, as one of the most common outcomes of untreated infection is meningo-encephalitis. Another frequent manifestation of systemic cryptococcosis is the appearance of cutaneous lesions in the guise of eruptions or abscesses. Diagnosis is achieved by the observation of fungi in clinical samples such as sputum (lung infections), pus from abscesses, and cerebrospinal fluid. Fungi may be observed directly in these samples using negative staining or by culturing. Organisms produce large mucoid colonies which reveal budding cells but no hyphae under the microscope. Antimicrobial agents such as amphotericin B may be used with limited success in treating infections but should be used with extreme caution as they are toxic.

Cryptosporidium

Protozoan parasite of the intestinal tracts of many different animals and often associated with veterinary gastrointestinal disease. Human infections, although rare in healthy individuals, have been seen with increasing frequency and severity among immuno-compromised individuals. These parasites were first identified over 100 years ago as inhabitants of the gastric epithelium in mice, but they did not garner much attention from scien-

tists until after the mid-1950s, when they were found to be associated with gastrointestinal infections of humans. The representative species and the most common human pathogen is called *Cryptosporidium parvum*.

The cryptosporidia are classified within the same group as *Plasmodium* and *Babesia*, but within a different order along with another pathogen, *Toxoplasma*. *Cryptosporidium* has a complex life cycle with many different motile and non-motile forms in both asexual and sexual phases, but unlike the malarial parasite, it does not require a second host. Virtually the entire life cycle of this parasite is lived out in the vicinity of the microvilli of the intestinal wall—either within the epithelial cells or in the intestinal lumen. All together, six distinct developmental stages of this organism have been observed. The infective phase of this organism is an environmentally resistant oocyst, which consists of several haploid cells called sporozoites within a cyst wall. Oocysts are the form which are discarded from one host along with the feces and transmitted to new hosts via contaminated water or soil. Upon entry into a new host, or upon maturation within the same host in which the oocyst was formed, the sporozoites are released into the intestinal lumen where they enter the epithelial cells and multiply into trophozoites by a process of binary fission within special protective vacuoles. At this stage, the parasite may undergo several cycles of multiplication and release into the intestinal lumen, which corresponds to the phase of infection when the clinical disease may manifest itself.

Trophozoites that are released into the lumen are termed merozoites. Merozoites may either infect new intestinal cells as described or may differentiate into sexual forms (gametes), which then fertilize within a host cell to form a diploid zygote. This zygote then develops into the oocyst with a protective wall, within which the cell undergoes meiosis and further division to form sporozoites. The oocyst may develop a very thick and resistant wall—which enables it to survive outside the host and infect new animals—or remain as a thin-walled structure that will germinate within the same host, i.e., maintain the parasite in the same host via a process of auto-infection.

The propagation of the cryptosporidia may induce a self-limiting enteritis with symptoms of watery stools (diarrhea), nausea, vomiting, and low-grade fever that may last 3–12 days. The same infection in AIDS patients results in far more drastic consequences because the body is unable to clear the infection, causing a protracted illness that may lead to dehydration and even death. Primary diagnosis is by the observation of the parasites in the stools, using a modification of an acid-fast stain to differentiate them from yeasts. Specific immunofluorescent dyes may also be used to confirm diagnoses. While the disease is self-limiting in normal individuals, the prognosis is poor in immuno-compromised patients. Antibiotic drugs have been used with limited success to clear infections, and the only consistent treatment appears to be supportive in nature—for example, reconstitution of body fluids via saline drips and drinking plenty of fluids.

curing

Method of food preservation that relies on desiccation, using heat (smoke) and high salt concentrations to discourage the growth of various spoilage microbes. This method is most commonly used for preserving meat and fish. *See also* FOOD PRESERVATION.

cutaneous leishmaniasis

Parasitic disease of skin and the mucous membrane caused by the protozoan *Leishmania*. This disease goes by various popular names, which often reflect the geographical location of occurrence, e.g., Baghdad or Delhi boil, or Oriental sore. Other names for leishmaniasis in the Americas (mainly in the regions from Texas to South America and also the Caribbean) include Uta and Chiclero ulcer.

The chief culprits in the Asian countries are *L. tropica, L. major,* and *L. aethiopica,* while the New World pathogens are *L. braziliensis* and *L. mexicana.* The disease starts with one or more papules at the site of insect bites that develop into ulcerated pus-filled lesions that are self-limiting and heal spontaneously for the most part. The lesion is formed as a result of the accumulation of macrophages and other immune cells at the site of parasitic infection, and the severity of the lesion formed is a function of the immune status of the infected individual. In rural areas in Asia, where the pathogen is *L. major,* the skin lesions tend to stay moist and typically disappear in 3–6 months. *L. tropica,* which is the main pathogen found in urban areas, induces the development of dry sores, which ulcerate after months and only heal 12–18 months after the initial infection.

The diseases caused by the South American counterparts of these parasites often involve the mucosal layers underlying the skin, causing what is called espundia, which may appear years after the original cutaneous sores have healed. The involvement of nasopharyngeal tissues—the face is most exposed to flies—often leads to erosion of tissue and disfigurement (not unlike leprosy) if the disease is not diagnosed and treated properly. *L. braziliensis* is often involved in causing a diffuse type of leishmaniasis associated with a lack of cell-mediated immune activity and in which adjacent sores or ulcers fuse to form larger diffuse lesions. Depending on the immune status of the individual, the cutaneous lesion might serve as the source of parasites for a systemic or visceral disease resembling Kala-azar.

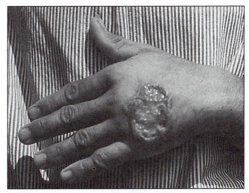

Skin ulcer due to cutaneous leishmaniasis. *CDC/Dr. D. S. Martin.*

Diagnosis of these diseases is achieved by identification of the amastigote stages of the parasite in either the macrophages (in case of Old World cutaneous lesion) or other immune cells found at the lesion. Treatment is mostly with either antimony compounds or amphotericin B. *See also* KALA-AZAR; LEISH-MANIA.

Cyanobacteria

Collective name for the photosynthetic prokaryotic organisms. This group of microbes were earlier classified with the algae—were called blue-green algae—and, even today, they are studied by phycologists, scientists who consider the algae their special domain. However, the cyanobacteria are true prokaryotes with naked DNA (no nucleus), ribosomes characteristic of prokaryotes, and chlorophyll present in thylakoids rather than separate membrane-bound organelles (chloroplasts). They use the sun's radiant energy via the classical photosynthetic pathway—namely, by converting atmospheric carbon dioxide to oxygen using chlorophyll. The cyanobacteria are widely distributed throughout the world, with their only constraint being the presence of moisture. They may be found in a number of different marine and freshwater ecosystems—as free-living phytoplankton, attached to rocks in damp environs, or in running streams or tidal pools, where they may be seen forming a blackish crust on rocks when they dry out. They are also seen in a variety of soils in terrestrial environments.

With the exception of their formation of toxic blooms, these autotrophic, photosynthetic organisms have a relatively low importance in terms of human disease, but a tremendous impact upon human life in a myriad of other ways. As the largest contributors to the earth's photosynthetic activity, they are our principal source of energy. Terrestrial cyanobacteria play a major role in establishing the microflora of soil because they are the primary colonizers that hold soil particles together by means of their extracellular products. They are also important in main-taining soil fertility by dint of their role in accumulating humus. Finally, different cyanobacterial species are used in certain communities as food. Examples include *Nostoc commune* use by certain Chinese and Japanese cultures and *Spirulina* species among Mexicans. Some of these species are also found among the ingredients of certain health food products found in North America.

Various species of the cyanobacteria occur in different morphological forms and show different modes of reproduction, which form the basis for the subdivision of this group into a number of different orders. The simplest types are independent, single-celled organisms that may or may not be held together in loose colonies by extracellular mucilage or slime. Examples include such genera as *Synechococcus* and *Synechocystis,* which represent the major portion of the total photosynthetic capacity in open oceans; *Microcyctis,* implicated in the occurrence of toxic blooms in lakes (freshwater); and *Gloeothece.* By far the largest group of cyanobacteria is that represented by genera such as *Nostoc, Anabaena,* and *Spirulina*, organisms that exist as unbranched filaments or trichomes of prokaryotic cells, which may or may not be encased in a common sheath. Individual cells are capable of communicating with adjacent cells via pores in the intracellular membranes or septa. The filaments of some species may contain a single differentiated heterocyst, a protected cell in which nitrogen-fixation occurs under anaerobic conditions. Different species are capable of forming akinetes or resting spores that are resistant to unfavorable conditions and allow the cyanobacteria to survive harsh conditions by shutting off most of their metabolic activity. *See also* ALGAE; BLOOM; BLUE-GREEN ALGAE; CHLOROPHYLL; PHOTOSYNTHESIS.

cyclosporins

A group of closely related peptides produced by fungi, such as *Trichoderma polysporum* and *Cylindocarpon lucidum,* which have significant immuno-suppressive properties, as well as some antifungal and possible anticancer effects. A suggested application of these compounds is as a mechanism to prevent or suppress graft rejection in tissue and organ transplantation. Cyclosporine, a specific drug in this class, has been shown to target its suppressive activities against T lymphocytes.

cyst

Non-motile, non-reproducing, metabolically inert and infective stage of protozoan parasites such as the amoebae. *See also* PROTOZOA; TROPHOZOITE.

cytomegalovirus (CMV)

A member of the herpesvirus family, cytomegalovirus (CMV) is perhaps best known to us as the salivary gland virus, which is the name given to the strain that infects human hosts. It ranks high among those viral infections whose incidence has risen drastically with the advent of AIDS. The characteristic feature of these viruses is their ability to induce their host cells to form enlarged masses called cytomegalia, both in whole animals as well as in tissue culture. The virus is widespread in nature, but although different strains may infect a large variety of animals, the host range for individual strains is restricted. They tend to establish a long-term coexistence within their host cells; once established, CMV cannot be eliminated from the cell for the lifetime of the host. Infections may remain silent for long periods of time, and the development of disease depends largely upon the age and immune status of the infected individual.

The most severe form of CMV disease occurs in congenitally infected infants, who acquire the virus from pregnant mothers unable to produce antibodies against the virus. Babies may exhibit a number of symptoms of developmental damage, including jaundice, liver and spleen enlargement, chronic lung disease, retarded growth, lack of motor coordination, and abnormalities in hearing or vision. In adults, CMV is one of the most common problems in recipients of bone-marrow and organ transplants (who are given high doses of immunosuppressants to pre-

vent graft rejection) and, like so may other infections today, in AIDS patients. Among the common symptoms in these people are prolonged fever, intestinal disorders, hepatitis, and retinitis (especially in AIDS patients). Symptoms have been known to manifest themselves in apparently healthy hosts as much as 10 years after the initial exposure to the virus. Because viruses are shed in various parts of the body, including in saliva, urine, and the nasal cavity, infection is easily spread via droplets and contact. There are no known drugs against this virus, and treatment is aimed largely at alleviating symptoms, reducing damage, and preventing secondary complications due to infection. *See also* HERPES VIRUS.

cytopathic effect (CPE)

Observable morphological changes in cells that occur due to the growth of viruses. These changes include the formation of inclusion bodies due to the intracellular accumulation of various viral components, the fusion of cell membranes to form giant polynuclear cells or syncytia, and the destruction of various intracellular components. These changes occur due to the shutdown of various cellular processes such as RNA and protein synthesis, the release of digestive enzymes within the cell, and the interference of the virus with RNA. The induction of various types of cytopathic effect (CPE) in tissue culture is used to identify and monitor viruses growing in cells. Cellular transformation by viruses is not regarded as a true CPE. *See also* INCLUSION BODY; NEGRI BODIES; SYNCYTIUM; TISSUE CULTURE; TRANSFORMATION.

cytoplasm

The matrix of any cell regardless of whether it is a prokaryote or eukaryote. The cytoplasm of eukaryotic cells is distinguished by the presence of different membranous structures called organelles, e.g., mitochondria and chloroplasts, which form the site for different metabolic and other activities. Prokaryotic cells lack organelles but contain different kinds of granules for nutrition. Metabolic and digestive activities typically occur in different vacuoles. Ribosomes, albeit of different types, are present in the cytoplasm of both prokaryotes and eukaryotes. The cytoplasm is the site for all cellular activity except DNA replication and RNA transcription. In the nucleus-free prokaryotic cells, these activities also take place in the cytoplasm. *See also* CELL; CHLOROPLAST; ENDOPLASMIC RETICULUM; MITOCHONDRIA; NUCLEUS.

D

darkfield microscopy

A microscopic technique that allows the viewing of nearly transparent specimens, such as bacteria, under an optical microscope without the use of stains or dyes. Ordinarily such visualization is difficult because of the lack of contrast between the organism and its surrounding medium. Darkfield microscopy uses a condenser between the specimen and light source to eliminate all direct light on the slide and to ensure that light rays not reflected by the organisms are deflected away from the microscopic lenses. This technique results in the formation of bright images on a dark background or field. This technique is especially useful for observing specimens that are sensitive to heat or stains and for observing organisms while they are still living. *See also* OPTICAL MICROSCOPE; STAINING.

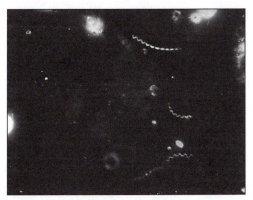

Micrograph of *Treponema pallidum,* darkfield preparation. *CDC/Susan Lindsley.*

Darwin, Charles (1809–1882)

This famous scientist's ideas about evolution and natural selection revolutionized the field of modern biology. Darwin's work, represented in his book *The Origin of Species,* might have concerned the evolution of larger organisms, but the principles upon which he based his ideas form the basis of evolution at all levels, including microbes, genes, and DNA.

defective virus

A virus that lacks a complete genome and is therefore unable to replicate properly without the help of some external genes. Satellite viruses require the presence of a helper virus to complement their replication, while other defective viruses are host-dependent and require some of the genetic functions encoded in the host cell.

dehydration

Clinical condition in humans characterized by the excessive loss of bodily fluids due to heavy diarrhea or vomiting, usually induced by toxic substances produced in infections. Timely treatment—namely, rehydration with adequate electrolytes—is of tantamount importance, otherwise the outcome of dehydration can be death. *See also* DIARRHEA.

delayed type hypersensitivity

Hypersensitivity reaction of the cellular arm of the immune system, this type of response is also known as a Type IV hypersensitivity reaction. Symptoms typically appear a couple of days or so after exposure to an antigen (disregarding the initial priming), reflecting the time it takes for the T cells and tissue or circulating macrophages to respond to the antigens. As these cells accumulate at the site of the antigen, they release various chemical substances that are normally intended to help clear the agent, usually some pathogenic organism. However, due to various factors, such as the ability of the organism to resist phagocytosis, these chemicals may fail to perform their normal functions, and attack the host tissues instead. In contrast to the antibody-mediated hypersensitivity reactions, the delayed reactions become apparent after 18–24 hours of antigen exposure, and reach maximum intensity after 1–2 days. The reaction takes the form of erythema (redness) and a raised thickening of the skin around the region where the antigen is introduced. There is no edema or accumulation of fluid, as is evident in case of an immediate reaction. A typical example of this reaction is seen when the tuberculin antigens are injected into the skin, and in fact, this procedure is used to test for prior exposure of an individual to the tuberculosis bacilli.

Delayed type reactions typically resolve slowly over time, causing little tissue damage except when the antigens are protected or persistent, e.g., when *Mycobacterium tuberculosis* persists in the lung tissue. The macrophages that gather at the infection site are unable to get rid of the bacilli but nevertheless continue to accumulate because of the perceived challenge. Over time, these cells fuse together to form granulomas, which displace and interfere with the functioning of normal tissue, and lead to the formation of necrotic lesions around the area where the antigen made its first contact with the body. *See also* HYPERSENSITIVITY; IMMUNE SYSTEM.

Delbrück, Max (1906–1981)

This scientist was one of the key figures in the famed "phage group" that was responsible for working out the details of the genetic structure of the bacteriophage and developing it as a quantitative model for research in the life sciences. Most of our modern understanding about the working and replication of viruses rests on fundamental knowledge derived from this system. Delbrück, a recipient of the 1969 Nobel Prize in Medicine/Physiology, was originally trained as a physicist and moved into biology only later in his career, inspired by a suggestion from the Nobel–winning physicist Niels Bohr. *See also* BACTERIOPHAGE.

dengue fever

Also known as breakbone fever in certain parts of the world, dengue fever is an acute febrile disease caused by one of four serotypes of a flavivirus called the dengue virus. Although it appears as though the original natural reservoirs for these viruses were monkeys in Africa and possibly Asia, the dengue viruses have since become endemic in human populations in different regions of the world, including southern China, Southeast Asia, the Indian subcontinent, and parts of Australia, Africa, and South and Central America. Millions of cases are reported annually, and some 2 billion people are at risk of contracting the disease.

The transmission of dengue fever, both from primates to humans and from human to human, is via infective mosquitoes of the genus *Aedes*, primarily *A. aegypti*, but also *A. albopictus* and *A. scutellaris*. Following an incubation period of about a week (3–15 days), the disease appears suddenly with high fever, intense headaches, muscle and joint pains, loss of appetite, gastrointestinal upsets, and the appearance of a rash. Infection with a single strain of dengue virus induces a permanent immunity to that specific strain, but unfortunately has the negative impact of exacerbating infections by the other strains. Indeed, scientists attribute the emergence in the 1970s of a more serious dengue virus infection, known as dengue hemorrhagic fever

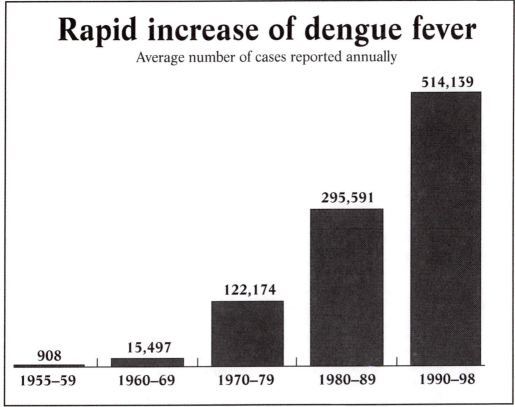

Rapid increase of dengue fever
Average number of cases reported annually

Figure 4. Rapid Increase of Dengue Fever, 1955–1998. Reproduced by permission of WHO, from *Report on Infectious Diseases 1999,* © World Health Organization.

or dengue shock syndrome, to the circulation of multiple serotypes of the virus. This form of the disease is characterized by excessive permeability of the blood vessels and abnormalities in the ability of blood to clot, leading to spontaneous bleeding or petechiae all over the body, bleeding gums, internal hemorrhage in the gastrointestinal tract, and a tendency to bruise easily. Excessive blood loss often leads to shock, and even death, if not treated in a timely and proper manner.

The disease is diagnosed by means of various serological tests and treatment is supportive rather than specific, aimed at alleviating symptoms and forestalling complications. Control measures are poor at best, aimed at minimizing exposure to infective mosquitoes, because anti–dengue fever vaccines are as yet only in the research stages. In view of the exponential increase in incidence in the past three decades (see the chart "Rapid Increase of Dengue Fever," accom-

panying this entry), scientists and public health officials are justified in viewing dengue viral diseases as potential time bombs of epidemics. *See also* FLAVIVIRUS.

denitrification
Energy-yielding process through which various bacteria reduce nitrates to nitrogen or nitrous oxide. These reactions, which are forms of anaerobic respiration, are called denitrification because they result in the net release of nitrogen from the earth to the atmosphere. A wide number of heterotrophic bacteria—e.g., common soil organisms such as *Pseudomonas, Bacillus,* and *Flavobacterium*—perform these conversions with the help of enzymes called dissimilatory nitrate and nitrite reductases. In addition to these organisms, a few chemoautotrophs, such as *Thiobacillus* and *Vibrio,* will act as denitrifiers under appropriate conditions.

Whether denitrification is seen as useful or detrimental depends on the context in which we view the process. In the nitrogen cycle, it serves as the main avenue for restoring nitrogen to the atmosphere. Environmentalists view it as an important mechanism to prevent the accumulation of nitrate pollutants in the soil and water. In agriculture, however, denitrification reduces crop yields by depleting the soil of the primary source of nitrogen directly usable by plants. *See also* ANAEROBIC RESPIRATION; NITRATE ASSIMILATION; NITRATE REDUCTION; NITROGEN CYCLE.

density gradient centrifugation

A specialized technique of centrifugation, usually used in biochemistry and microbiology laboratories for separating cells, cellular components, viruses, microbes, or certain macromolecules from suspensions that may contain one or more of these entities. The basic technique is to subject the sample, along with a supersaturated solution of some solute (usually a sugar or high-molecular-weight salt) in appropriate solvent, to very high-speed centrifugation, or ultracentrifugation. This treatment results in the formation of a density gradient of the solute, with the heaviest material at the bottom. During the centrifugation, different components of the sample settle along the gradient at densities corresponding to their own, and they may be recovered after centrifugation is complete. Such operations are generally carried out under refrigeration to prevent damage by the heat generated during centrifugation. Examples of solutes and samples include cesium chloride for the separation of DNA molecules and sucrose for cell organelles. *See also* CENTRIFUGATION.

dental caries

The medical term for tooth decay, caries encompasses the destruction of the hard dental tissues as well as gums and other tissues surrounding the teeth. A variety of microbes may be involved in this process, typically beginning with forming dental plaque on tooth enamel. The bacteria in the plaque metabo-

lize various sugars present in the foods that we eat and produce acids that act on tooth enamel and wear it away. Over time, this demineralization and erosion leads to the formation of dental cavities, and the acids may proceed to irritate and inflame underlying nerves and the surrounding gum tissue, leading to various forms of periodontal disease.

The usual site of initiation of caries in children and young adults is the crown of the tooth with lesions developing in pits or fissures or between the teeth. Very young infants are susceptible to a form of caries called "nursing bottle caries," which is usually correlated with the high use of sugars in baby foods, drinks, and with sweetened pacifiers. This form of decay spreads rapidly and causes the destruction of the crowns of all newly forming milk teeth of the children. As people grow older, the erosion of their gums due to gingivitis and other gum diseases often leads to the exposure of underlying surfaces of the teeth, such as the dentine and roots, to the bacterial acids, which leads to further inflammation and decay. The practice of good oral hygiene—regular brushing, eating fewer sweets, and periodic cleaning to remove plaque—is perhaps the most effective control measure against dental caries. Once disease sets in, it is relatively difficult to treat with antibiotics because as the structure of the biofilm restricts exposure of its inhabitants to these agents. *See also* BIOFILM; DENTAL PLAQUE.

dental plaque

This term refers to a structureless accumulation of a number of different types of bacteria, their extracellular products (notably bacterial polysaccharides), and protein material produced by the host on the enamel of the teeth. Among the organisms that are commonly found in dental plaque are species of *Actinomyces* including *A. viscosus* and *A. naeslundii*. Both species have fimbriae that help them attach themselves to the surface of the tooth. Once attached, these organisms also produce gummy polysaccharides that attract other bacteria such as streptococci to the site, which results in the formation of a

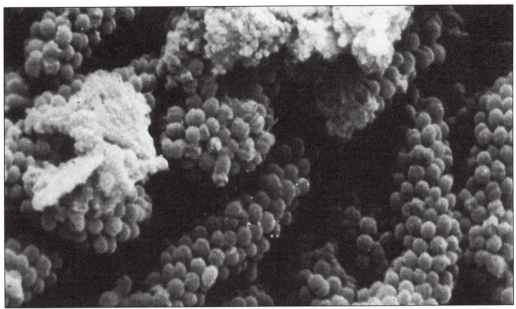

Micrograph of dental plaque showing corncob configurations, an example of bacterial coaggregation. The central cores of the cobs are filamentous bacteria surrounded by streptococcal "kernels." *Courtesy of the National Institute of Dental and Craniofacial Research.*

stable ecosystem of organisms, namely plaque. Dental plaque is one of the most common naturally existing biofilms. The formation of plaque is a prerequisite for the development of various forms of tooth decay and gum disease. *See also* BIOFILM; DENTAL CARIES.

dermatophytes

Species of fungi belonging to the genera *Epidermophyton, Microsporum,* and *Trichophyton,* present ubiquitously in nature, that infect the hair, nails, or skin of a living animal or human host and cause acute inflammatory conditions known as dermatophytosis. Generally known in lay terms as tinea or ringworm due to their clinical appearance, the dermatophytoses are classified according to the body area affected and not on the basis of the infecting species. Thus, tinea pedis refers to a foot infection (athlete's foot), tinea capitis affects the scalp, and tinea unguium refers to an infection of the nail plate. Generally, *Microsporum* species show a preference for hair and skin, *Epidermophyton* infects skin and occasionally nails, but not hair, while *Trichophyton* may infect all three

sites with equal facility. More than one species may be isolated from a single lesion.

All dermatophytes produce a number of proteolytic enzymes that enable them to degrade keratin, the main protein in surface tissues, and thus colonize the host. Normally, the chemicals produced by the sweat glands serve as natural inhibitors for the organism, but infections take hold under weakened circumstances or in areas where fewer sweat glands exist. Primary symptoms are due to inflammatory reactions, leading to the development of erythematous, scaly, or vesicular lesions at the site of infection.

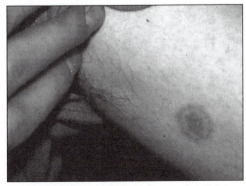

Ringwork lesion on the arm of a sheepworker. © *Science Source/Photo Researchers.*

desiccation

The removal of water or moisture from a substance. In the context of microbiology, desiccation is used as a technique for preserving foods, both by reducing chemical reactivity and discouraging the growth of microbes. Desiccation under vacuum is also the method used to preserve bacterial cultures over long periods of time. Readers should note the difference between desiccation and dehydration, the latter specifically referring to a clinical condition rather than a technique. *See also* DEHYDRATION; FOOD PRESERVATION; LYOPHILI—ZATION.

Desulfomonas

Gram-negative, chemoautotrophic, non-motile, rod-shaped bacteria that gain their energy by anaerobic respiration using sulfates as their primary oxidizing source and pyruvate as the electron donor. The net result is the production of hydrogen sulfide (H_2S). The GC content of this genus falls within a very narrow range: 66–67 percent. *Desulfomonas* are found along with other sulfate and sulfur reducers in anaerobic mud and soils, as well as in animal and human intestines where they are shed with the feces. Together, these bacteria are the major contributors to the H_2S of organically polluted waters. The type species is *D. pigra*.

diagnosis

The process by which a medical practitioner determines the cause and nature of a patient's illness before prescribing any course of treatment. A good diagnosis incorporates several elements, including taking a medical history, performing a physical examination, careful analysis and interpretation of signs and symptoms, as well as the use of a battery of biochemical, histological, microbiological, genetic, physiological, imaging, and psychiatric tests to pinpoint and rule out various disease possibilities. The tests used in any given diagnosis are usually based on the patient's oral report and an evaluation of the signs and symptoms. For example, a person suffering from an acute attack of diarrhea and chills would be subjected to certain biochemical and microbiological tests on blood, urine, and stools samples, but would not typically require a genetic or psychiatric test.

diarrhea

Symptoms of gastrointestinal upset characterized by frequent and often liquid stools. The condition is nonspecific, which means that it may arise from a number of causes such as intestinal infections by bacteria, protozoa, or fungi; disturbances in the natural chemical (especially electrolyte) composition and microflora of the intestines; and psychosomatic reasons. Diarrhea is often one symptom in a spectrum of several that make up a specific disease.

diatom

Group of mostly unicellular algae with cell walls made up of hydrated silica in an organic matrix. The silica gives these algae the sculpturing for which they are noted. After the death of the organisms, the silica is deposited to form what is known as "diatomaceous earth." Each diatom cell is split into two halves that fit together like the base and lid of a box. These organisms may be found in a variety of fresh, brackish, and saltwater environments and are especially abundant in plankton. *See also* ALGAE.

differential staining

Microscopic staining procedure for the visualization of microbes that simultaneously allows us to differentiate between two or more organisms, usually on the basis of their chemical composition. Common examples of bacteriological stains of this type are the Gram stain and the acid-fast stain. *See also* ACID-FAST STAIN; GRAM STAIN; STAINING.

diphtheria

Potentially fatal disease caused by an inflammation in the throat in response to an infection by the bacterium *C. diphtheriae*. The primary mode of transmission of these bacteria is via droplet infection. The bacilli themselves are minimally invasive and tend to confine themselves to the tonsils and upper respiratory tract. Under appropriate conditions—i.e., with adequate iron in the surroundings—*C. diphtheriae* produces an extremely potent exotoxin. This protein has the ability to block protein synthesis in cells by specifically inactivating the enzyme that builds up the chains of amino acids.

The diphtheria toxin is readily absorbed by the epithelium of the throat and tonsils. This causes an inflammation of the local tissues, characterized by the production of a fibrinous exudate due to the destruction of the cells and mucus membranes. This exudate coagulates to form a leathery pseudomembrane that coats the throat and windpipe and obstructs the passage of air. Unless the pseudomembrane is removed, or the air supply is restored mechanically by inserting a tube through the occlusion (by a procedure called tracheotomy), the patient will die of asphyxiation. In addition, other tissues in organs such as the heart, nerves, kidneys, and adrenal glands may absorb the exotoxin. About 10 to 25 percent of diphtheria patients, for instance, succumb to a fatal myocarditis,

a result of the inflammation of the heart muscles.

Diphtheria is likely to be most severe in young children, and a relatively high percentage of deaths occur in children under five years old. But the disease is by no means confined to childhood. Because many adult infections are often subclinical—causing no more than a sore throat—it is easy for the organism to pass virtually unnoticed from person to person.

Throat swabs and secretions, as well as pseudomembranes, are all suitable materials for isolating the organism to confirm diagnosis. Culturing the material on tellurite media is the quickest way to obtain pure cultures, as well as to confirm the presence of the diphtheria bacilli. It is important to confirm the virulence of the bacteria either by an animal virulence test or an *in vitro* procedure called the Elek diffusion test. Another test commonly used in association with diphtheria is the Schick test, which is not a diagnostic test as much as a method to monitor the status of a person's immune system vis-à-vis the disease and so assess susceptibility.

Corynebacteria are susceptible to the commonly used antibiotics, and treatment with

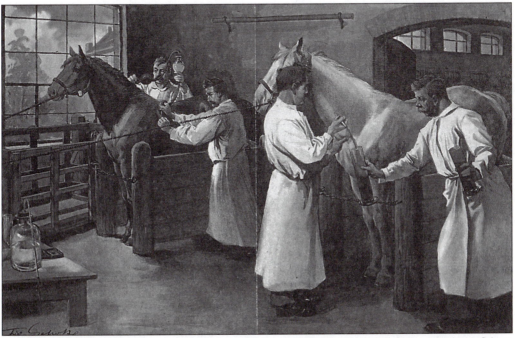

A painting showing inoculated horses being bled for their serum, which contains dyphtheria antitoxin. © *Science Source/Photo Researchers.*

these drugs will limit the infective phase of diphtheria. However, because exotoxin, rather than the physical presence of the bacteria, causes the majority of the symptoms, the main treatment is the administration of antitoxin to neutralize the damaging effects. This passive immunization is only useful if used early in the disease because it is inactive against the toxin once the latter is bound to the cells. Epidemiological studies have shown a dramatic difference in the outcome of the disease depending on when the antitoxin is administered. Fatality is virtually nil in patients treated on the first day of exposure, rising to about 5 percent on the second day and to more than 20 percent by day five. If the disease progresses to the point of the formation of the pseudomembrane, surgical measures may be necessary to enable the patient to breathe.

Diphtheria bacilli are easily transmitted via aerosols—generated by coughing and sneezing—and from contact such as kissing. Schoolrooms, where many children are in close proximity and are likely to exchange articles such as pencils and drinking cups, are especially infective environments. Limiting contact among children, especially those with colds, is a sensible precautionary measure against the rampant spread of infection. Once feared as one of the major killing diseases of children—at the beginning of this century, the annual incidence of diphtheria ran to several thousand in the United States—diphtheria was dramatically brought under control by the 1970s. To a large degree, this control is attributable to the success of immunization procedures. Active immunity against the disease may be induced with a nontoxic form of the diphtheria exotoxin, called a toxoid. Nowadays, children receive doses of the diphtheria toxoid as a matter of course, in combination with vaccines for tetanus and whooping cough (the DTP vaccine). *See also* ANIMAL VIRULENCE TEST; *CORYNEBACTERIUM DIPHTHERIAE*; DTP VACCINE; ELEK DIFFUSION TEST; SCHICK TEST.

diphtheroid

Term for the nonpathogenic species of *Corynebacterium* that are otherwise similar to *C. diphtheriae*. Examples include *C. pseudodiphtheriticum* and *C. xerosis*. These bacteria are gram-positive, pleomorphic, non-spore-forming, non-motile, and aerobic. They can reduce potassium tellurite and are often club-shaped due the accumulation of volutin granules. They differ from the pathogenic species in that they cannot produce the toxin that induces disease symptoms. *See also* CORYNEBACTERIUM DIPHTHERIAE.

diploid

State of being of living organisms characterized by the presence of two complete sets of genes/chromosomes in the cell or cells of the organism. The presence of pairs of genes allows for the existence of dominant and recessive characteristics or phenotypes. Diploidy is almost always a state of higher plants and animals, except for brief phases in their sexual reproduction. The converse is true for the lower eukaryotes such as fungi or algae, which exist in a diploid state for brief phases of their life cycles. Many protozoan parasites have life cycles that are roughly even in either phase. While prokaryotes are largely haploid in nature, some bacteria may exhibit a false diploidy due to the presence of the plasmid. *See also* CHROMOSOME; HAPLOID; ZYGOTE.

disease clustering

A noticeable pattern in the occurrence of a disease, such that there is a higher incidence than would be expected by chance. Clustering may be spatial (i.e., observable in a given geographic area or space) or temporal (discernible within a certain time frame). Typically the term is used in connection with noncontagious diseases such as cancer, although this classification is loose, particularly when dealing with diseases whose origins and etiology are uncertain. The observation of disease clusters provides epidemiologists with good clues to identify potential sources and causal links.

disease incidence

Term used by public health workers and epidemiologists to represent the number of new cases of a specified disease to occur in a given population over a given period of time. The incidence rate for the disease is the ratio of incidence to the population that was exposed to the disease. The rate is usually expressed in terms of number of cases per 1,000 or 100,000 people per annum.

disease notification

The notification or reporting of the incidence of a "notifiable disease"—typically a contagious or infectious disease such as tuberculosis or AIDS—by a physician or other healthcare provider to the appropriate public health surveillance body—e.g., county health center, hospital, or the CDC. *See also* NOTIFIABLE DISEASE.

disease prevalence

The total number of cases of a particular disease occurring in a specified population. The prevalence of a disease is related to, but not the same as, its incidence—the latter term applies only to new cases. Thus, a chronic disease may have high prevalence in a community due to the past exposure of members to the disease agent, even if the current incidence of the disease is close to nil. The prevalence rate of a disease is the ratio of prevalence to the total population at risk for the disease, expressed in terms of prevalence per 1,000 or 100,000 people. Prevalence rate may be measured with respect to a specific instance or episode in time (called point prevalence) or over an interval of time (period prevalence).

disease transmission

The transmission of an infectious disease from one person or organism to another. Horizontal transmission refers to the spread of a disease from one individual to another in the same generation. Vertical transmission, which is the spread of the disease from one generation to another, may occur *in utero* (i.e., the agent is passed from mother to fetus via the placenta), during birth, or after birth via breast milk. The transmission of a condition due to the passage of a defective gene, rather than an infective organism, from parent to offspring is not normally considered vertical disease transmission. *See also* CONTACT TRACING; DROPLET INFECTION; FOMITE; VECTOR.

disinfection

The treatment of an environment, usually with chemical or physical agents, to remove or inactivate microbes. These methods do not necessarily kill or remove all traces of the microbes, but they are usually effective in discouraging the growth of harmful organisms. One of the most common methods of disinfection in the laboratory is the use of chemical disinfectants to wipe surfaces and wash hands. Other means include heat (e.g., boiling water), filtration, irradiation, and drying (desiccation). *See also* ASEPTIC TECHNIQUE; FILTERATION.

DNA

Molecular form in which all living things, with the exception of some viruses and prions, store the information required for carrying out various processes of life. One can imagine DNA (deoxyribonucleic acid) as the central repository of instructions that must be read and then carried out to perform different functions properly. Chemically, it is a polymer of nucleotides, which in turn are composed of a deoxyribose sugar, a phosphate group, and one of four nitrogenous bases—adenine, guanine, cytosine, and thymine (designated A, G, C, and T, respectively). The information is stored in the DNA in the form of sequences of the four base pairs, packaged into discrete units known as genes. A molecule of DNA typically consists of a double-helix structure of two anti-parallel chains of nucleotides twisted around each other like a coiled rope ladder. The "rungs" of the DNA ladder are made of specific hydrogen bonds between A and T, and between G and C bases. Because of the specific nature of base-pairing, the two strands of DNA are complementary. The anti-parallel structure enables a single DNA molecule to contain different genes on the two different strands. In some instances, for example, in certain viruses, DNA may be present in a single-stranded form. Three-and four-stranded DNA forms may also exist during certain phases of replication and gene expression. *See also* CHROMOSOME; DNA REPLICATION; GENE; NUCLEUS.

DNA fingerprinting

A technique for evaluating the similarity of DNA from related organisms (or in the case of humans, between related individuals) without resorting to sequence determination and a comparison of complete DNA molecules. The DNA molecules are digested with specific restriction enzymes that recognize and cut these molecules at specific locations according to sequence. This procedure generates a mixture of pieces of DNA, which may be separated by electrophoresis and visualized by hybridization with probes. The pattern of bands or segments that emerges is specific for a given organism and is called its DNA fingerprint. In general, the degree of similarity between the DNA fingerprints of two organisms indicates a higher degree of relatedness between them. In the DNA fingerprints of humans and other higher organisms (used, for instance, in forensic medicine), the probes used are certain core sequences that are present in several stretches of tandem repeat sequences. Probes used for the detection of bacterial DNA fingerprints are usually species, genus, or family-specific gene sequences. *See also* DNA HYBRIDIZATION.

DNA hybridization

Because the natural form of existence for DNA molecules is a complementary, double-stranded configuration (i.e., the double helix), the tendency of most single-stranded DNA molecules is to find their complements and anneal to them by the formation of hydrogen bonds between the complementary bases. Hybridization refers to the laboratory technique by which single-stranded DNA molecules are made to anneal to their complements—usually also single-stranded DNA, but sometimes also RNA. Hybridization also takes place between sequences that are similar to one another without being exact complements. This property of hybridization is used in a variety of laboratory techniques for detecting specific sequences in a library of unknowns, for instance, or to find specific genes using a signature sequence as a probe. The probe DNA is tagged with a radioactive or fluorescent dye and allowed to react with a pool of molecules that have been previously denatured—made single stranded—and immobilized on a solid support. After exposure to the probe, the complementary sequence may be located by autoradiography or UV exposure, depending on the nature of the tag. *See also* AUTORADIOGRAPHY.

DNA replication

The synthesis of new molecules of DNA from a preexisting template. DNA replication is semi-conservative in nature because each so-called "new" DNA molecule actually possesses one old strand from the preexisting template and a freshly synthesized complement. The new strand in each case is synthesized by the stepwise addition of nucleotides on a primer, with the help of specific enzymes called DNA polymerases. In addition, some enzymes called topoisomerases and gyrases help the DNA molecule wind and unwind properly during the different stages of its synthesis. *See also* CELL DIVISION; DNA.

Domagk, Gerhard (1895–1964)

Scientist credited with the discovery of sulfonamide antibiotics ("sulfa" drugs) and their usefulness for treating human infections, particularly of the streptococci and staphylococci. His discovery came at a time when there were no adequate treatments for such infections, and it quickly led to the development of cures for several serious infections. He was awarded the Nobel Prize in 1939 but, because he was forbidden by the German government to accept the honor, had to wait until 1947 to receive the medal and diploma. *See also* ANTIBIOTIC.

droplet infection

Mode of transmission of infectious diseases, usually of the respiratory tract, in which the causative organisms (bacteria or viruses) are shed into the mucus and transported to the air via aerosols—droplets—created by sneezing, coughing, or sometimes even ordinary breathing. Depending on their hardiness, different organisms may survive in these droplets anywhere from hours to days and will infect another individual when he or she breathes in the droplet-contaminated air.

drug resistance

This term is often used interchangeably with the term antibiotic resistance but may be used more broadly to describe the resistance acquired by any infectious agent—bacterium, fungus, or virus—to any type of antimicrobial agent. *See also* ANTIBIOTIC RESISTANCE.

DTP vaccine

A combined vaccine for immunization against three bacterial diseases: diphtheria, tetanus, and pertussis (whooping cough). Protection against the first two diseases is achieved by the use of toxoids. Both toxoids are prepared by treating the exotoxin (produced by *Corynebacterium diphtheriae* and *Clostridium tetani,* respectively) with formaldehyde, which inactivates the toxic portion of the proteins without affecting their antigenic potential or specificity. The pertussis vaccine may be either a suspension of killed or inactivated cells of the organism, *Bordetella pertussis*, or an acellular preparation of the antigenic components of these bacteria. Of the two, the latter is preferable because it appears to cause fewer adverse reactions in the recipient. The combined vaccine is a mixture of the three products adsorbed on an aluminum salt, which is administered in an intramuscular injection.

DTP is a childhood vaccine, generally administered in a series of four primary shots at 2, 4, 6, and 15 months of age. The CDC recommends a booster shot sometime between the ages of 4 and 6 years, when the child begins to attend school. Additional boosters of the diphtheria and tetanus toxoids (DT)—with the former in lower doses—are recommended at 10-year intervals. At this stage, the pertussis vaccine is only recommended if necessary—e.g., in high-risk groups or in instances of epidemics. DT vaccines may also be used for the primary vaccines if pertussis appears to be contraindicated, i.e., if it elicits harmful reactions such as anaphylaxis or encephalopathy. In addition to scheduled boosters, the tetanus toxoid—available by itself in a soluble, nonadsorbed form—may be given in cases where bacterial exposure is suspected, e.g., cuts and punctures from rusty or dirty objects, particularly when the bleeding is minimal. *See also* BORDETELLA PERTUSSIS; CLOSTRIDIUM TETANI; CORYNEBACTERIUM DIPHTHERIAE; DIPHTHERIA; TETANUS; TOXOID; VACCINE; WHOOPING COUGH.

Dulbecco, Renato (1914–)

Italian-born American scientist who won the 1975 Nobel Prize in Physiology, along with David Baltimore and Howard Temin, for "discoveries concerning the interaction between tumor viruses and the genetic material of the cell." Dulbecco's contribution was to furnish proof that the genes of oncogenic DNA viruses—specifically the polyoma virus and SV40—became a physical part of the genome of the cells that they transformed, so that the host cells acquired inheritable properties derived from the virus. *See also* BALTIMORE, DAVID; POLYOMA VIRUS; ONCOGENE; TEMIN, HOWARD M.; TRANSFORMATION (CELLULAR).

dumdum fever. *See* KALA-AZAR

dysentery

Condition caused by bacteria such as *Shigella* or protozoan parasites, marked by symptoms of bloody diarrhea.

E

E. coli. *See* ESCHERICHIA COLI

Ebola virus

One of the notorious "new" viral infections to have broken out during the last few decades, the Ebola virus grabbed headlines because of the suddenness with which the first epidemic occurred and the severity of the symptoms. The first reported outbreak of hemorrhagic fever caused by this virus—which occurred in 1976 in Zaire, Africa, near a small river called Ebola—infected only 300 people and killed around 50 of them within a few days. Fortunately, the Ebola virus never spread to pandemic proportions.

Members of the filovirus family, the Ebola viruses are known to comprise at least four antigenically distinct strains. Three of these strains have been found confined to specific regions of Africa, namely Zaire, Ivory Coast, and Sudan. The fourth strain, called the Reston virus, appears to be restricted to non-human primates. The Ebola viruses are structurally identical but antigenically distinct from one another as well as from the Marburg virus, which also belongs in the same family. *See also* EMERGING PATHOGEN/INFECTION; FILOVIRUS; HEMORRHAGIC FEVER; MARBURG VIRUS.

ecology

Branch of the biological sciences that deals with the study of interactions between living organisms and their environment.

electron microscope

A microscope that uses beams of electrons to form enlarged images of miniscule objects. Because an electron is much smaller than the average wavelength of visible light, the resolving power of electron microscopes is correspondingly higher than that of optical microscopes. Both types of microscopes function on the same underlying principle; they produce images by focusing rays—either light or electron beams—reflected from the specimens. However, because electrons behave differently than light, the construction of the instrument and lens system is different.

In the case of the optical microscope, the original source of the light rays to be focused is the object itself, although a light bulb and condenser system are often used to provide uniform intensity and some measure of consistency. However, with the exception of certain radioactive materials, most objects do not emit electrons. Thus, the first order of business in fashioning an electron microscope is to have a steady source of electrons at fixed speed and intensity to hit the specimen and be reflected off it for further processing (i.e.,

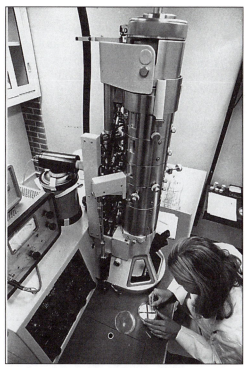

Electron microscope. © *Science Source/Photo Researchers.*

Nowadays, images are captured by an image processing system, which is transmitted to a screen for immediate viewing. Finally, the entire assembly of the electron gun and lens system needs to be operated under vacuum to enable the proper flow of electrons.

Because electron microscopic operations take place in a vacuum, specimens must be fixed so that they do not collapse during observation. Furthermore, they must be present in extremely thin sections, or else different layers will scatter reflected electrons at different rates and intensities, leading to multiple images that will show up as blurs on the screen. This ultrathin sectioning is accomplished by embedding the specimen in some medium such as epoxy resin or paraffin, which will not interfere with its natural structure but will provide support. Prior to embedding, the material must be "fixed" or preserved for viewing—this is usually done with osmium compounds (which also help to enhance contrast) and glutaraldehyde. Once embedded, the material is cut into extremely thin sections of less than 0.1μm thickness using a device called a microtome. These ultrathin sections are then placed on a support medium—usually a very thin film of carbon or carbon-coated plastic—on a metal support grid. Just to give an idea of the scale, about three or four of these grids can fit in the space of a single fingernail.

Once a section is prepared, the material must be treated to provide contrast in the reflecting abilities of the specimen and the medium. The application of materials containing heavy atoms, such as uranium or lead, provides a surface from which electrons can bounce off with greater ease. These types of "stains" have rather limited usage because the chemical reactions between biological materials and such compounds are relatively rare. Negative staining techniques by the use of chemicals containing such heavy atoms as phosphotungstate or molybdenum to coat the surroundings and leave the specimen unaltered are more widely applicable in electron microscopy. Another approach to viewing particulate specimens is by metal decorating.

image formation). The most commonly used source is a metal such as tungsten, which emits electrons in response to heating its surface, in a process called thermionic emission. Alternatively, certain metal surfaces may be induced to release electrons in response to a high-voltage electrical potential. The electrons are then focused into an intense beam using a positively charged electrode (anode), which is functionally equivalent to the condenser system of the light microscope. The device containing the metal surface and focusing anode is called an electron gun.

Once the incident radiation is reflected from the object, it needs to be focused through some sort of lens system to form the image. In the case of the electron microscope, this lens is provided by an electromagnetic field, which directs the path of the reflected electrons. The image thus produced is not detectable by the eye and must be captured on some radiation-sensitive medium such as photographic surfaces or fluorescent screens.

Here, a very thin layer of electron-opaque material such as tungsten is deposited under high vacuum at an angle to the supporting medium. These atoms will accumulate on the surface features of the object and produce a high level of topographic contrast, which can be detected by the differential reflection of the electrons.

Instruments that rely on the transmission of electrons from the object via the lens to the screen are called transmission electron microscopes or TEM. Techniques such as metal decorating make use of an adaptation of this basic design called a scanning electron microscope (SEM), which provides images of the external surfaces of the objects only. *See also* CONTACT MICRORADIOGRAPHY; CONTRAST; METAL DECORATING; MICROSCOPE; NEGATIVE STAINING; SCANNING ELECTRON MICROSCOPE; SCANNING TRANSMISSION ELECTRON MICROSCOPE; TRANSMISSION ELECTRON MICROSCOPE.

electron transfer chain

This phrase refers to the various components of a biochemical pathway—versions of which are found across the board in the living kingdom—that allow metabolic reactions such as respiration and photosynthesis to proceed smoothly within cells. Specifically, the electron transfer chain is made up of a system of specialized molecules and enzymes that form a path or channel to move the electrons that are generated during metabolism. These molecules are typically found attached to the external or mitochondrial membranes and are arranged in a gradient of gradually increasing (electrically) positive potential. Electrons move down this gradient through a series of "redox" reactions, i.e., reactions in which one molecule is oxidized by the loss of its electrons and the second becomes reduced by acquiring them. The first step of the chain is the reduction of the compound NAD (nicotinamide adenine dinucleotide) or NADP (NAD phosphate) to NADH or NADPH, respectively. Other molecules that make up electron transfer chains include proteins such as cytochromes and iron-sulfur proteins and

aromatic compounds such as quinones. The final step in electron transport involves the transfer of the electrons to an external oxidizing agent, which happens to be oxygen in the case of aerobic respiration, and water in photosynthesis. Because the electrons move down a gradient from higher to lower energy locations, there is a net energy gain in the system. This energy is used to pump protons in the reverse direction to the flow of electrons, resulting in the generation of what is called proton motive force. The proton-motive force, in turn, may be channeled towards functions like flagellar movement, ion transport across channels, and in the creation of high-energy phosphate bonds in ATP and GTP molecules. *See also* ATP; METABOLISM; PHOSPHORYLATION; PHOTOSYNTHESIS; RESPIRATION.

electrophoresis

A laboratory technique to separate macromolecules such as proteins and nucleic acids on the basis of their mobility in an electric field. A mixture of molecules is allowed to separate out on a solid support medium, such as a gel, paper, or a glass slide, while the electric field is maintained by the use of some electrolytic solvent. The use of support materials such as polyacrylamide or agar/agarose adds a second dimension to the separation, so that the molecules are separated on the basis of size (i.e., length of a molecule) in addition to charge. The electrolyte system being used may be either continuous—i.e., using just a single solvent—or discontinuous. Electrophoresis is mostly used for analytical purposes, e.g., to determine the different proteins present in an immune serum or to obtain DNA fingerprints after restriction digestion, but it is also used for the preparation and purification of small amounts of material such as plasmids. Many adaptations have been introduced to the basic technique to impart sensitivity or specificity to the electrophoretic system. *See also* DNA FINGERPRINTING.

Elek diffusion test

An *in vitro* diagnostic test for diphtheria that looks for toxin production by the infecting organisms. Strips of filter paper soaked in diphtheria antitoxin are placed on a special medium, and the bacterial isolate from the infected individual is inoculated onto the plate by streaking it perpendicular to the filter paper strips. The antitoxin from the strips diffuses into the medium, and as the organism grows, it produces the toxin—if it is a virulent strain—which also diffuses into the medium. The reaction of the toxin with antitoxin will produce a precipitate, which will be seen as a thin white precipitate line. A number of cultures may be tested on a single plate. *See also* DIAGNOSIS; DIPHTHERIA.

Embden-Meyerhof-Parnas pathway.

See GLYCOLYSIS

emerging pathogen/infection

Term used by health professionals to describe an infectious agent or disease that is totally new to both their current experience and their historical knowledge. Different kinds of emergent pathogens are typically identified in terms of their origins. For instance, neither AIDS nor its causative agent HIV had been described by anyone, until the 1980s. A vast group of previously unknown viruses causing different types of hemorrhagic fevers are also emerging as serious new pathogens, and are causing diseases for which there are as yet no cures. Another type of condition, namely the development of peptic ulcers, had been recognized, and treated symptomatically for well over a century before Barry Marshall discovered the role of *Helicobacter pylori* in their formation. Finally, there are known organisms, which have developed new pathogenic strategies through mutation and evolution, resulting in the creation of virtu-

---

Table 3. New and Emerging Infectious Diseases and Pathogens

1997	Avian flu virus, H5N1
1996	Australian Bat lyssavirus (new rhabdovirus isolate)
1995	Human herpes virus * (HHV8); identified as cause of Kaposi's sarcoma
1993	Hantavirus pulmonary syndrome caused by the Sin Nombre virus
1992	A new strain of *Vibrio cholerae* (O139)
1991	Guanarito virus; found causing Venezuelan hemorrhagic fever
1989	Hepatitis C virus
1988	Hepatitis E
	Human herpresvirus 6
1983	HIV
1982	*Escherichia coli* strain O157:H7 found in gastrointestinal infections
	Lyme disease
	HTLV-2 virus
1980	HTLV
1977	*Campylobacter jejuni*
1976	*Cryptosporidium parvum*
	Legionnaire's disease
	Ebola

Source: Adapted from the WHO's *Infectious Disease Report 1999.* © Copyright World Health Organization 1999.

ally new diseases and agents. *Streptococcus pyogenes*, for example—long known to cause human infections such as strep throat and scarlet fever—reappeared in a new guise during the 1980s, causing a serious new form of toxic shock syndrome, a disease earlier associated mostly with staphylococci. *Escherichia coli*, long considered innocuous or only an occasional agent of disease, has been increasingly identified in serious outbreaks. At least part of the emergence of this organism as a pathogenic threat is the development of antibiotic resistance among different strains.

EMP pathway. *See* GLYCOLYRIS

encephalitis

Generally, encephalitis refers to the infection or inflammation of brain tissue, which may occur as a complication in infections by any number of pathogens, including bacteria, viruses, and protozoa. Rabies, for instance, is a type of encephalitis with almost always fatal consequences. This section, however, deals specifically with the group of encephalitides that typically break out in epidemic fashion and are caused by RNA viruses in the bunyavirus, flavivirus, and togavirus families. The clinical manifestations of various encephalitides are similar. Onset is marked by symptoms not unlike meningitis (in which the inflammation occurs in the meninges or the brain covering), namely fevers, headaches, rigidity of the neck muscles, and vomiting. These symptoms are followed quickly by alterations in behavior, which is the hallmark of tissue involvement. The patient becomes lethargic and gradually shows increasing signs of confusion, eventually lapsing into a coma. Death is a frequent outcome, and survivors are often left with permanent effects such as paralysis, mental retardation, epilepsy, blindness, or deafness.

The flaviviruses are perhaps the most important group of encephalitis viruses in terms of their human impact. They are often enzootic in small animals such as swine, dogs, and rodents and are transmitted to humans via specific arthropod vectors such as mosquitoes or ticks. The epidemiology of these diseases is often seen to correlate closely with the life cycles of the vectors and behavior of reservoirs. Thus, outbreaks of the mosquito-borne viruses, such as the Japanese encephalitis and St. Louis encephalitis viruses (whose names reflect their geographic distribution), occur most frequently during the warm seasons when there is usually a sudden increase in the population of these insects due to the hatching of the larvae. On the other hand, tick-borne viruses such as Powassan, Omsk, Russian, and Central Europe viruses are less seasonal in their attack patterns. The togaviruses, specifically members of the genus alphavirus, are the etiologic agents for equine encephalitides, an epizootic disease of horses, which may also break out in humans due to transmission of the virus via mosquito vectors. Mosquitoes are also the primary vector for the La Crosse virus, a member of the bunyavirus family that causes California encephalitis. The geographic distribution, vectors, and reservoirs of various encephalitides are summarized in Table 4 "Arboviral Encephalitides," accompanying this entry. *See also* ARBOVIRUS; BUNYAVIRUS; FLAVIVIRUS; TOGAVIRUS.

endemic

The constant presence of a disease, usually but not necessarily infectious in nature, in a portion of a given population group or geographic area. Cholera, for example, is endemic in certain parts of India. Endemic diseases may be either holoendemic or hyperendemic in nature. Holoendemic diseases predominantly infect the children of a population in such a manner that the disease reaches a state of equilibrium before they attain adulthood. Thus, at any given time, a proportion of adults of the population will have the disease, albeit in a relatively mild form (e.g., malaria in some cases). Hyperendemic diseases affect all age groups more or less equally, at high intensity. The difference in patterns of occurrence as well as in

Table 4. Arboviral Encephalitides

Virus (Disease)	Geographical Distribution	Reservoir Host	Arthropod Vector	Epidemiological Notes
I. Bunyavirus				
La Cross virus (California encephalitis)	Western United States	Chipmunks, squirrels	Mosquitoes	Endemic in wooded areas of the US; Annual incidence is about 100 although an estimated 300,000 humans are infected annually; Children are affected more often than adults.
II. Flavivirus				
Japanese encephalitis virus	Asia, particularly	Swine, birds	Mosquitoes	Endemic in Southeast Asia and occurs in summer epidemics in Northern and central parts of Asia (i.e. outbreaks related to breeding cycle of mosquitoes).
Kunjin virus	Australia	Birds	Mosquitoes, Ticks	Causes a lethal encephalitis in horses and only occasional encephalitis in humans.
Kyasanur forest virus[1]	India	Rodents	Ticks	Primarily a cause of hemorrhagic fevers and causes encephalitis only occasionally.
Louping ill virus	British Isles	Birds (Grouse)	Ticks	Causes lethal disease in sheep and sporadic outbreaks among humans.
Murray valley encephalitis virus	Australia, New Guinea	Birds (Waterfowl)	Mosquitoes	Causes periodic summer epidemics
Omsk hemorrhagic fever virus[1]	Central Russia	Rodents	Ticks	Primarily a cause of hemorrhagic fevers but causes sporadic outbreaks of encephalitis also; Both epidemics have been correlated with winter musk-trapping season.
Powassan virus	North America, Russia	Small Mammals	Ticks	Sporadic outbreaks;
Rocio virus	Brazil	Birds	Mosquitoes	Sporadic outbreaks;
Russian/Eastern Europe encephalitis virus[2]	Russia, Eastern Europe, Scandinavia	Rodents, birds, Domestic/Farm Animals	Ticks	Mostly sporadic outbreaks; Some epidemics have been traced to the ingestion of raw (unpasteurized) milk.
St. Louis encephalitis virus	Americas	Birds	Mosquitoes	Identified in periodic epidemics in North America and sporadic outbreaks in South America.
West Nile virus	Africa, Mediterranean and tropical Asia	Small birds	Mosquitoes, Ticks	Endemic in tropical countries with sporadic outbreaks in summer; Causes mostly a dengue-like syndrome with only occasional encephalitis;
III. Togavirus[3]				
Eastern equine encephalitis virus	Americas and the Caribbean	Birds	Mosquitoes	This virus is maintained in the US in freshwater marshes; The insect vector can transmit virus and disease to both equine and human hosts.
Venezuelan equine encephalitis virus	Tropical Americas including Florida Everglades	Rodents (Enzootic)	Mosquitoes	Epizootic outbreaks in horses; Human disease is relatively mild.
Western equine encephalitis virus	Americas	Small Birds	Mosquitoes	Can cause encephalitis in both horses and humans.

Notes:

[1] Also see Table 5.
[2] Collectively called the tick-borne encephalitis viruses.
[3] All the encephalitis viruses of the togavirus family are classified in the alphavirus genus.

the epidemiology of endemic and epidemic diseases entails different approaches to the management of these diseases. *See also* EPIDEMIC.

Enders, John F. (1897–1985)

Head of the team of scientists who were the first to successfully cultivate the poliomyelitis virus in *in vitro* tissue cultures. This finding had a direct and dramatic impact on Salk's ability to grow sufficient quantities of the virus for the preparation and testing of his vaccine against the disease. Enders and his co-workers, Frederick Robbins and Thomas Weller, won the Nobel Prize in 1954 for their discoveries. *See also* POLIOVIRUS; SALK, JONAS; TISSUE CULTURE.

endocytosis

One method by which a cell takes up materials from its external environment, i.e., phagocytosis. Endocytosis is often initiated by the adherence of a molecule to the cell surface, either by specific attachment to a receptor or simple proximity, e.g., with nutrients. To begin with, the particle is drawn deeper into the cell by the invagination of the membrane. Ultimately, the membrane pinches off to form an intracellular vesicle or vacuole that keeps the material sequestered away from the rest of the cytoplasm.

endoplasmic reticulum

System of membranous tubes or cisternae found throughout the cytoplasm of eukaryotic cells, continuous with both the inner surface of the plasma membrane and the outer membrane of the nuclear envelope. Membranes are typically 50–200 nm apart, creating distinct inner and outer spaces. The cisternae play a role in transporting materials from the nucleus or other organelles, such as the Golgi apparatus, to the cell membrane and beyond. The outer surface of the endoplasmic reticulum (ER) may be studded with ribosomes at locations where active synthesis of extracellular and membrane proteins is taking place. This surface is called the rough

endoplasmic reticulum. Smooth ER lacks the capacity to bind ribosomes but is otherwise biochemically similar to the rough ER. These structures are more abundant in cells concerned with lipid metabolism, such as the liver cells of mammals. *See also* CYTOPLASM.

endosymbiont

An organism that resides within a second organism in a symbiotic relationship. *See also* SYMBIOSIS.

endosymbiont hypothesis

A widely accepted (albeit unproven) hypothesis that various organelles present in eukaryotic cells, such as mitochondria and chloroplasts, are actually the descendants of endosymbiotic respiratory bacteria and cyanobacteria, respectively. Among the arguments offered in favor of this hypothesis are (1) they have their own DNA; (2) they contain their own ribosomes, which are more akin to those found in prokaryotic organisms than to those in the cytoplasm of the cells in which they are present; and (3) they are more or less autonomous with respect to replication. *See also* CHLOROPLAST; CYTOPLASM; MITOCHONDRIA; SYMBIOSIS.

endotoxin

Surface or intracellular components of bacterial cells, which have harmful effects when they come in contact with the immune systems of humans or animals. An example of a bacterial endotoxin is the lipopolysaccharide (LPS) component of gram-negative cell walls, which elicits an inflammatory reaction when these organisms enter the bloodstream and leads to symptoms of shock.

Entamoeba

This genus contains the most abundant and widespread members of the parasitic amoebae that colonize the intestinal tracts of humans and animals such as dogs, cats, and rats. The most important human pathogen is *E. histolytica*, the causative agent of amebic dysentery and other forms of amebiasis. All other species of this genus, namely *E. coli, E. hartmanni, E. polecki,* and *E. gingivalis,* are commensals that do not normally cause any disease.

The differentiation between various species is crucial for administering proper therapy to affected individuals and to prevent unnecessary drug treatment. *E. gingivalis,* which is often found in the mouth, tartar, and tonsils, is different from all the other species on two important counts: it does not have a cyst stage, and it is the only species that can ingest white blood cells. One of the main distinguishing features of the trophozoites (active, vegetative form) of *E. histolytica* is the presence of whole or partially digested erythrocytes inside the cytoplasm. Nonpathogenic commensals do not derive their nutrition from this source. Another characteristic feature is the presence of a dense body of condensed chromatin material (called the karyosome) in the center of the nucleus and uniform granules of peripheral chromatin lining the nuclear membrane. In contrast, the nucleus of *E. coli*—the species that is most often confused with *E. histolytica*—has a large eccentric karyosome and irregularly arranged chromatin granules. Another commonly mistaken species is *E. hartmanni,* which forms small cysts of a maximum diameter of 10 mm, in contrast to the larger cysts of *E. histolytica.* The pathogen may be differentiated from *E. polecki*—which is usually isolated from animals and only occasionally found in humans—on the basis of erythrocyte ingestion. In addition, the cysts of *E. polecki* have a single nucleus and give rise to a single trophozoite, while *E. histolytica* cysts typically germinate to release four new trophozoites. *See also* AMEBIASIS.

Enterobacter aerogenes

A normal inhabitant of the intestinal tracts of humans and various animals, *E. aerogenes* has been implicated as an opportunistic pathogen in a number of different respiratory, urinary, and gastrointestinal tract infections, usually following the inadvertent transfer of bacteria during surgery. Earlier references to this bacterium may identify it as the genus *Aerobacter.* The bacteria are gram-negative, motile rods with several flagella located all over the cell. Metabolically, *Enterobacter* requires low amounts of oxygen (microaerophilic) and can ferment common sugars such as glucose and lactose. Proper identification and differentiation from other members of the Enterobacteriaceae family is required for proper therapeutic and epidemiological management. *E. aerogenes* may be differentiated from other coliforms on the basis of the IMViC tests. The organism possesses three major types of antigens—somatic (O), flagellar (H), and capsular (K) antigens—which provide the basis for strain differentiation. Antibiotics have been used with some success to deal with infections, although the development of multiple drug resistance in this and other coliforms is a fast-growing problem. The use of careful, aseptic surgical techniques is the best preventive measure for the inadvertent transfer of organisms from the intestinal tract to other locations. *See also* COLIFORM.

enterotoxin

General term for toxic substances produced by bacteria and other pathogens, which affect the water absorption and other properties of the cells lining the intestinal tract and which give rise to symptoms of gastrointestinal distress including vomiting, diarrhea, and stomach cramps. Both endotoxins and exotoxins may have enterotoxic effects. *See also* DIARRHEA.

enterovirus

Small RNA viruses constituting the largest group within the picornavirus family. Common examples of human enterovirus species are the poliovirus, Coxsackie viruses (types A and B), and the enterocytopathic human orphan (echo) virus. There are some 70 individual serotypes based on differences in the capsid proteins. Enteroviruses show a special preference for gastrointestinal tissue, where infection may be non-apparent or result in mild gastrointestinal upsets. However, they may spread to other organs and tissues and cause disorders of varying degrees of severity. Scientists have suggested that at least part of the reason for the infection of specific tissues is the expression of specific viral receptors, although this is not the only reason. The most severe disease caused by picornaviruses is poliomyelitis. Other manifestations of enteroviral infection include febrile illnesses, aseptic meningitis, rashes, conjunctivitis, and myopericariditis. Enteroviruses have also been suggested as a possible cause of chronic fatigue syndrome. *See also* CHRONIC FATIGUE SYNDROME; COXASACKIE VIRUS; ECHO VIRUS; PICRONAVIRUS; POLIOMYELITIS; POLIOVIRUS.

enzootic disease

A zoonotic disease that is endemic in an animal population. *See also* ENDEMIC; EPIZOOTIC DISEASE.

enzyme

A protein molecule that mediates biochemical reactions in living cells. Most of these reactions would not take place without the enzyme, although the protein molecule itself does not undergo any net change as a result of the reaction. Thus, an enzyme may be said to be a catalyst in biological systems.

epidemic

This term is used, especially in the public health context, as a synonym for a disease outbreak—namely, the sudden, unexpected occurrence of a disease amid a large population in a given geographical area over a certain period of time. The use of this term is often restricted to infectious diseases that can spread from person to person. In contrast to an endemic disease, which is always present in a proportion of the population, an epidemic disease is typically a "visitor," with some external source.

Epidemics may be classified into three main groups based on the mode of disease transmission, the duration of the disease, and the conditions of exposure to the infectious agent. In a common source epidemic, for instance, several people are more or less simultaneously exposed to the same source, e.g., a town's drinking water contaminated at the source or contaminated food at a picnic or other large gathering. Propagative epidemics occur due to the transmission of the disease agent from person to person, either by direct contact, such as with smallpox and influenza, or via insect or other vectors, such as with malaria and rickettsial diseases. Possibly the most widespread epidemics are of a mixed nature—they start out by affecting multiple people from one common source but are quickly spread to others as well. The cholera pandemics of the nineteenth century, as well as the infamous Black Death in fourteenth-century Europe, were likely mixed epidemics. Food infections, but *not* toxin-mediated poisoning outbreaks, are also examples of this latter type. *See also* PANDEMIC.

epidemiology

The study of human health at the level of populations, i.e., the *public's* health, rather than that of individuals, the final aim of which is to promote good health and prevent diseases. This definition encompasses a vast terrain of study areas, including such basic medical sciences as microbiology and pathology (to look into the causes of various diseases), and applied disciplines such as psychology, history, and human ecology (to understand the behavioral patterns among populations), so that effective measures might be instituted to prevent or control a specific disease.

epizootic disease

A sudden outbreak (epidemic) of an infectious disease in an animal population. The term is most often used in referring to disease outbreaks among livestock. *See also* ENZOOTIC DISEASE; EPIDEMIC.

Epstein-Barr virus

A member of the gamma subfamily of the double-stranded DNA herpesvirus family, Epstein-Barr virus (EBV) is perhaps best known as the causative agent of infectious mononucleosis. The same virus has also been found associated with two different cancers, Burkitt's lymphoma and a nasopharyngeal carcinoma. This virus may be distinguished from other herpes viruses by its preference for B-lymphocytes and epithelial cells of primate origin. It may also be recognized by its ability to transform or immortalize B cells both *in vivo* and *in vitro*. *See also* BURKITT LYMPHOMA; CANCER; HERPESVIRUS; INFECTIOUS MONONUCLEOSIS.

Erwinia

In a family of bacteria dominated by organisms that prefer to reside in the intestinal tracts of animals and may cause a variety of intestinal infections, *Erwinia* is a unique genus in its choice of host organisms. These bacteria reside in various green plants rather than in animals, and they are associated with a number of wilt and rot diseases of plants. The main determinant of plant pathogenicity of various species of *Erwinia* is an enzyme called protopectinase, which functions by breaking down pectins, the material that makes up the cementing medium of various plant tissues. In addition, these bacteria have capsules that enhance pathogenicity and are motile. Different species of *Erwinia* are capable of fermenting a wide variety of sugars and sugar-alcohols found in plants, producing acid but little or no gas. The ambient temperature for growth is 27–30°C, although some species may grow at body temperature, while certain other strains are psychrophilic and can grow at temperatures as low as 1°C. These latter types are implicated in refrigerator-associated food spoilage. Over 30 species have been identified—*E. amylovora* is the most abundant species, and *E. carotovora* has been found associated with a number of rots, including the "black leg" of potatoes. *See also* FOOD SPOILAGE.

Erysipelothrix

These bacteria were first isolated and identified as the causative agent of "rotlauf," or swine erysipelas, during the final decades of the nineteenth century by Koch and Pasteur. Currently, there are two known species: *E. rhusiopathiae*, which is naturally pathogenic in a wide range of bird and animals and has been associated with human infections on occasion; and *E. tonsillarum*, a normal, non-pathogenic inhabitant of the tonsils in pigs. Strangely, however, this species has been implicated in certain canine diseases.

The bacteria are slender, non-motile rods with a tendency to form filaments. They are gram-positive by virtue of their peptidoglycan cell walls, but may exhibit gram-variable characteristics because they decolorize easily. They are facultative anaerobes that weakly ferment sugars such as glucose to form acid but no gas. Although they may be isolated from the soil on occasion, they typically reside in various animals. Human infections of *Erysipelothrix* may manifest in a number of different ways, including skin lesions with sequelae of septicemia, arthritis, or endocarditis. Primary infection most likely occurs via scratches or other open skin wounds and the diseases are seen most commonly among people likely to contact diseased animals in their profession, e.g., butchers, fish handlers, farmers, and veterinarians. Diagnosis is straightforward and involves the isolation of the bacteria from clinical samples appropriate to the nature of infection. *Erysipelothrix* is susceptible to penicillin, erythromycin, and cephalosporin, and infections may be treated accordingly.

Escherichia coli

A normal and abundant inhabitant of our intestinal tracts, *Escherichia coli* may be aptly regarded as the bacteriologist's guinea pig. Discovered well over a century ago, this bacterium has been used since the 1940s as the model for studying various aspects of bacterial physiology and genetics. The most recent increment to this knowledge was the determination and publication of its entire genome sequence in 1997. Although *E. coli* was not the first microbe to be sequenced in totality, the data obtained is arguably the most informative simply because this is the microbe about which the most is known.

Classified in the family Enterobacteriaceae, *E. coli* is a gram-negative, non-sporeforming, usually motile rod. Most strains lack capsules but have outer membranes made of lipopolysaccharide. Organisms have fimbriae for adhesion to the intestinal wall. Motile strains have peritrichous flagella, and some also have pili, which allow genetic material to pass from one organism to another during conjugation. *E. coli* rods show differences in their antigenic structure based on their somatic (O), capsular (K), and flagellar (H) antigens. These bacteria are microaerophilic and facultatively anaerobic. The organisms draw energy by fermenting glucose but are also capable of fermenting lactose and mannitol. The average size of the genome is about 4.6 million base pairs. There are some 4,300 genes present, although only about two-thirds of these have known functions. In addition to the chromosome, the organism contains plasmids, which it can exchange with other enteric bacilli. Some strains contain the F factor, either in a plasmid or in the chromosome, and are thus capable of transferring genes by conjugation.

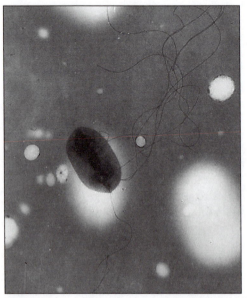

Micrograph of *Escherichia coli* O157:H7. *CDC/ Elizabeth H. White, M.S.*

E. coli is part of the normal intestinal microflora of several animals (including humans), and as a rule, it does not cause disease. In fact, it serves an important function in manufacturing large quantities of the B-complex vitamins and Vitamin K, which are then absorbed into the body via the intestinal lining. It also inhibits the growth of potentially harmful proteolytic organisms in the intestinal tract. Because the organisms are routinely excreted from the body in the stools, public health officials use them as an index for the fecal contamination of water.

Pathogenic strains of *E. coli* do exist, however, and are mostly responsible for different types of intestinal disorders. These strains—collectively referred to as the enterovirulent *E. coli* (EEC)—owe their pathogenicity to their ability to produce exotoxins. One of these, veratoxin, resembles the toxin that causes cholera and is easily destroyed by heat. The second, *E. coli* enterotoxin, is heat stable. Among the first diseases found to be associated with *E. coli* was infantile diarrhea, but there has been an increasing incidence of adult diarrhea associated with this organism. This organism can also cause disease when displaced from the intestine to other locations in the body such as the urinary tract, liver, and gall bladder. In some instances, *E. coli* invades the bloodstream and releases endotoxins, which produce symptoms of the chills, fever, nausea, and vomiting. This is known as gram-negative sepsis or endotoxic shock. In 1996, for the first time, a group of scientists from Israel reported the incidence of an intestinal infection by an *E. coli* strain (O144:NM) complicated by encephalopathy.

Because large numbers of the nonpathogenic organisms are excreted with the stools as a matter of course, diagnoses of *E. coli* infections require immunological tests for pinpointing the pathogen. Concomitantly, it is also important to examine the water supply and all suspected foods—such as unpasteurized milk and raw foods like meats and eggs—for the presence of the organism via routine culturing and biochemical tests. A number of broad-spectrum antibiotics, such as ampicillin, cephalosporin, and tetracycline, have been used successfully in the past, although antibiotic resistance is fast becoming a problem. Fortunately, most of the conditions are self-limiting, and the primary concern in any of the disorders is to control diarrheal symptoms and prevent dehydration. Possibly the single most important preventive measure is to ensure the proper processing of potentially infected foods. In the past few years, some of the most notorious culprits for *E. coli* outbreaks have been ground beef (specifically hamburgers) and apple juice. Cooking the meat and pasteurizing the juice will ensure the elimination of *E. coli* and other pathogens that may be present in the foods.

Based on disease mechanisms and symptoms, there are four classes of enterovirulent *E. coli*.

- Infantile diarrhea is caused by the enteropathogenic group (EPEC) of organisms. These organisms do not typically cause their effects by the enterotoxin, but rather through active infection that disturbs the integrity of the intestine and affects its normal functions. The disease is spread via what is known as the oral-fecal route—the feces of infected children introduce organisms into the sewage system, and any contamination of the drinking-water supply by sewage can lead to outbreaks.

- Enterotoxigenic strains of *E. coli* (ETEC) are responsible for a diarrheal condition not unlike cholera which goes by the common name of "traveler's diarrhea." This syndrome is more common in underdeveloped countries, and there have only been four known outbreaks in the United States. ETEC strains cause their effects mainly due to the production of the veratoxin, which interferes with the water-absorbing ability of the intestinal wall.

- The enteroinvasive *E. coli* (EIEC) cause a form of bacillary dysentery characterized by the appearance of blood and mucus in the stools. The illness is often mistaken for the more widespread shigellosis, which is caused by a closely related member of the Enterobacteriaceae family. EIEC strains are extremely virulent and as few as 10 individuals can induce disease symptoms. Outbreaks have been associated with foods such as hamburger and unpasteurized milk but it is often difficult to pinpoint the source of these infections because the bacteria may be at undetectable levels and still cause disease.

- Finally, there is *E. coli* serotype O157:H7, which is enterohem–orrhagic (and therefore also called EHEC). It is capable of causing sufficient damage to the intestinal wall to induce bleeding due to the production of veratoxin. First discovered as a pathogen in 1982, it was the cause of an epidemic of hemorrhagic colitis in 1985. It has since been implicated in the United States alone as the culprit responsible for outbreaks related to infected drinking water (1990), apple cider (1991 and 1996), fresh produce (1995), and undercooked or raw hamburgers (1993). *See also* CONJUGATION; ENTEROTOXIN; SEPTICEMIA.

eukaryote

Living organisms whose cells have a special, membrane-bound compartment called the nucleus that houses the organism's DNA material (i.e., genes) separately from the rest of the cellular matrix (known as the cytoplasm). The nucleus and cytoplasm of eukaryotic cells may be distinguished under an ordinary light microscope with the use of selective dyes. In addition to the nucleus, eukaryotic cells also possess special membrane-limited structures called organelles (little organs) to perform various functions that the cell needs to carry on living, such as respiration, metabolism, and synthesis. The eukaryotic world encompasses a wide variety of organisms, from unicellular microbes such as yeasts and algae, to multi-cellular macroscopic beings, including plants, fish, corals, sponges, birds, and humans. *See also* CELL; CYTOPLASM; DIPLOID; HAPLOID; NUCLEUS.

eutrophication

An increase in the nutrient content—primarily nitrogen in the form of nitrates and ammonium compounds, but also phosphorus and sulfur (in compound forms)—in bodies of water such as lakes, rivers, and oceans, due to an imbalance in the normal cycles of production and consumption that exist in nature. These imbalances are created naturally or by human activities, e.g., indiscriminate disposal of sewage, chemicals, and other industrial wastes into the water. The presence of excess nutrients encourages a profuse growth of microbes in the water, often leading to the formation of toxic "blooms." *See also* BLOOMS

Evans, Alice C. (1881–1975)

In 1928, Alice Evans became the first woman president of the American Society for Microbiology (then the Society for American Bacteriologists). She was also the first woman to hold a permanent appointment with the U.S. Dairy Division of the Bureau of Animal Industry. In 1917, Evans demonstrated the presence of the pathogen *Brucella* in raw milk. Her personal experience with brucellosis made her a particularly strong advocate of the importance of pasteurization, and it is thanks largely to her efforts that routine pasteurization of milk became mandatory in the dairy industry in the U.S.

exocytosis

A method used by cells to release material from within their cytoplasm to their surroundings. The material to be exported is typically contained within a membrane-bound vesicle so that it is released externally when the vesicular membrane fuses with the cell membrane.

exotoxin

Extracellular proteins—i.e., proteins released into the environment after synthesis—that are produced by bacteria and that are known to adversely affect specific cells or sites in the body. Examples include the diphtheria toxin of *Corynebacterium diphtheriae*, responsible for inducing the formation of the pseudomembrane in the throat, and the botulism and tetanus neurotoxins produced by different *Clostridium* species. Extracellular enzymes that bacteria produce for the specific purpose of breaking down and using materials in the environment are not generally classified as exotoxins.

F

facultative aerobe

A normally aerobic organism that is capable of living under anaerobic conditions for extended periods of time, either by slowing down its metabolism or by turning to sources besides oxygen, such as formate and ferric ions, as the primary electron acceptors. *See also* AEROBE; ANAEROBE; RESPIRATION.

facultative anaerobe

Living organisms that are normally anaerobic but that can survive and grow even in the presence of oxygen. This is a common metabolic type among the bacteria. *See also* ANAEROBE.

fatal familial insomnia

A relatively recent discovery, fatal familial insomnia (FFI) is a prion disease, marked by difficulty sleeping, combined with a dysfunction of the autonomic nervous system (the part of the nervous system regulating activities such as heart rate and perspiration). The major symptoms appear to be related to the specific degeneration of the thalamus in the brain, leading to the formation of the characteristic spongiform lesions associated with these diseases. Dementia usually follows the insomniac phase, and death invariably results. As its name indicates, FFI appears to be inherited rather than acquired via an infection. *See also* PRION; SPONGIFORM ENCEPHALOPATHY.

fermentation

General term for energy-converting metabolic processes that involve the conversion of a substrate without the involvement of an external oxidizing agent. Typically these reactions involve the breakdown of some carbohydrate source into smaller molecules, with the net release of energy in the form of ATP for the organism to store or use in other energy-requiring processes. Many, but not all, fermentation reactions proceed anaerobically. Fermentation begins with glycolysis, or the Embden-Meyerhof-Parnas (EMP) pathway, which converts glucose to pyruvic acid. The pyruvic acid (or pyruvate) is then converted into different end products such as ethanol, lactic acid, and butanol in such a manner that the reducing agent used up during glycolysis is regenerated. The specific final products depend on a combination of factors, including the type of organism, the available substrates, and the enzymes present in the organism.

It should be noted that the definition above corresponds to the biochemical or microbiological concept of fermentation. In the industrial context, however, the term fermentation is used to describe any chemical process that is mediated by microbes, regardless of whether these are actually fermentative or not. For instance, the production of acetic acid or vinegar, which is actually an oxidation of ethanol to acetic acid conducted by nonfermentative aerobic bacteria, is also called fermentation. *See also* ATP; GLYCOLYSIS; METABOLISM.

filovirus

This family of viruses, which consists of agents of certain viral hemorrhagic fevers, are among the newly emerging pathogens to have gained prominence in the past few decades. Well-known entities within this group include the Ebola and Marburg viruses. Structurally, these viruses are identical and appear as long filamentous structures consisting of a helical nucleocapsid within a tightly bound lipid envelope with viral proteins on the surface. The genome is a linear, single-stranded RNA molecule of negative polarity with complimentary ends. Individual filoviruses differ with respect to their antigenic structure.

Disease symptoms are typical of hemorrhagic fevers, with acute onset of fevers, muscular weakness and pains, and headaches some 2–21 days after infection, followed by systemic signs such as diarrhea, vomiting, pharyngitis, and a rash. The hemorrhagic manifestations of the disease—namely pinpoint bleeding (petechiae) and bleeding along the gastrointestinal tract—become apparent rather quickly (around the third day of infection). Of all the viruses known to cause hemorrhagic fevers, Ebola and Marburg have the highest fatality rate as well as the severest symptoms. Diagnosis may be achieved by virus isolation or by serological tests, although there are no specific drugs or vaccines against these viruses. Treatment is therefore mainly supportive.

Exactly how humans first acquired Ebola virus infections is not clear because the natural reservoirs have not yet been ascertained. The first outbreak of the Marburg virus was traced to the handling of infected tissues from African green monkeys. Infection is spread through close contact, including sexual intercourse, and control methods are aimed at minimizing close contact between infected and infection-free individuals. Careful handling of clinical and other laboratory specimens and proper treatment and disposal of laboratory equipment and materials (particularly in primate labs), are some of the most important precautionary measures against unexpected outbreaks. *See also* EBOLA VIRUS; EMERGENT INFECTION/PATHOGEN; HEMORRHAGIC FEVER; MARBURG VIRUS.

filtration

Technique of passing fluid materials—either liquids or gases—through a physical barrier so as to separate some components from the others. The smallest sized particle that can pass through the barrier or filter is determined by the screening material and the size of the spaces or pores present in the filter. Examples of the applications of filtration of fluids include the purification of drinking water and the sterilization of heat-sensitive fluids, such as serum or antibiotic solutions. The most common bacteria-proof filters are made of cellulose membranes with pores less than 0.75μm in diameter, which hold back most bacteria. Today, technology has advanced to a point where it is possible to make filters with pores of 10 nm diameter, which are also capable of screening out many viruses (ultrafiltration). Filtration is also the most widely used approach to sterilize air and other gases used in safety cabinets and air conditioners. In those cases, the filtering medium is typically a glass-asbestos screen.

fimbria

Extracellular protein appendages of bacteria, usually gram-negative organisms, which may

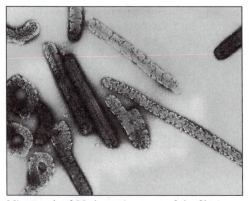

Micrograph of Marburg virus, one of the filoviruses. *CDC/Dr. Erskine Palmer.*

be present all over the cell surface or localized in one spot. Fimbriae that are distributed all over the bacterial cell help the organisms adhere to other cells, a particularly useful property for pathogens. Localized fimbriae impart a twitching type of mobility to the cells.

flagella

Thread-like structure protruding from a cell, that functions as the organ of motility in a single-celled organism or in the mobile cells (e.g., sperm) of a multicellular organism. Although the basic function is the same—they act like propellers to enable the organism to swim in liquid environments—the flagella of prokaryotes and eukaryotes have markedly distinct structures.

A bacterial flagellum is made up primarily of a protein called flagellin. It originates inside the cell and is powered by an electrical motor–like apparatus at its base. The number and arrangement of flagella is characteristic for a given bacterial species. Bacteria may be monotrichous (with a single flagellum) or possess several flagella. These may be regularly distributed all over (peritrichous), present in tufts (lophotrichous), or in polar arrangements (amphitrichous). The visualization of flagella presents particular difficulties because they are slender and do not take up ordinary stains. Therefore, these structures are viewed with the help of mordents, such as tannic acid, which coat the flagella and make them wide enough to be perceived under an ordinary light microscope. *See also* MOTILITY; STAINING.

flavivirus

Family of mostly arthropod-borne (tick-borne) RNA viruses associated with such human diseases as yellow fever, dengue, and viral encephalitis, which are among the most important viral diseases in the developing world. Originally classified under the arbovirus label as the Group B viruses, the flaviviruses were separated into an independent family based on genomic characteristics, antigenicity, and replication strategies. As it stands today, this family contains not only the original arbovirus members, but also certain non-arthropod-borne members, including the hepatitis C virus and certain veterinary pathogens called pestiviruses. Viruses are spherical, enveloped particles, consisting of an inner core made of a single protein surrounded by a tightly attached lipid envelope derived from the host. The flavivirus genome is a single-stranded, positive-sense RNA molecule of about 11 kilobases, capable of infecting cells on its own, i.e., when naked. Replication occurs in the cytoplasm. *See also* ARBOVIRUS; DENGUE FEVER; ENCEPHALITIS; HEPATITIS C VIRUS; YELLOW FEVER.

Fleming, Alexander (1881–1955)

British scientist who is best known for his almost accidental discovery of penicillin, the first antibiotic substance known to humans. A chance contamination of certain staphylococcal cultures with the fungus *Penicillium notatum* led to the observation that colonies of bacteria were killed (or inhibited) by the mold. This piqued Fleming's interest because of his previous interest in naturally produced bacteriostatic and bactericidal agents. Earlier in 1921, Fleming had also discovered lysozyme in human saliva and tears. After discovering the killing properties of the mold, he proceeded to culture it in the lab and purify the material responsible for inhibiting the bacteria, which he named penicillin. His seminal discoveries earned him several honors throughout the world, the most noteworthy being knighthood in his native England and the Nobel Prize in 1945. *See also* CHAIN, ERNST; FLOREY, HOWARD WALTER; PENICILLIN.

Florey, Lord Howard Walter (1898–1968)

Leader of the scientific group that investigated the usefulness of the antibiotic penicillin, and developed standards for the required dosage and safety levels of this drug. For this work, Lord Florey shared the 1945 Nobel Prize with his colleague, Ernst Chain, as well as with Sir Alexander Fleming, the scientist who discovered penicillin. *See also* ANTIBIOTIC; CHAIN, ERNST; FLEMING, ALEXANDER; PENICILLIN.

fluorescent microscopy

A technique in optical microscopy that relies on the use of fluorescent —i.e., glow-in-the-dark—dyes or stains to achieve contrast between an object and its surroundings. This technique is also a good method to observe organisms that are themselves fluorescent. The light source in this case is an ultraviolet lamp, which excites the fluorescence of either the dye or the organism. Objects are thus seen as bright spots against a black background. These microscopes are typically operated in the dark.

There is no particular advantage in using fluorescent dyes as a substitute for ordinary stains in bacteriology, and fluorescent microscopy is most widely used in immunology. Using antibodies coupled with fluorescent markers, scientists may be able to identify specific organisms or even localize the presence of specific antigenic markers on a single organism. *See also* MICROSCOPY; OPTICAL MICROSCOPY.

fomite

An inanimate object that functions as the vehicle for transmitting infections from different sources—including people, animals, or the environment—to the (typically unsuspecting) patient. Virtually any object that comes in frequent contact with the source of infection might play the role of a fomite. Common examples include eating utensils, clothing, bed linens, doorknobs, and hospital equipment. *See also* DISEASE TRANSMISSION.

food-borne diseases

In the context of microbiology, the food-borne diseases are a large group of disorders that are acquired by the consumption of foods contaminated with microbes. The causative organisms may be bacteria, viruses, protozoa, or fungal parasites. The term *food poisoning* is often used to describe these diseases, but it should be noted that this term also cn-compasses a number of illnesses caused by nonmicrobial contaminants such as organic toxins and heavy metals.

Food-borne diseases of microbial origin may be divided into two main classes. In the case of food infections, the contaminating organisms establish themselves in the body of the host and cause the disease. A number of diarrheal infections by organisms such as *Escherichia coli*, *Shigella*, typhoid fever, cholera, trichinosis, and hepatitis A are the result of food-borne infections. The second category of food-borne diseases, called food intoxication, contains diseases caused by toxins that are produced by the microbes in the food material prior to their ingestion. Common examples include botulism, food poisoning due to staphylococcal toxins, and fungal aflatoxins. The presence of the organisms in the food during consumption is not a requirement in these diseases.

A typical feature of food-borne infections is the sudden occurrence of symptoms among a number of people within a short period of time, traceable to a common food source. Picnic foods, particularly those containing cream or raw eggs, are notorious sources of *Salmonella* infections. Many food-borne outbreaks are closely linked to water-borne infections, because contaminated water used for washing or preparing foods is often the source of the offending microbes. This is particularly true in the less developed countries of the world. Among the most common sources of infections such as trichinosis and brucellosis are raw and undercooked meats, especially in tropical countries. Shellfish that grow in waters contaminated with improperly treated sewage tend to accumulate and harbor pathogens and have been associated with a number of sporatic outbreaks of diseases such as cholera.

Regardless of the specific cause of a disease, the prevention and control measures for various food-borne infections rely on the same basic principles: pinpointing the source, avoiding consumption of the contaminated foods, and destruction or treatment of the implicated foods. A prompt laboratory evaluation of the disease cases and suspected foods is essential both for determining the best course of treatment and for preventing the spread of the disease. Techniques such as pasteurization should be used to process foods such as dairy products that often serve as sources of infections. Public education regarding the common sources of infection, and proper methods of food handling and storing, are important elements in controlling food-borne diseases. The World Health Organization (WHO) has developed a set of common-sense guidelines for safe preparation which include such measures as ensuring food is properly cooked, storing foods carefully using safe water, reheating of stored foods to ensure the destruction of any microbes that may have survived or even multiplied during storage, maintaining hygienic food-preparation practices, minimizing contact between raw and cooked foods, and protecting foods from exposure to insects, rodents and, other potential vectors.

food chain/food web

The patterns of transfer of energy and matter—mostly carbon, but also other essential elements including oxygen, nitrogen, and sulfur—among various inhabitants of the biosphere. There is some hierarchy in this transfer, which gives the appearance of the formation of a linear "chain." However, because many organisms consume more than one type of food source, the food chains become interconnected. Thus, the term "food web" is a more accurate description of the consumption patterns. Plants, cyanobacteria, phytoplankton, and certain chemolithotrophs form the basis of the web and are the primary food sources because they are the only members of the biosphere that are able to harness energy from nonliving things, using sunlight and chemical energy. Microbes are also responsible for converting various inorganic forms of the essential elements and incorporating these into different biomolecules.

The food web consists of many processes with energy and matter moving in different directions of evolution. In a predator chain,

for instance, plants are consumed by plant eaters, which in turn are consumed by larger carnivorous animals (or fish, in a marine environment). A parasite chain is one in which energy moves downward in evolution—e.g., from a mammalian host to a bacterium or fungus, which in turn is colonized by a still smaller parasite. The maintenance of balance between various compartments of the food web is important for sustaining the earth's ecosystems.

food inspection

The examination of saleable food by certified officials so as to assure the production of clean and wholesome foods, free from microbial contamination, chemicals, decomposition, or physical damage that may have occurred during the production, processing, packaging, or transport of the foods.

food irradiation

Exposure of food products to relatively high-energy radiation as a means of preservation from microbial spoilage. The most widely used types of radiation for this purpose are gamma radiation (from radioactive cobalt or cesium) and electron beams from synchrotrons. Irradiation is often used to treat processed foods prior to canning so as to destroy the spores of potential pathogens such as *Clostridium botulinum*. Fresh produce, meats, grains, and other foods that are to be transported or held in markets for extended periods of time are also irradiated to reduce the number of spoilage organisms on their surfaces. This approach is limited to the treatment of foods that require only surface sterilization. Similar technology is applied for the sterilization of medical supplies such as syringes, intravenous fluids, and surgical gowns and gloves.

food preservation

Nearly all types of natural foods, as well as many processed edible products, are susceptible to spoilage by the growth and activity of various microbes. Depending on the source of the food, its chemical properties, and the way it is handled and stored until consumption, the changes in the food may range from alterations in taste and texture (organoleptic qualities) to more drastic consequences, such as food infections or poisoning. Thus, various techniques of preservation are aimed at altering the microenvironment of the food products so that harmful microbes are either destroyed outright or discouraged from growing. Sterilization techniques include high heat and irradiation, although these have limited value in food preservation because of the significant changes they cause in the organoleptic qualities of the original food. Of considerably more significance in the food industry are processes such as pickling, salting, canning, and pasteurization, which alter one or more of the environmental factors—e.g., temperature, pH, moisture content, and salinity—in such a way as to create unfavorable conditions for the growth of microbes in the food product. For instance, the pickling of vegetables in a mixture of vinegar and brine both lowers the pH of the food and increases salinity, creating conditions unfavorable for the proliferation of most contaminants. Another simple method of food preservation is refrigeration, which is sufficient to stall the growth of most microbes (except the psychrophiles).

food spoilage

Any process that results in damage or injury to food products in such a manner as to render the food undesirable for human consumption. Food may spoil for any number of reasons, including mechanical or physical injury, changes in chemical composition due to temperature changes or enzymes, and the presence of biological agents such as insects and microorganisms.

The basic mechanism by which microbes induce food spoilage is by metabolizing different substances in the food for their own growth. With the breakdown of its contents, different physical and organoleptic properties of the food product, such as texture, consistency, taste, and aroma, are changed.

Bacteria such as *Erwinia carotovora* and *Pseudomonas marginalis*, for instance, cause the rotting of fresh vegetables or fruit—called bacterial soft rot—by breaking down the pectins in these foods, which gives rise to a mushy, watery consistency, often with a bad odor due to some gaseous metabolites. The black leg of potatoes is a specific example of bacterial soft rot. Some examples of fungal rots are a gray mold rot of several summer vegetables caused by *Botrytis cinerea* and the downy mildew of lettuce by a species of *Bremia*.

In addition to plant products, bacteria and fungi also spoil a variety of fresh and preserved meats (including poultry and seafood) and dairy products. Both the specific spoilage organisms, as well as the nature of spoilage, depend on a host of influences, such as the pH, moisture content, nutrient content, storage temperatures, and the nature of the infecting organisms—namely, their tolerance for various conditions, specific enzymes they produce, and the action of these enzymes on the food material. Often there is a breakdown in the protein content of the food, which gives rise to putrefying odors, as well as changes in texture. Fermentation in foods such as milk drastically changes the pH of the medium, which causes the milk to curdle. A complete list of the organisms implicated in spoilage, and the nature of the damages they cause, is well beyond the scope of this volume; students are therefore advised to consult a book on food microbiology for more detailed information. *See also* ERWINIA.

foot-and-mouth disease

This highly infectious, usually nonfatal disease in cloven-hoofed animals, such as pigs, goats, sheep, and cows, is caused by aphthoviruses. It is mostly prevalent in parts of Africa, Asia, and South America. Recently eradicated in Europe, it figures as high priority among the reportable veterinary diseases in North America because of the heavy toll it exacts in terms of lost production of meat and milk, loss of animals during eradication, and loss of export markets.

The virus is highly contagious and spread easily over large distances due to the movement of infected animals or contaminated food and fodder. The most commonly affected animals are pigs, which ingest the viruses in contaminated foods. These animals shed large amounts of viruses via aerosols, which are the main means of spread to cattle and other animals. This shedding begins long before the clinical signs of the disease become apparent, often resulting in the spread of the disease even before it manifests itself in the source animal. Foot-and-mouth disease symptoms typically appear some 3–8 days after primary infection (or up to 21 days later) and include the formation of painful fluid-filled blisters in the areas around mouth and feet, resulting in excessive salivation, depression, appetite loss, and lameness. Fevers and decreased milk production may precede these blisters. The blisters may rupture to leave large unprotected areas that are susceptible to secondary infections. Foot-and-mouth disease is most severe in cattle. Efforts are currently underway to eradicate this disease worldwide. *See also* APHTHOVIRUS.

Fournier's disease

Also known as Fournier's gangrene, this disease is an acute inflammatory condition of the genital area, typically the scrotum, penis, and perineum, caused by any number of gram-positive organisms, enteric bacteria, and anaerobic bacteria not normally implicated in genital infections. This gangrenous infection often presents as a complication of local trauma, surgery, or urinary infections. Drug treatment should be administered based on the identification and antibiotic sensitivities of the specific causative organisms.

Fracastoro, Girolamo (ca.1478–1553)

Described variously in different sources as a physician, astronomer, geographer, or poet, this Italian was one of the first individuals to have advanced insights into the microbial nature of infectious disease. Fracastoro is recognized today both for his work on syphilis

and his ideas on "contagion." In his written work, he spoke of disease agents in terms of "seminaria," which may be roughly translated in English as "seeds" or "germs." Although not explicitly stated in his writings, some historians believe that he considered these invisible seeds to be living entities.

Francisella tularensis

This highly contagious bacterium is the causative agent of a systemic disease in humans known as rabbit fever or tularemia. In nature, these bacteria are found in a variety of mammals such as rabbits, rodents, dogs, and deer and are usually transferred from one host to another via fleas that infest these animals. The bacteria are gram-negative, non-motile, extremely tiny rods that are strictly aerobic in nature. *F. tularensis* has a specific requirement for the amino acid cysteine in the growth medium, in addition to proteins from blood or serum. Due to its highly contagious nature, the cultivation of this organism, even for the purposes of diagnosis, is restricted to laboratories with special safety equipment. *See also* TULAREMIA.

fumigation

The application of different types of fumes—smoke, vapor, or gas—for disinfecting areas such as houses and public buildings. Fumigation is often used as a public health measure to rid areas of microbial or insect (termite) pests.

fungi

A dozen different images spring to mind when one thinks about the fungi—mushrooms and truffles, blue cheese, the black and white molds that grow on bread and fruit, yeast granules, and the patchy lesions of athlete's foot. The various life forms that we classify together as the fungi share the basic properties of a eukaryotic cellular form and a heterotrophic mode of nutrition, (they derive their energy and carbon from organic molecules in the environment). These organisms are important not only because they cause a number of disease conditions (collectively referred to as mycoses) in humans, but also because they produce a number of antibiotics that have proven useful in treating bacterial and fungal infections. In addition, various fermentations performed by the fungi are central to the making of edible products such as bread, beer, wine, and cheese.

There are over 200,000 individual organisms in the fungal kingdom, displaying a wide variety of morphological forms and reproductive modes but showing little metabolic diversity. Fungi resemble higher plants in that they are not capable of movement or locomotism from one place to another—and so live fixed in one spot for the duration of their lives—and in that they possess cell walls outside the cell membrane. Unlike plants, however, fungi are haploid; lack differentiated parts such as roots, stems, and leaves; and, most importantly, do not perform photosynthesis. They are completely heterotrophic, and for the most part saprophytic—i.e., living off dead organic matter, although a few parasitic forms do exist. Thus, their preferred habitats are damp environments with an abundance of organic material; they grow best in soil amid dead and decaying plants and animals. Morphologically, the fungi may consist of both unicellular and multicellular forms. Yeasts typically exist as single cells that appear to give a positive reaction when treated with the Gram stain, although the cell wall does not contain peptidoglycan or resemble the bacterial cell wall in any other manner. The basic structural unit of a multicellular fungus (commonly known as a mold) is a slender filament called a hypha, which may or may not be divided into compartments/cells by septa or walls. Multiple hyphae form an interconnected network or mat called a mycelium, which constitutes the vegetative part of the fungus and is responsible for feeding. The unicellular and multicellular modes of fungal life are not always mutually exclusive. For instance, some single-celled yeasts are capable of changing into cotton-like masses under certain environmental conditions. Such fungi are said to be dimorphic in nature.

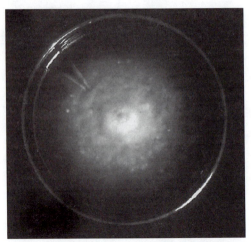

Culture of *Histoplasma capsulatum* showing the typical fuzzy appearance of a mold colony. *CDC.*

Fungi display both asexual and sexual modes of reproduction. The simplest mode of replication is budding, typically seen in the single-celled yeasts. A small spore begins to bud from the parent cell, matures while still attached, and then separates to form a new individual, which in time will also begin to form budding spores. During the maturation phase, the parent cell steps up the production of various cellular materials and replicates its DNA. Vegetative or asexual reproduction in multicellular fungi typically takes place in aerial hyphae that grow vertically from the mycelium to form the fuzzy, cottony growth that we typically associate with fungal growth on stale bread, cheese, and other food products. The basic unit of propagation is called a spore, which is the smallest structural unit capable of being separated from the parent cell or mycelium and developing into a new organism. The simplest spores are formed directly from vegetative hyphae, either by budding from the tips or by the enlargement of certain hyphae, accompanied by a thickening of the cell wall. Some spores are formed only in special hyphae called the conidiophores.

Both single-celled yeasts and multicellular fungi can undergo sexual reproduction. In the former case, two cells merely fuse to form a diploid spore, which then undergoes meiosis to create new organisms that contain genetic material from both parent cells.

In molds, sexual reproduction is initiated by the differentiation of vegetative cells into special female and male structures. The cytoplasm of these two structures fuse together and form dikaryotic (with two nuclei) hyphae, which give rise to fruiting bodies. Nuclei from the two different cells then fuse together inside the fruiting body. Once the nuclei fuse, the cells undergo meiosis to form haploid spores, which contain genetic information from both parent nuclei. These spores are released into the environment and initiate the vegetative phase of the fungal life cycle in a new location. The major criteria used to classify different fungi into families include their morphology, modes of reproduction, and type of spores. *See also* YEAST.

fungus ball

Mass of fungal hyphae—of organisms such as *Aspergillus*—that forms in cavities of the lung, due to the germination of inhaled spores. Individuals with fungus balls in their lungs may be completely asymptomatic or may develop nonspecific symptoms such as coughing. The fungus ball—which is also known as a mycetoma— may also serve as a site for secondary infection with bacteria, which can lead to the development of severe pulmonary and respiratory symptoms.

G

Gajdusek, D. Carleton (1923–)

Working during the 1960s, this scientist was the first to attempt to delve into the mechanisms underlying the kuru disease of the Australian Aboriginals, and to draw connections between the pathogenesis of kuru and those of sheep scrapie and Creutzfeld-Jacob disease. Until Gajdusek conducted his investigations, these diseases were not ascribed to any specific causative agents, but, with meticulous experimentation, he provided evidence for the infectious nature of the disease agent. In a tangible sense, Gajdusek's work set the stage for Stanley Prusiner's discovery of the prions. In 1976, Gajdusek was a co-recipient of the Nobel Prize along with Baruch Blumberg, whose work on hepatitis B, though unrelated, shared a common thread with Gajdusek in that both scientists suggested "new mechanisms for the origin and dissemination of infectious diseases." *See also* CREUTZFELD-JACOB DISEASE; KURU; PRION; PRUSINER, STANLEY; SCRAPIE; SPONGIFORM ENCEPHALOPATHY.

Gallo, Robert C. (1937–)

Probably best known for his role in determining the causative agent of AIDS, Gallo also conducted pioneering research that led the way to the discovery of HTLV, the first RNA virus shown to be associated with cancer (in this case, leukemia) in human beings. He has twice been awarded the prestigious Lasker Awards: the first in the basic research category in 1982 for his work on leukemia viruses and the second in 1987 in the clinical research category for his HIV/AIDS research. *See also* AIDS; HIV; HTLV; MONTAGNIER.

gamete

Haploid cell that represents the mature sexual stage of an organism. Gametes may be found in those fungi, algae, and protozoa capable of sexual reproduction. Prokaryotes do not form true gametes. The fusion of two gametes results in the formation of a diploid zygote. *See also* EUKARYOTE; HAPLOID; ZYGOTE.

gametocyte

A cell that gives rise to gametes or sex cells in organisms capable of sexual reproduction. *See also* GAMETE.

gas gangrene

Rotting disease of soft and muscular tissues associated with infection of *Clostridium* species, most often *C. perfringens*, but also *C. septicum* and *C. novyi*. The two main requirements for this disease to take hold are a clostridial infection and an anaerobic environment, which stimulates the growth of these bacteria. A common route of infection is via open wounds exposed to contaminated soil— for example, during World War I, gas gangrene claimed the lives of many soldiers because shrapnel wounds dirtied with contaminated soil were left untreated for long periods. Gas gangrene is also seen as a postoperative complication, particularly in patients whose natural skin and intestinal microflora includes these organisms. Diabetic patients undergoing surgery are particularly susceptible to this disease because the decreased blood flow and oxygenation of tissues provides the anaerobic conditions conducive to the clostridia.

The main symptoms of gas gangrene are the destruction of tissue proteins (gangrene) accompanied by swelling in the infected area with pockets of gas produced by the vigorous fermentation of tissue components such as glycogen by the infecting organisms. The gas tension within the tissues reduces blood flow and oxygenation, further improving conditions for the growth of the anaerobic clostridia. Tissue destruction is worsened by the action of the various exotoxins produced by the clostridia. The disease may be controlled by a combination of antibiotic therapy and surgery (to remove necrotic tissue). In recent times, these measures have been augmented with hyperbaric oxygen therapy to increase oxygen tension in the tissues, which has the dual benefit of discouraging the proliferation of bacteria and enhancing the activity of the patient's immune system. *See also* CLOSTRIDIUM PERFRINGENS; HYPERBARIC OXYGEN THERAPY.

gas vacuole

Cluster of tiny, gas-filled vesicles found in the cytoplasm of certain aquatic, usually photosynthetic, bacterial species, e.g., *Halobacterium*. The main function of these vacuoles is to provide buoyancy to the cells. This property is especially important in photosynthetic organisms because it plays a direct role in determining the intensity of sunlight received. The walls of individual vesicles of the gas vacuoles are made of protein rather than lipids. *See also* VACUOLE.

GC percent. *See* GC RATIO

GC ratio

A numerical value denoting the total amount of guanosine (G) and cytosine (C) bases in a DNA molecule as the percentage of its total nitrogen-base content. Mathematically, GC ratio = $[(G + C) / (A + T + G + C)] \times 100$. It is expressed in units of moles percent. In microbiology, the GC ratios of the DNAs of various prokaryotic organisms are used to compare their taxonomic or evolutionary relatedness. Generally, two bacteria with similar GC ratios are considered more closely related than those with values that are further apart. However, this is not a consistent relationship because two organisms with similar GC ratios may actually have different sequences and genes and not be related at all. The converse, however, is not true, and two organisms with large differences in GC ratios are likely unrelated.

gene

Long before the discovery of DNA, the term gene was coined to represent a functional and physical unit of heredity, i.e., a physical entity that determined the presence of a particular observable trait in an organism and passed this trait on to the next generation. In a sense, this definition still holds true because the physical entity we call a gene today is, in fact, a discrete linear segment of DNA, which functions as the set of instructions for the production of a single protein product in living cells, and which also passes on this information to the progeny cells during cell division (except perhaps when mutations occur). However, the one-to-one relationship between a gene and a trait—implied by the original definition—is not always present. For example, the idea of a gene for "eye color," as such, is incorrect. Rather, the gene or genes that are responsible for this trait encode one or more protein enzymes, which act on specific substances within the cell to produce a pigment with the specific eye color in question. The information in a gene is stored in the form of specific sequences of nucleotides; this genetic information is then translated into the cognate protein with the help of appropriate enzymes via an intermediary RNA molecule called the messenger RNA. In the case of the RNA viruses, genes are encoded into the RNA sequence rather than into DNA. *See also* CHROMOSOME; DNA; GENETIC CODE; PROTEIN; RNA.

gene expression

The synthesis of the product encoded by a gene. Although this term might conceivably refer to either RNA or protein products, it is generally used in the context of protein synthesis. The expression of a gene in a unicellular organism thus determines its phenotype. Multicellular organisms will typically exhibit differential gene expression. For example, despite the fact that all cells in a human body have the same set of chromosomes and genes, only specific cells in the irises express the genes for eye color, while the expression of the insulin genes is localized in the pancreas. *See also* GENE; GENOTYPE; PHENOTYPE; TRANSCRIPTION; TRANSLATION.

genetic code

Codified form in which DNA molecules store their information. This code is specified by the linear arrangement of the four nitrogen bases—adenine, thymine, guanine, and cytosine (A, T, G, and C)—read in a single direction determined by the orientation of the sugar backbone. Because of the directional nature of the genetic code, the two complementary strands of a single piece of DNA encode different genes.

The sequence of nucleotides in DNA (or its complement in RNA) also functions as the template for the sequence of amino acids in proteins. However, decoding the language of nucleic acids into that of proteins is not simple because the latter have 20 basic building blocks instead of just 4. To deal with this disparity, nature devised a specific relationship between nucleotide and amino acid sequence, called the triplet code. By this scheme, each amino acid is encoded by a sequence of three nucleotides called a codon—for instance, a triplet of TTT in DNA codes for lysine and CAT codes for methionine. There are 64 (4^3) possible codons but only 20 amino acids. Therefore, an amino acid may be specified by more than one codon; i.e., the genetic code is redundant. In fact, only methionine and tryptophan (UGG) have unique codons; all the others display a redundancy varying from two to five. The codons UAA, UAG, and UGA constitute a special group of what are called the terminator or stop codons. They function as punctuation marks in the DNA sequence, signaling the end of the instructions for one protein in a continuous strand of nucleic acid. Triplet sequences that are neither amino acid specific nor stop codons are called nonsense codons. With one or two exceptions, the genetic code is universal among all living things. *See also* AMINO ACID; DNA; GENE; PROTEIN; TRANSCRIPTION; TRANSLATION.

genetic engineering

The *in vitro* introduction of genes from one organism into another so that the recipient organism acquires certain new properties upon gene expression. The main vehicles used to transfer the genes are relatively short pieces of DNA, such as viruses and plasmids. Using a combination of restriction endonucleases and other enzymes, the specific gene is "spliced" into the viral genome or plasmid—not unlike the splicing of an audio cassette tape—and introduced into the cell of the target host (or cells, when dealing with diploid organisms). The earliest attempts at genetic engineering involved the transfer of genes between species; for example, the "superbug" of the 1960s was a single bacterium that contained spliced genes for various oil-degrading enzymes that were normally present in other bacteria. Since then, technology has advanced sufficiently to allow the transfer of genes across species, and it is now possible for us to express the genes of one organism in another unrelated species. *See also* RECOMBINANT DNA; RESTRICTION ENZYMES; TRANSDUCTION.

genome

The complete set of genes present in any living organism. The genome of a prokaryote is typically contained within a single chromosome (with a few traits occasionally specified in plasmids), while eukaryotic genomes are much larger. For example, the complete human genome has 23 chromosomes, each of which is several orders of magnitude larger than the average bacterial chromosome. *See also* CHROMOSOME; DNA; GENE.

genotype

The genotype of an organism is the complete set of genes present in its genome. The genotype represents the potential of an organism to produce various proteins, whether or not it may be doing so at any given instant in time. *See also* GENE; GENE EXPRESSION; PHENOTYPE.

genus

A taxonomic category that groups together organisms that are essentially alike in all the broad features and differ only with respect to a few details. Genus is the first part of the identification label for any living organism in the binomial system of nomenclature. Hierarchically, the genus is subordinate to the family and superior to the species category. *See also* BINOMIAL NOMENCLATURE; LINNEAUS, CARL; SPECIES.

Germ Theory of Disease

Historically speaking, the idea that infectious diseases are caused by living entities—microbes or "germs" that are invisible to the naked eye—is a relatively recent development. Perhaps the earliest references to the possibility of a "seed" or "germ" for initiating disease was made by Girolamo Fracastoro during the early part of the sixteenth century, but this was before the invention of the microscope, and his conjectures were not verifiable in any way. Nearly a century later, Antoni van Leeuwenhoek began his now-famous correspondence with the Royal Society by describing his observations on all

manner of living "animalcules," but even then, scientists did not make the connection between his findings and the agents of various diseases. Two centuries later, Agostino Bassi conducted investigations into the cause of a disease in silkworms and pinpointed a microscopic fungus as the culprit. He published his findings in an 1835 monograph, which is believed to be the first documented evidence forwarding the idea of the microbial cause for disease. A few decades later, Louis Pasteur and Robert Koch made their grand entrance into the field of microbiology and laid the foundations of our modern understanding of infectious disease. *See also* BASSI, AGOSTINO; FRACASTORO, GIROLAMO; KOCH, ROBERT; LEEUWENHOEK, ANTONI VAN; PASTEUR, LOUIS.

German measles

Human disease caused by the rubella virus. Perhaps its most common guise is a mild syndrome characterized by a measles-like rash, low-grade fever, and lymphatic involvement in children and young adults. Rubella (German measles) rashes are more erratic than rashes associated with measles, and may or may not appear, which makes the clinical diagnosis of this disease difficult. Adults who contract the infection may suffer more severe symptoms, and a significant proportion of affected individuals develop arthritis. Rubella infections in pregnant mothers in the first trimester have particularly disastrous effects due to the teratogenic potential of the virus.

The rubella virus enters the body via the respiratory route and travels from the mucosal cells of the nose and pharynx to local lymph nodes where it multiplies and causes viremia some 2–3 weeks after infection. Patients become infective after about one week and begin to shed virus into respiratory secretions. The appearance of the rash corresponds to the viremia stage. Diagnosis may be achieved by virus isolation from nasal secretions and through serological means. No specific treatment is available, although patient isolation is recommended to prevent spreading the virus. The development of vaccines (now administered as MMR, in combination with vaccines for measles and mumps) has greatly reduced the incidence of the disease in the United States. *See also* MMR VACCINE; RUBELLA VIRUS.

Gerstmann-Straussler-Scheinker syndrome

A rare neurodegenerative prion disease in humans, which occurs at a rate of 2 percent of the frequency of the Creutzfeld-Jacob disease (CJD). This disease, whose onset is somewhat earlier (appearing in individuals in their 40s or 50s), is characterized by the development of spongiform lesions in the cerebellum rather than in the cerebrum, leading to motor problems, rather than the dementia characteristic of CJD. Probably because of the location of damage and age of incidence, Gerstmann-Straussler-Scheinker (GSS) syndrome appears to run a longer course before resulting in death. It appears to be a familial disease, although sporadic cases have also been noted. *See also* CREUTZFELD-JACOB DISEASE; PRION; SPONGIFORM ENCEPHALOPATHY.

Giardia

This intestinal protozoan holds a special place in the history of parasitology because it was the first organism of its category—i.e., protozoan—to be observed under a microscope. Antoni van Leeuwenhoek discovered the

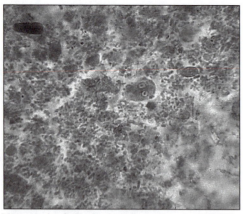

Micrograph of *Giardia lamblia* trophozoite. *CDC/Dr. George R. Healy.*

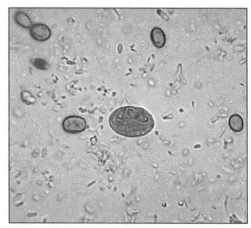

Micrograph of *Giardia lamblia* cyst. *CDC.*

motile stage of *Giardia* while examining his own stools under the microscope sometime around 1681, and wrote a detailed description of the microorganism in one of his many letters to the Royal Society. Although Leeuwenhoek was suffering from a bout of diarrhea at the time, the connection between his illness and the presence of the *Giardia* parasites was not one the scientific community made until after the World War II.

One of the most common and widely distributed intestinal protozoa in the world, *Giardia* is a flagellate organism with a relatively simple life cycle alternating between an active trophozoite and a metabolically inert cyst stage. The trophozoite is a flagellated cell with four pairs of flagella and two nuclei and an external sucking disc. This form of the parasite has a distinctive teardrop front view and a concave spoon-like appearance when seen sideways. The sucking disc is present in the concave side. Multiplication is by simple binary fission. The trophozoite is the stage that resides in the intestinal tracts of animals, where the parasite attaches to the epithelium by means of its disc. The most common location appears to be the duodenum, and for reasons not entirely understood, cyst formation occurs as the organisms move down the colon. The cysts are rounded or oval cells formed by the retraction of the flagella. They are passed out along with the feces and may reside for long periods in the soil. Infection occurs via the ingestion of contaminated water or of foods washed in this water. As a cyst undergoes maturation, its internal structures duplicate themselves so that it gives rise to two new trophozoites upon excystation. The representative species of this parasite is *G. lamblia*, also sometimes known as *G. intestinalis* or *G. duodenalis*.

Giardia infections may be asymptomatic or cause an acute illness (typically within 7–10 days of infection) with a variety of gastrointestinal symptoms, notably diarrhea with pale, greasy, foul smelling stools accompanied by flatulence, cramps, and low-grade fever. The growth of this organism appears to affect the fat-absorption properties of the gastrointestinal epithelium, which accounts for the quality of the stools. The infection remains localized to the intestinal tract. The acute disease may progress to a chronic form wherein the patient suffers from intermittent bouts of diarrhea alternating with normal stools or constipation—presumably mirroring the life cycle of the organisms. The disease is diagnosed by the observation of the parasites from the stools or aspirated material from the duodenum. It is particularly important to differentiate chronic giardiasis from other types of intestinal infections because the treatments are different. A negative diagnosis depends upon at least three (if not more) consecutive examinations because trophozoites may often be so tightly attached to the duodenal/intestinal walls that they are not shed with the feces. More recently, serological tests have become available for detecting anti-*Giardia* antibodies in the blood, and such tests are considered more sensitive. The drug of choice in treating infections is metronidazole. Control measures against these infections are aimed at promoting good hygiene and sanitation and the adequate purification of drinking water. A general trend in the distribution of giardiasis is the apparent preference for warmer climates, and for this reason special precautions should be taken during the summer months. *See also* DIARRHEA; LEEUWENHOEK, ANTONI VAN.

global warming

Postulated phenomenon of the gradual, overall increase in the earth's surface temperature, believed to be occurring due to the rise in proportion of atmospheric carbon dioxide, i.e., the "greenhouse effect," on the Earth. The immediate manifestations of this warming phenomenon are not experienced in warmer climates with each passing year but, rather, in a greater incidence of temperature extremes, as well as other climactic changes. The rate of atmospheric carbon dioxide has risen especially sharply in the past two or three centuries due to a variety of human activities, including population growth (which, too, has experienced a dramatic increase in the past few centuries), a high rate of fossil-fuel burning, and widespread deforestation. *See also* GREENHOUSE EFFECT.

glucose

A six-carbon sugar/carbohydrate molecule that is one of the most easily usable substrates for carbon metabolism among living things. Different heterotrophic organisms break glucose down via glycolysis and other pathways to produce energy as well as a multitude of useful chemical end products such as alcohol and acids. Glucose is also a monomer in a vast number of different structural components of living things, e.g., the cell wall, cellulose, and starch in plants and algae, and glycogen in higher animals.

glycolysis

Metabolic pathway through which a vast number of bacteria and fungi conduct fermentation, also known as the Embden-Meyerhof-Parnas (EMP) pathway. The overall reaction may be viewed as the breakdown of glucose (most sugars are typically converted to this form first) to pyruvic acid, with the net generation of energy in the form of ATP molecules. The reaction actually takes place in a series of six enzyme-mediated steps and uses an internal reducing agent in the cell called NAD, which is converted to NADH.

Glucose ($C_6H_{12}O_6$) + 2 ADP + 2 NAD + 2 phosphates → 2 pyruvate (COOH–CO–CH$_3$) + 2 ATP + 2 NADH

Although this equation shows an overall gain in ATP, a look at the individual steps of the pathway in sequence reveals that some ATP molecules are also needed for glycolysis to begin, i.e., to earn some, the cell needs to spend some first. Glycolysis seldom represents the end of fermentation, and might be more justifiably expressed as the metabolic pathway by which organisms *begin* sugar metabolism. The conversion of pyruvic acid (or rather pyruvate) to further products almost invariably regenerates the NAD molecule. In the most complete possible breakdown, pyruvate is converted to water and carbon dioxide, as happens in the Krebs cycle. Many bacteria, however, convert pyruvate to other end products such as ethanol and lactic acid. *See also* FERMENTATION; METABOLISM.

glycoprotein

Class of macromolecules in which a protein is combined with a carbohydrate group. The carbohydrate moiety is usually added onto the protein after translation; in eukaryotes, this glycosylation occurs in the Golgi apparatus. The composition of these molecules endows them with strongly antigenic properties. Cell-surface components such as the receptors for viruses, hormones, and other substances are frequently composed of glycoproteins. Many of the envelope proteins of viruses, such as the hemagglutinins of the influenza viruses, are also made of glycoprotein molecules.

Golgi apparatus

This organelle is the site in eukaryotic cells where newly synthesized proteins acquire modifications—called post-translational modifications—such as the addition of carbohydrate molecules, intramolecular bonding, and folding into three-dimensional shapes. The apparatus consists of a stack of flattened membrane-bound vesicles that re-

ceive the newly made protein chains from the rough endoplasmic reticulum on one end and direct the modified proteins to different destinations, such as the secretory vesicles, lysosomes, or the cell membrane, at the other end. *See also* CYTOPLASM.

gonorrhea

A genital disease caused by *Niesseria gonorrhoeae*. Recognized in humans for several centuries before the discovery of the bacteria, gonorrhea has a long history and has been described in various ancient Hebrew and Chinese medical texts. In the modern context, gonorrhea remains an important disease, affecting an estimated 2 million people annually in North America alone. Gonorrhea is marked by an initial inflammation of the mucus membranes in the genito-urinary tracts, causing vaginitis in women and urethritis in men, 4–10 days after infection. The primary mode of infection is through sexual contact. Bacteria invade and damage the epidermis, which results in the copious discharge of pus in the early (acute) stages of disease. Men find urination painful and often develop scar tissue in the area, sometimes leading to complete blockage or stricture of the urinary passage. Among females, preadolescent girls who have more delicate vaginal lining tissue show a much greater susceptibility to the gonococci. Many infected adult women may be asymptomatic but nevertheless capable of transmitting the gonococci. If not treated, the initial infection may spread to surrounding tissue, extending to the uterus and ovaries, or the seminal vesicles and prostate, as well as the bladder and even the rectum. A frequent outcome is the development of sterility due to damage to the fallopian tubes or vas deferens caused by the gonococci. Pregnant women who harbor gonococci can pass on the infection to the baby at the time of birth. The most common manifestation of infection in newborns is a conjunctivitis due to the inflammation of the mucus membranes of the eyes; in fact, gonococcal infections were once one of the major causes of congenital blindness. *See also* NIESSERIA GONORRHOEAE.

Gram stain

Staining procedure used to differentiate among different types of bacteria on the basis of cell wall structure. Developed in 1890 by Danish pathologist Christian Gram, this technique takes advantage of the differential permeability of different types of cell walls, which is a function of the structure and amount of peptidoglycan in different organisms. Bacterial cells are heat-fixed on a glass slide, stained with a crystal violet-iodine complex and then washed with alcohol. Gram-positive organisms have a thick layer of peptidoglycan, which retains the dye complex, and provides a barrier to the alcohol. Thus they appear as blue cells when viewed under the microscope. Gram-negative organisms, which have only a thin layer of peptidoglycan, are more permeable and are thus decolorized by the alcohol wash. These bleached cells may be visualized by the use of a second dye, such as safranin, which provides a contrasting pink color. The organism *Legionella,* as well as a few anaerobic rods, e.g., *Brucella,* are unable to retain safranin and require stronger counterstain dyes such as carbolfuchsin. As an organism grows old or gets ready to divide, the bonds between the peptidoglycan layers in the cell wall are loosened so that new material may be added. During this time, the walls become more permeable to alcohol and may thus give a false indication of being gram-negative—such organisms are called gram-variable cells. Typically, they regain their gram-positive character after the cell wall is restored. The Gram reaction is one of the most important criteria for the classification of the bacteria, and almost always the first test performed in a laboratory to identify a new or unknown isolate. *See also* CELL WALL; DIFFERENTIAL STAINING; PEPTIDOGLYCAN; STAINING.

granuloma

A chronic inflammatory lesion, often in the area of the skin (but also at other locations such as the lungs, connective tissue, and eyes), arising from a delayed type hypersensitivity response to a persistent antigen. It is characterized by the accumulation and subsequent fusion of various immune cells—mostly macrophages, but also some T lymphocytes—at the site of inflammation. These granulomas cause further damage to the surrounding tissues, e.g., necrosis, in part due to the displacement of the normal cells, and in part due to the destructive activities of the various immune cells summoned to the site. A large number of agents are capable of triggering this granulomatous reaction, including infectious organisms such as fungi, bacteria, and protozoa, as well as other materials, including pollen and animal dander. *See also* IMMUNE RESPONSE; INFLAMMATION.

greenhouse effect

The phenomenon of the gradual increase in proportion of atmospheric levels of carbon dioxide on earth, due to increased activity of living organisms. For several centuries—millennia, even—the cycling of different elements such as carbon, oxygen, and nitrogen through the biosphere has maintained a balance ideal for sustaining life as we know it. However, the proportion of living things, especially humans, has risen greatly in recent times, resulting in higher rates of respiration (and hence CO_2 generation). In addition, the past few centuries have seen an accelerated rate of CO_2 release due to the burning of fossil fuels. The term "greenhouse effect" was derived from observations of a similar phenomenon—i.e., carbon dioxide increase—at a small-scale level when green plants grew rapidly within the confines of a greenhouse. Some of the predicted outcomes of the greenhouse effect include a decrease in available oxygen due to the imbalance created by carbon dioxide, a gradual increase in the earth's temperature (global warming), and the possible increase in the incidence of respiratory, water-borne, and vector-borne diseases. *See also* GLOBAL WARMING.

GSS syndrome. *See* GERSTMANN-STRAUSSLER-SCHEINKER SYNDROME

GTP

Guanosine triphosphate (GTP) is sometimes used in lieu of ATP as the energy currency in some living cells because it carries the same amount of energy in its phosphate bond as ATP. It is, however, far less abundant than ATP in most cells. Like ATP, GTP is a monomer used for the synthesis of RNA chains. *See also* ATP.

Guillain-Barré Syndrome

Acute disease of the nervous system that may develop after a number of different viral infections. The main symptom is a gradual paralysis of the limbs, brought about by the inflammation of the peripheral nerves followed by the removal of the myelin sheaths of the nerve cells (demyelination). Most cases recover spontaneously within a few weeks, although the damage may be permanent in about 15 percent of cases. Perhaps the most usual causative agent in Guillain-Barré syndrome is the Epstein-Barr virus, and symptoms often develop some 1–4 weeks after an episode of infectious mononucleosis. Other viruses implicated as possible causes include the Cytomegalovirus, the enteroviruses, and HIV. The exact mechanisms that trigger the demyelination are not completely understood, although scientists suspect that the pathogenesis might have an immunological basis. This conjecture is supported by the fact that a 1976 outbreak of the syndrome in the United States was linked to the use of an influenza vaccine using formalin-inactivated virus.

H

Haemophilus

Genus of gram-negative, non-motile, encapsulated, coccobacillary bacteria, which form the normal indigenous microflora of the upper respiratory tract and mouth of humans. These bacteria are facultative anaerobes capable of both respiration and glucose fermentation, but not, typically, of lactose. Growth in the laboratory requires an enriched media such as chocolate agar. The name of this genus was derived from the requirement of all species for one or both of two blood factors: factor X or hemin, normally required for the synthesis of respiratory enzymes, and the V factor, which is actually the coenzyme NAD. There are at least 16 *Haemophilus* species, which exhibit various degrees of virulence in humans. The most important pathogen, *H. influenzae*, is also the species about which the most is known. Examples of other pathogens include *H. aegyptius*, *H. ducreyi*, and *H. haemolyticus*.

Haemophilus aegyptius

The exact taxonomic position of this organism is not clear because it appears very similar to *H. influenzae* in virtually all phenotypic and genotypic characteristics, save pathogenicity. This organism is known to cause an acute and contagious form of conjunctivitis, which is typically not a serious problem with the influenza organisms. More recently, it was identified as the cause of an invasive disease

called Brazilian purpuric fever, which has raised further confusion as to its taxonomic status.

Haemophilus ducreyi

This organism was originally classified under the genus *Haemophilus* due to its requirements for blood factors X and V, but it is likely to be reclassified as a separate genus or even in a separate family. It has been associated with an endemic chancroid in different tropical and subtropical parts of the world. Susceptibility is highest in uncircumcised, non-white males, and prostitutes are believed to be a major reservoir of the infectious bacteria. The primary lesion develops as a small papule with surrounding erythema some 4–7 days after initial infection. It rapidly becomes ulcerated, and bacteria spread to the local lymphoid tissue to cause swelling in that region. There is no systemic involvement. This infection is also associated with HIV infections.

Haemophilus influenzae

This species, which is restricted to the human host, causes a wide spectrum of diseases in humans, ranging from localized respiratory infections to serious systemic conditions. Perhaps the most common disease associated with *H. influenzae* is otitis media or middle ear infection, seen mostly in children. However, if left untreated, these initial conditions can develop into serious complications involving the nervous system. Indeed, until the development of a protective vaccine, the b strain of *H. influenzae* was the single most important cause of meningitis in children in the United States. It is also a major cause of pneumonia. Bacteria typically gain entry through inhaled aerosols. Ear infection is the result of the proliferation of the bacteria through the Eustachian tubes. Meningitis and other systemic infections arise as a consequence of bacterial entry into the bloodstream. This occurs mostly in children and is characterized by symptoms such as high fever accompanied by severe headache and a stiff neck. Pneumonia is often the outcome in people already suffering from chronic respiratory conditions.

Early diagnosis of *H. influenzae* infections is very important, especially in the case of meningitis, because the damage, once caused, is irreversible. Organisms may be identified on the basis of cultural and serological characteristics. There are six different serotypes—designated a through f—identified on the basis of differences in their capsular antigens. Vaccines are now available against this organism, and the U.S. Centers for Disease Control recommends three doses at two-month intervals until the age of six months, followed by a booster between 12 and 18 months. Different antibiotics may be used to treat infections, the choice depending on the resistance developed by the particular strains being treated.

In addition to its clinical importance, *H. influenzae* is also of general interest to scientists because it was the first organism from which restriction enzymes were purified, and because it was one of the first bacterial genomes to be sequenced in its entirety.

Halobacterium salinarium

This member of the archaebacterial domain is a photosynthetic organism that requires high concentrations of salt for its growth— i.e., it is extremely halophilic. It has been implicated in the spoilage of foods such as processed meats and fish, which typically rely on high salinity for protection from microbes. *H. salinarium* exhibits some peculiar properties, such as phototaxis and chemotaxis, namely, directed movement in response to light and chemicals, respectively. It moves toward nutrient compounds such as glucose, asparigine, and histidine, and away from phenol. It carries out a unique mode of photosynthesis using a pigment called bacteriorhodopsin to harvest light. It can also ferment organic compounds such as the amino acid arginine. *H. salinarium* has a 40 kb genome (being sequenced in early 2000), which is divided into two distinct fractions according to GC content.

haloduric

A microbe that can tolerate, i.e., continue to live, under conditions of high salt concentration in the environment, but which is not metabolically active under such conditions.

halophile

A microorganisms that requires high concentrations of salt in its environment, e.g., *Halobacterium* species. A rough estimate of the salinity required by these organisms is about 10 times the concentration of seawater. As a rule, these organisms are chemoheterotrophs and contain bacteriorhodopsin, which they use in a unique type of photosynthesis.

hand, foot, and mouth disease

Common childhood disease caused by an infection with the Coxsackie A virus, usually the serotype A16. Infection with the virus can result in the development of painful, fluid-filled blisters in the regions of the mouth (lips, gums, and tongue), hands, and feet, which are usually the points of contact with infective materials such as food, fomites, and other infected persons. Symptoms may last for 7–10 days, but the disease is usually self-limiting and there are no serious complications. The fluid of the blisters contains a large number of viruses and is the principal source of infections, although the virus may be shed for weeks after the disappearance of symptoms. This disease occurs most commonly in children under the age of 10 (though adults may also be infected) and outbreaks often occur in childcare facilities. The U.S. Centers for Disease Control advises frequent hand-washing to minimize spread but does not strongly recommend isolation of infected children. There is no noticeable immunity to the disease, and vaccines are not available. *See also* COXSACKIE VIRUS.

hantavirus

This virus is unique among the members of the bunyavirus family in that it is harbored in, and transmitted by, rodents rather than arthropods. The first hantavirus was discovered during the Korean War in both humans and rats. It was found to be associated with different forms of viral hemorrhagic fevers, including an especially virulent form that was complicated by acute renal failure (known as HFRS or hemorrhagic fever with renal syndrome). Then, in 1993, there was a series of unexpected deaths of young Native Americans, following a mysterious malady that started as a fever but quickly progressed to symptoms of acute respiratory distress. Within six months of beginning the search for the culprit, investigators had identified the agent as a new hantavirus, which they named the Sin Nombre Virus (SNV). The disease was called hantavirus pulmonary syndrome (HPS). Other hantavirus species, causing a similar malady, have since been found harbored in such reservoirs as deer mice, voles, rats, and mice. Strains of hantavirus causing two different conditions differ with respect to the target organ for attack (i.e., kidneys or lungs), as well as mortality rates—the renal syndrome has a 1–15 percent mortality rate, while nearly 50 percent of those affected by HPS succumb to the disease. Physically, these viruses consist of a single-stranded RNA genome with three unequal segments bound to a capsid that is enclosed within a lipid envelope. They multiply exclusively in the cytoplasm of the host cells. *See also* BUNYAVIRUS; HEMORRHAGIC FEVERS.

haploid

Genetic state of living organisms characterized by the presence of a single set of the complete genome in each cell of the organism. Haploidy is the normal state of prokaryotic organisms, as well as the vegetative state of eukaryotes such as fungi and algae. The only haploid cells in a human being, however, are the reproductive cells—namely, the sperm and ova. *See also* CHROMOSOME; EUKARYOTE; GAMETE.

Helicobacter pylori

This organism was catapulted to fame during the 1980s, when the Australian physician Barry Marshall demonstrated a causal link between it and peptic ulcers in humans. A curved rod with gram-negative staining characteristics, *H. pylori* was initially identified as a new species of the genus *Campylobacter*, on account of the close similarities in microscopic appearance and biochemistry between the two genera. Upon further investigation, however, the scientists found sufficient differences in the structure of the fimbriae to justify the creation of a new genus. In addition to *H. pylori,* there are two other species, *H. cinaidi* and *H. fennelliae,* which possess single polar flagella instead of the polar tufts present in *H. pylori.*

H. pylori is a habitat-specific organism found almost exclusively in the lining of the human stomach and duodenum. It is microaerophilic and requires slightly higher concentrations of carbon dioxide (10 percent) and lower concentrations of oxygen than are found in the atmosphere. Surprisingly, given the highly acidic conditions in the stomach cavity, *H. pylori* is not very acid-resistant and multiplies best at near-neutral pH. It protects itself from constant exposure to the acidic gastric juices in the stomach with the help of the enzymes mucinase and urease. Mucinase enables the bacterium to burrow into mucus layers of the stomach wall (but not the cells) while urease provides buffering capacity by converting urea—breakdown product of proteins—into carbon dioxide and ammonia. In laboratory cultures, *H. pylori* grows best on fresh moist media in a humid environment. *See also* ULCER.

hemadsorption

The adsorption of a substance or agent to the surface of a red blood corpuscle (RBC). Hemadsorption may come about as a result of the specific binding between molecules or receptors on the RBC surface and the adsorbed substance, or it may be nonspecific in nature. One outcome of hemadsorption is the acquisition of some of the properties of the adsorbed substance, e.g., antigenicity and binding capacity, by the blood cells. Viral and bacterial proteins absorbed by RBCs attract antibodies and complement, which, in turn, leads to such consequences as hemolysis or hemagglutination.

hemagglutinin

Glycoprotein antigens on the surfaces of viruses (e.g., the influenza virus) that have the property of being able to induce the clumping of erythrocytes, namely, hemagglutination. The main function of the hemagglutinins is to enable intact viruses to attach to the surface of the host cell. Thus, these proteins aid in viral pathogenicity, and their presence in the blood may be taken as an indication of viral infection. Hemagglutinins may be detected by certain hemagglutination tests that check the ability of viral suspensions—e.g., serum from viremic individuals—to agglutinate sheep red blood corpuscles *in vitro*. *See also* GLYCOPROTEIN; INFLUENZA VIRUS.

hemolysin

Extracellular protein products of such pathogenic bacteria as *Streptococcus* and *Staphylococcus*, identified on the basis of their ability to induce hemolysis, i.e., the lysis of erythrocytes (red blood corpuscles or RBCs). Different types of hemolysins lyse erthrocytes to varying degrees. Presumably these proteins aid in the virulence of the pathogens because they are not produced by commensals or by nonpathogenic species of bacteria in the same genus. When grown on blood agar plates, the

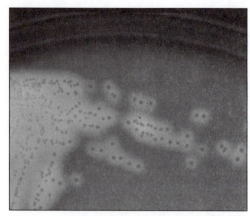

Streptococcus pyogenes beta hemolysis on sheep blood agar. © *Science Source/Photo Researchers.*

hemolysin-producing bacteria form colonies with zones of complete or partial clearing around them, due to the lysis of RBCs in that area. The nature of lysis is a reflection of the type of hemolysin produced by the organism on the plate. *See also* STREPTOCOCCUS.

hemorrhagic fevers

The viral hemorrhagic fevers comprise a diverse group of viral infections that share the common characteristic of widespread bleeding from the epithelial tissue on both internal and external surfaces of the body. Featured in this group are yellow fever and dengue, which have a long history of devastation, as well as fevers caused by such deadly newly emerging pathogens as Ebola and Marburg. The exact mechanisms underlying the hemorrhages are still not understood completely, but they are a consistent feature of these diseases. The different viruses causing these diseases also differ with respect to their mode of spread—for example, members of the bunyavirus family, which cause diseases such as Rift Valley fever, sandfly fever, and the Crimean-Congo hemorrhagic fever, normally reside in nonhuman animal hosts such as cattle and rodents and are transmitted to humans via such insect vectors as mosquitoes, sandflies, and ticks. Some others, such as the hantaviruses, which belong to the same family, are found in rodent hosts but do not require insects for their transmission to humans. The first outbreak of a Marburg virus

Table 5. Etiology and Distribution of Viral Hemorrhagic Fevers

Disease (HF[1])	Virus[2]	Virus classification[3]	Distribution
Argentine HF	Junin virus	Arenavirus	Argentina[4]
Brazilian HF	Sabia virus	Arenavirus	Brazil
Bolivian HF	Machupo virus	Arenavirus	Bolivia
Colorado tick fever[5]	Coltivirus	Reovirus	North America
Crimean-Congo HF	*epon*	Bunyavirus	Africa, Eastern Europe, Asia
Dengue HF	Dengue virus	Flavivirus	Worldwide
Hemorrhagic fever[6]			
	Ebola virus (3 separate strains)	Filovirus	Zaire, Ivory Coast, and Sudan
	Marburg virus	Filovirus	Africa[7]
Hemorrhagic fever with renal syndrome (HFRS)		Bunyavirus (genus Hantavirus)	
	Belgrade virus		Balkans
	Hantaan virus		Asia, Europe
	Puumala virus		Scandinavia
	Seoul virus		Worldwide
Kyasanur Forest disease	*epon*	Flavivirus	India
Lassa fever	Lassa virus	Arenavirus	Africa
Omsk HF	*epon*	Flavivirus	Russia
Rift Valley fever	*epon*	Bunyavirus (genus Phlebovirus)	Africa
Venezuelan HF	Guanarito virus	Arenavirus	Venezuela
Yellow fever	*epon*	Flavivirus	Africa, South America, and Central America

Notes:

[1]HF = Hemorrhagic fevers

[2] Individual viruses that are named for the disease they cause are identified by the abbreviation *epon* for eponymous, (i.e. having the same name).

[3]In the majority of the cases, this column identifies the virus **family** to which the disease agent belongs. Exceptions (genera etc.) are noted in parentheses.

[4]Please note that the distribution of the virus is not limited by the boundaries of the country/countries named but gives an idea of the region of the continent (in this case South America) where the virus/disease predominates.

[5]The hemorrhagic form of this disease is a relatively uncommon complication of viral infection seen mainly in children.

[6]These fevers are sometimes referred to as the African hemorrhagic fevers and identified by their causative agents (e.g. Ebola virus fever).

[7]Although the Marburg virus is endemic in Uganda and nearby parts of Africa, the first outbreak occured in Germany and Yugoslavia as a result of importing of infected monkeys.

infection was traced to a batch of infected African green monkeys, and its subsequent spread was linked to direct contact, not insects. Table 5, "Etiology and Distribution of Viral Hemorrhagic Fevers," lists the causative agents, along with the classification and geographic distribution of various diseases in this category.

hepatitis

Class of liver diseases characterized by inflammation in regions of the liver and either degeneration or necrosis of the liver cells (hepatocytes). Hepatitis has a varied etiology and may originate due to liver damage by alcohol, drugs, or other toxins, as well as by viruses. Due to the scope of this book, only the viral forms of hepatitis will be discussed in detail. The basic symptoms are similar in different types of hepatitis; however, knowledge of the etiology is important for administering proper therapeutic and public health measures. Patients suffer from generalized symptoms of fever, nausea, and lethargy, as well as localized symptoms of liver dysfunction such as the enlargement of the organ and associated pain and tenderness, jaundice, bilirubinuria (appearance of bile pigments in the urine), and pale feces. Treatment is two-pronged, aimed at eliminating the causative agent on one hand, and limiting damage and restoring function to the liver on the other. In general, precautions should be taken to not overburden the liver during the early stages of recovery from the disease. Patients are advised to avoid alcohol and fatty foods for several weeks after the infection recedes. Many apparent relapses of the disease may be attributed to a premature bout of alcohol consumption or very heavy exercise, both of which place undue stress on the liver.

Viral hepatitis may be distinguished from nonviral forms of the disease by the detection of elevated levels of the liver enzymes, alanine aminotransferase and aspartate aminotransferase, in the serum. There are at least five different types of viral disease caused by distinct viruses from different families, which are classified into two main groups based on mode of transmission. These viruses are designated A through E, with the A and E viruses causing enteric infections, while B, C, D, and miscellaneous non-A non-B viruses are implicated primarily in blood-borne infections. Other infectious agents such as the Epstein-Barr viruses, cytomegalovirus, and herpes simplex virus, as well as nonviral entities such as the malarial parasite, might present with hepatitis as one of the symptoms of a more broadly defined disease.

hepatitis A

A highly infectious hepatitis caused by the hepatitis A virus (HAV). Known for several years before the isolation and identification of the causative agent, this disease was termed "infectious hepatitis" to differentiate it from the blood-borne versions of hepatitis. The primary route of entry for HAV infection is via food contaminated with the virus. Blood-related transmission is rare, although it is possible during the viremic phase of infection. The cardinal sign of the disease is the appearance of dark-colored urine some 2–6

weeks after initial infection, followed by the typical signs and symptoms. Infections in children may be subclinical (i.e., without jaundice), and the severity of symptoms appears to increase with the age of onset. The infection usually subsides within 4 weeks, although the convalescence period is much longer to allow the liver time to recover fully. Because there is only one serotype, infection with HAV endows lifelong immunity. Patients will either recover fully from this disease or die; there is no chronic carrier state. Passive immunization using nonspecific immunoglobulins is routinely administered in endemic areas, and the jaundice is treated symptomatically. Because hepatitis A is transmitted mainly via the fecal-oral route, good hygiene practices, proper regulation of drinking water, and proper sewage disposal are targets for improvement. Because most of these conditions are bigger problems in underdeveloped parts of the world, hepatitis A is more prevalent in those regions. *See also* HEPATITIS A VIRUS.

hepatitis B

Blood-borne viral hepatitis caused by the hepatitis B virus (HBV), which is spread only when there is a transfer of bodily fluids (e.g., blood transfusions, the use of shared hypodermic needles, and sexual intercourse). *In utero* infections and mother-to-infant infection via breast milk have also been reported. The earliest outbreaks of this disease, seen around the time of World War I, were associated with vaccines later found contaminated with human serum, which gave it the name "serum hepatitis." The disease may be asymptomatic or symptomatic (30–40 percent of the total cases), and with or without jaundice, with the outcome depending on both the virus and the immune status of the host. The disease has a long incubation period ranging from three weeks to six months, and the clinical disease may last from four to six weeks after symptoms first become apparent. Acute infections become chronic in about 10–15 percent of the cases, with viral surface antigens found circulating in the blood.

A chronic infection may be either asymptomatic or develop into chronic hepatitis. A fulminant variation of hepatitis B, which occurs in 1–3 percent of the total infected adults, gives rise to generalized symptoms, including encephalopathy, which progresses to coma and death. Chronic HBV-hepatitis may also be associated with primary carcinoma of the liver cells or with cirrhosis.

With an estimated infection rate of about 300,000 people per year in the United States alone, hepatitis B is one of the major challenges facing physicians and public health officials, along with AIDS and other venereal diseases, which share the same mode of transmission. In addition, hepatitis B is endemic in certain underdeveloped areas of the world. The disease is controlled primarily by prevention, e.g., the careful screening of the blood of potential donors, testing before use in transfusions, and safe sexual practices. A synthetic intramuscular vaccine administered in a series of three injections is now available. Interferon-a is being used in the treatment of chronic infections, but only with limited success. *See also* HEPATITIS B VIRUS.

Hepatitis A virus (HAV)

The causative agent of hepatitis A or infectious hepatitis, hepatitis A virus (HAV) is a picornavirus, consisting of a single m-RNA genome in an icosahedral protein coat, without any envelope. HAV was originally identified as an atypical enterovirus (72), but significant differences in genome sequence and replication cycle led to its classification in a separate genus, hepatovirus, so named because of the virus's predilection for the liver. The virus is resistant to antiviral agents that inhibit the enteroviruses. It is antigenically conserved with just a single serotype, but three genotypes are based on differences in the genetic sequence. Different strains are capable of infecting human and simian hosts. HAV grows slowly and poorly in tissue culture. *See also* ENTEROVIRUS; HEPATITIS A; PICORNAVIRUS.

Hepatitis B virus (HBV)

A complex DNA virus that causes hepatitis B in humans, hepatitis B virus (HBV) is classified along with certain animal-infecting species—found in woodchucks, ground squirrels, and Chinese ducks—in a single-membered family called Hepadnavirus (for hepatic DNA virus). The virus is hardy and can survive at room temperature for about 6 months without losing its infectivity. The human HBV is known to exist in three distinct morphological forms. The fully infective form, called a Dane particle, is about 42 nm in diameter, and is composed of a dense nucleocapsid enclosed in a lipid envelope. The capsid comprises an antigenic protein called the HBV core antigen found in association with the genome, a DNA polymerase, and a kinase enzyme. The genome is unusual among viruses in that the double-stranded structure is interspersed with a gap of single-stranded DNA, representing about one-third of the total size. A complex lipoprotein known as the HBs antigen (for surface antigen), composed of lipid and several polypeptides, is associated with the envelope. Its presence in the serum is an indication of the patient's infective state. The two other forms of HBV are spherical or filamentous forms made up of incomplete HBs antigen and are not infectious. In addition, HBV has a defective form made of single-stranded RNA associated with an incomplete surface antigen. This form is called the Hepatitis D virus or delta antigen and is capable of superinfecting cells/organisms already infected with HBV. Various vaccines against HBV are directed at eliciting immune responses to the different antigenic components. *See also* AUSTRALIA ANTIGEN; BLUMBERG, BARUCH; HEPATITIS B.

Hepatitis C virus (HCV)

An RNA virus, tentatively classified in the flavivirus family, responsible for the majority of non-A, non-B hepatitis infections that are transmitted via the parenteral route, namely, via blood, serum, or other bodily fluids. The existence of this and other non-A, non-B viruses was first suspected during the 1970s and 1980s, when it became obvious that a large proportion of the so-called "serum hepatitis" did not demonstrate the serological specificity and reactivity expected of the hepatitis B virus. Although scientists have the complete genomic sequence for hepatitis C virus (HCV) and can thus diagnose infections with ease, clinical treatment has remained elusive. One of the main reasons for this is that the HCV genome mutates at an extremely rapid rate and thus quickly develops resistance to any drugs that might be used. In addition, the virus is extremely difficult to cultivate, both in tissue cultures and in nonhuman hosts. The HCV consists a single-stranded linear gonome approximately 9,500 nucleotides in length. *See also* FLAVIVIRUS.

hepatitis E virus

This virus is a spherical, non-enveloped, single-stranded RNA virus, which is the major cause of food-borne or water-borne non-A, non-B hepatitis. The classification of this virus is uncertain because, while it resembles the Calicivirus family (Norwalk virus) in some respects, its genome organization is substantially different.

Herpes Simplex virus (HSV)

Viruses of the herpesvirus family known to cause a number of human infections—e.g. of the gums, lips, tongue, eyes, genitalia, skin, and brain tissue—collectively called herpes. Like all other members of this family, the herpes simplex viruses (HSVs) have the ability to enter a latent state in the host. Specifically, these viruses achieve latency in neurons. Other characteristics that distinguish this group of viruses from other herpes viruses (except the Varicella-Zoster virus) include its rapid growth in cell cultures and its ability, when in the active state, to induce lysis of the infected host cells.

There are two main types of human herpes simplex viruses, designated HSV-1 and HSV-2, which are differentiable on the basis of antigenicity. Either virus can cause any of the aforementioned infections, although one or the other appears to predominate in a particular location or tissue. Clinically, it is important to distinguish between primary and recurrent infections because therapeutic approaches may be different. Primary infection with the herpes simplex virus usually takes place in the epithelial cells of the oral cavity or genitalia, usually by contact with another infected person. Exactly how an infectious outbreak begins is not fully understood by epidemiologists because these viruses are typically host-specific and are not transmissible via contaminated food or water. One possibility is that the virus is already present in a person in a latent form—a characteristic of the herpes viruses is their ability to exist in vertebrate hosts over generations—and is triggered to proliferate and cause disease by some external, as-yet-unidentified stimulus. A primary epithelial infection is often inapparent or subclinical, but when apparent, it is considerably more severe than recurrent episodes in the same tissue.

One of the most common sites for primary infection, especially with HSV-1, is the mouth. In young children, this manifests as gingivitis, with the formation of fluid-filled vesicles that rupture to form ulcers in the gums, tongue, and lips. Other than a mild fever and bleeding gums, children tend to recover spontaneously with few systemic complications. In adults, this infection more often manifests in the form of pharyngitis or tonsillitis. Following this infection, the virus often travels to the trigeminal nerve, where it remains latent and periodically makes its presence known by the formation of fever blisters or cold sores, otherwise known as herpes labialis or herpes facialis. In contrast to facial infections, primary genital herpes is caused by HSV-2 about 80 percent of the time. Genital herpes is also highest among those age groups that are most active sexually and in individuals with multiple sexual partners. As in the case of oral infections, primary genital infections may be either subclinical or manifest via the formation of ulcerating vesicles around the genital regions. Primary infections typically occur with greater severity and frequency among females. Other symptoms include pain, itching, redness, swelling, discharge, and systemic manifestations such as fever and malaise. Recurrent infections are milder than the primary attack.

In addition to mouth and genitalia, primary infections with herpes simplex virus may also occur in one or both eyes—often with the fingers as fomites—and in any location of the skin. Eye infections usually take the form of keratoconjunctivitis, an inflammation of the cornea and conjunctival membrane, and the formation of a ulcer. Repeated recurrences in this region lead to corneal scarring and, eventually, blindness. Although a relatively rare occurrence, encephalitis from herpesvirus is one of the commonest forms of nonepidemic viral encephalitis. Both types of infection are caused more often with HSV-1. Neonatal herpes is a serious infection of newborns that occurs due to viral transmission from infected mothers to babies during delivery. As such, these infections are more often due to HSV-2. Such infections may result in systemic infections that prove fatal in about 80 percent of cases. Survivors are often left with serious neurological or ocular disorders; encephalitis, also with high fatality rates; and skin or mucus membrane infections, that must be treated promptly to avoid systemic spread. Heavily immuno-compromised individuals, such as AIDS patients and transplant recipients, are also at great risk for severe systemic or disseminated HSV infections.

HSV infections are best diagnosed by culturing clinical specimens in tissue cultures for detecting the virus and with serological tests to identify the culprit. These infections are amenable to drug therapy with acyclovir. Vaccine development has proven difficult, especially due to the ability of these viruses to become latent. *See also* HERPESVIRUS.

herpesvirus

Family of large, enveloped, DNA viruses associated with a variety of human diseases, such as chickenpox, shingles, infectious mononucleosis, certain types of cancer, and various herpes infections, including genital herpes. Under an electron microscope, these viruses appear as spherical particles some 120–200 nm in diameter, which actually consist of four concentric layers—the outer lipid envelope, surrounding an amorphous layer called the tegument, within which is an icosahedral capsid that encloses a doughnut-shaped core that bears the viral genome in association with proteins. The herpesvirus genome consists of a linear piece of double-stranded DNA, ranging in length from 125–230 base pairs, depending on the genus. These viruses replicate in the nucleus of the host cell, and the envelope is acquired by budding through the nuclear rather than the plasma membrane.

A characteristic feature of the herpes viruses is their ability to persist indefinitely in their hosts in the form of extranuclear DNA. Indeed, virtually every known vertebrate has been found to harbor at least one host-specific herpes virus that may or may not announce its presence in the form of disease. This ability to remain silent within the host is also the reason that so many human herpes viruses are associated with latent diseases that manifest themselves years, even decades, after initial infection.

The herpes viruses classified into three categories based on biological properties. The herpes simplex and the Varicella-Zoster (chickenpox/shingles) viruses are classified together as the alpha herpes viruses. They share the properties of growing rapidly in cell cultures, inducing the lysis of the host cells,

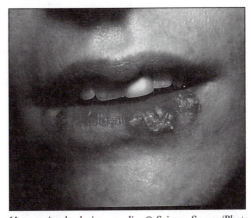

Herpes simplex lesion on a lip. © *Science Source/Photo Researchers.*

and achieving latency in neurons. Beta herpes viruses are slow growing, and instead of lysing host cells, induce the fusion of their membranes to form large multinucleated cells (cytomegalia). They become latent in a variety of sites including the salivary glands, kidneys, macrophages, and lymphocytes. A well known example of a member of this subfamily is the cytomegalovirus. The third group, called the gamma herpesvirus, is represented by the Epstein-Barr virus. It is characterized by its ability to either become latent in B lymphocytes or induce the proliferation of these cells rather than lyse them. These viruses are not amenable to cell culture. Different viruses and the diseases they cause are discussed individually. *See also* CANCER; CHICKENPOX; CYTOMEGALOVIRUS; EPSTEIN-BARR VIRUS; HERPES SIMPLEX VIRUSES; INFECTIOUS MONONUCLEOSIS; KAPOSI'S SARCOMA; POSTHERPETIC NEURALGIA; SHINGLES; VARICELLA-ZOSTER VIRUS.

Hershey, Alfred D. (1908–)

An American microbiologist whose work on mutations in the bacteriophage provided the first line of evidence (circa 1946) for the presence of more than one gene in viruses. His collaboration with the research groups of Max Delbrück and Salvador Luria led to the development of the phage as the first standardized system for studying virology and genetics at the molecular level. The three scientists shared the 1969 Nobel Prize in Physiology/Medicine.

heterocyst

Specialized cell found in cyanobacterial species, which functions as the site of nitrogen fixation. *See also* CYANOBACTERIA.

heterotroph

An organism that derives energy from chemical sources, and whose main source of carbon is organic material.

Histoplasma capsulatum

Originally identified in the early 1900s as a protozoan parasite possibly related to *Leishmania*, this organism is actually a fungus that causes human disease. It is a dimorphic organism found in highly organic soils (rich in bird or animal droppings, for instance) as a saprophytic mold with septate hyphae and in mammalian hosts as parasitic, unicellular yeasts. The saprophytic molds are capable of both asexual and sexual reproduction, while the infective forms multiply only by simple budding. Human infection occurs via the inhalation of airborne, vegetative spores, and because the yeast form does not produce spores, person-to-person transmission of these fungi is only possible when infected tissue is implanted into the healthy individual.

Scientists have identified three distinct types of *H. capsulatum* that are differentiable on the basis of their clinical manifestations and geographic distribution, but similar with respect to structure and life cycle. The three types are classified as varieties of the same species rather than as individual species. *H. capsulatum* var. *capsulatum* is the most widely distributed of these fungi and causes a systemic infection (histoplasmosis). *H. capsulatum* var. *duboisii* is endemic to central Africa and Madagascar, where it has been associated with granulomatous disease of the skin and bones. The third variety, called var. *farciminosum*, has been found in Africa, Asia, and Europe but never in the New World; it causes an epizootic infection in horses but has thus far not been isolated from humans.

The clinical presentation of *H. capsulatum* infections varies greatly according to the infectious dose of the organisms, as well as the immune status of the infected individual. The most common manifestation of primary infection is an acute respiratory/pulmonary disorder with fever and coughing. Usually these infections resolve themselves with such consequences, but if the immune system is weakened, organisms persist at the site of infection and lead to lesions characteristic of delayed-type hypersensitivity reactions in the lungs. The clinical presentation of such a chronic infection is not unlike tuberculosis.

Patients infected with a heavy dose of spores are more likely to develop a disseminated infection due to fungal invasion of the bloodstream. This usually occurs in immuno-suppressed individuals such as AIDS patients, transplant recipients, and in the very young or old. Clinical presentation includes such nonspecific symptoms as fever, weight loss, spleen and liver enlargement, and bone-marrow suppression, and will terminate in death unless treated properly. Diagnosis is by the isolation of fungi from clinical samples and by serological evidence of exposure to the fungal antigen called histoplasmin. Antifungal drugs should be administered to treat apparent infections.

The subacute, granulomatous disease caused by *H. capsulatum* var. *duboisii* is a disseminated inflammatory reaction to the fungal antigens, resulting in the formation of abscesses in the subcutaneous tissues and bones and spreading skin lesions. A definitive diagnosis is made by examining pus from the lesions for the presence of the budding yeasts—the cells of the *duboisii* variety are typically larger than other *Histoplasma* types—and culturing the material. Antifungal agents such as amphotericin B may be used to treat severe cases.

Ho, David (1952–)

Named by *Time* magazine as "Man of the Year" in 1996, this AIDS researcher is perhaps best known for developing the multidrug approach, popularly known as "cocktail therapy," for treating HIV infections, which has proven to be one of the most effective strategies for clearing the virus from the blood. The basis for this method of treatment lies in Ho's observations that, contrary to earlier speculation, HIV is actively multiplying inside the host's lymphocytes even during the phase when it appears dormant or inactive in that its physiological and clinical manifestations (e.g., the lymphocyte population) appear normal. Ho's approach also deals with the problem presented by the ability of HIV to mutate rapidly and thus acquire resistance to an antiviral drug. The use of multiple drugs at once presents HIV with so many challenges in a single instance that it is unable to develop resistance to all of them before it is inactivated. *See also* AIDS; HIV.

Hooke, Robert (1635–1703)

This British scientist secured a permanent place in the history of biology for coining the term "cell." While he used the term to describe the microscopic structural units that he observed in cork tissue, the word is now universally used to represent the organizational unit of living things. An assistant in the laboratory of Robert Boyle, Hooke was also one of the early members of the Royal Society, where he worked as a microscopist and demonstrator of biological specimens. He was the author of *Micrographia*, a treatise published in 1665, in which he reported various microscopic principles and observations. *See also* CELL.

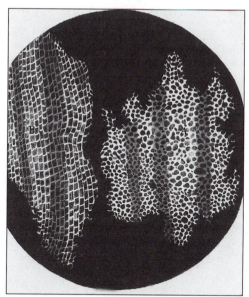

Hooke's drawing of cork under magnification. © *Science Source/Photo Researchers*

host

Any living organism that functions as the natural habitat for another organism (usually a parasite or commensal). Certain protozoa and helminths may require more than one host in the course of their life cycle. In such cases, the organism in which the parasite undergoes its sexual phase and reaches maturity is the primary or definitive host, and the host in which the parasite is present in the larval stages is the secondary or intermediary host. In addition, there may be vectors or transport hosts, which mediate the transfer of the parasite from one host to another without supporting any developmental changes. *See also* PARASITISM; VECTOR.

HTLV. *See* HUMAN T CELL LEUKEMIA VIRUS.

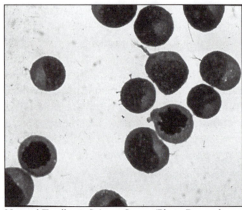

Normal T-cells. © *Science Source/Photo Researchers.*

human immunodeficiency virus (HIV)

Better known by its acronym, HIV, this retrovirus, classified in the lentivirus family, is widely held to be the primary cause of acquired immunodeficiency syndrome (AIDS). Morphologically, HIV consists of a dense core containing the diploid, single-stranded RNA genome associated with various enzymes, including reverse transcriptase, encased in a lipid envelope studded with glycoproteins that endow the virus particles with a high degree of antigenicity. A notable feature of the HIV genome is its high rate of mutation during replication. One of the major reasons for this is the high error rate of the reverse transcriptase enzyme, which incorporates mismatched nucleotides at about the rate of 1 in 2,000 bases. With a genome of about 9,000 nucleotides, this translates to roughly 8 or 9 point mutations per generation, due to the enzymatic error alone, over and above the random mutations that occur in any replicating genome. Among the consequences of the high rate of mutation of the viral genome is the ability of the antigens to change structure (and therefore evade the host's immune system) and the easy adaptability to various antiviral drugs. Still another result of this mutability is the presence of different variants of HIV in different cells and tissues of the same individual.

The surface glycoproteins of HIV can recognize and bind to specific receptor molecules called CD4, which are especially abundant on the surface of the T lymphocytes. HIV can also bind to other receptors such as those for complement or the F_c portion of antibodies. This enables the viruses to target and bind to a variety of immune cells, including blood monocytes, tissue macrophages, B and T lymphocytes, dendritic cells, the epithelial cells of the digestive tract, microglial cells in the brain, and the lymphatic tissue that gives rise to various blood cells. Once the virus enters the cell, it takes over the host cell machinery for its own purposes, quickly leading to the lysis or other destruction of the cells. An infection may be latent for a long period of time because the body's own powers of regeneration are activated upon the destruction of the lymphocytes—the first, and most commonly attacked cell population—and replenish what

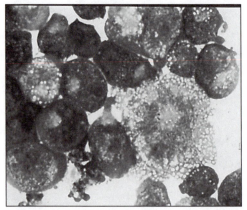

T-cells infected with HIV virus. © *Science Source/Photo Researchers.*

is lost to the virus. Over time, however, the body is unable to withstand the sustained attack, and the production of new cells is unable to keep up with destruction. The end result of this is an increased susceptibility to all manner of infections—especially opportunistic infections by organisms that do not normally infect humans—and rare cancers, also usually associated with immuno-compromised individuals.

The virus is transmitted from one infected person to another via the parenteral route (e.g., sexual contact, blood transfusions, sharing hypodermic needles), via the placenta, or via breast milk. Thus far, two main types of the virus have been isolated. By far, the most widespread human pathogen is designated as HIV-1 and is responsible for most of the AIDS cases reported worldwide. The second variant, HIV-2, is found predominantly in regions of West Africa and appears only sporadically in other parts of the world. HIV-2 appears to share more genetic features with the simian (monkey) immunodeficiency virus than with HIV-1. It causes a less aggressive form of AIDS, with slower progression and a lower overall mortality rate.

Despite the considerable evidence linking HIV to AIDS, there is a strong minority with different levels of scientific training that sees any association between the two as spurious. Indeed, some deny the very concept of a "retrovirus." Readers are referred to <http://www.tulane.edu/~dmsander/garryfavwebaids4.html#altaids> for links to different viewpoints about the subject of HIV and AIDS.

The U.S. Centers for Disease Control upholds the belief that HIV is responsible for AIDS based on three main lines of evidence. First, evidence of viral infection, obtained either directly by culturing or indirectly through antibody tests, is easily obtainable. Second, in keeping with the guidelines of Koch's postulates for determining the cause of a disease, HIV has been isolated from AIDS patients and grown in pure culture in the laboratory. Last, but not least, documented cases of AIDS resulting from blood transfusions have shown a correlation between the trans-

mission of HIV to previously uninfected individuals with the subsequent development of AIDS. *See also* AIDS; ANTIVIRAL DRUGS; GALLO, ROBERT; HO, DAVID; KOCH'S POSTULATES; MONTAGNIER, LUC; RETROVIRUS.

human T cell leukemia virus (HTLV)

The first human retrovirus to be discovered, human T cell leukemia virus (HTLV) is also the only RNA virus positively known to cause a human cancer. Specifically, HTLV has been linked with an unusual form of T cell leukemia in adult humans. There are two main types of this virus: HTLV-1, which is the confirmed cause of the leukemia; and HTLV-2, which has been found among intravenous drug users but not been demonstrated to cause cancer. Because of this, a more correct—and now more widely accepted—form of the acronym HTLV is the human T cell lymphotropic virus.

Exactly how HTLV first became integrated into the genome of host cells is not clear from our present understanding of retroviruses and oncogenesis, but the virus may be transmitted from one person to another via the transfer of blood or semen containing infected T cells. Vertical transmission from mother to newborn through the placenta or breast milk has also been observed. In the latter case, the virus may remain latent for years and sometimes even for life. The mechanism by which HTLV-1 induces cancer is unique among the oncogenic viruses in humans and depends on the activity of a regulatory gene called *tax*. This gene is required for the proper transcription of viral genes and is capable of stimulating the action of certain host genes, which leads to cancer. The long latency of the virus indicates that an additional factor may be required for oncogenesis, but this has not been confirmed.

When an HTLV-1 infection does become apparent, it can take one of two distinct forms, with little or no overlap in the symptoms. The first is the eponymous leukemia, which usually occurs in adults and which is typified by the development of malignant T cells. T cells become enlarged and pleomorphic (changing shapes) with misshapen nuclei, and lose their normal functions, resulting in a severe immunodeficiency. Lymphatic glands, as well as the liver and spleen, become enlarged and tender. Death usually occurs within a year of onset. The second manifestation of infection—called tropical spastic paraparesis—seems to occur mainly in women between the ages of 20 and 50. The main symptom is a progressive weakening and paralysis of the muscles, beginning with pain in the lumbar (back) region and progressing to the lower limbs, brought about by a demyelination of the motor nerves in the spinal cord.

HTLV infections may be detected by antibody tests during the subclinical phase of infection (such as in infants of infected mothers). Once infection is apparent, the appearance of the malignant cells is fairly recognizable. There are, however, no known cures for the diseases, aside from supportive therapy. Control measures are aimed at educating the public about safe sex, the careful screening of blood and organ donations for the virus, and discouraging infected mothers from breast feeding. *See also* CANCER; GALLO, ROBERT; ONCOGENIC VIRUS

human T cell lymphotropic virus. *See* HUMAN T CELL LEUKEMIA VIRUS.

hydrolysis

A type of chemical reaction that involves the participation of water to degrade or break (lyse) a chemical bond. Depending on the substrate molecule, hydrolysis may or may not require a catalyst or enzyme. Hydrolysis is a fundamental mechanism in many important biological reactions. *See also* ENZYME.

hyperbaric oxygen therapy

Therapeutic procedure in which a patient is exposed for short periods (1–2 hours) to higher-than-normal concentrations of oxygen gas under increased pressure. This technique increases the oxygen tension in the

patient's serum and interstitial tissues, thus making this vital substance more readily available for various metabolic activities. These conditions are also believed to inhibit the growth of anaerobic bacteria such as the clostridia. *See also* GAS GANGRENE.

hypersensitivity

An immune response to a previously exposed antigen, in which the immune system responds so vigorously that it harms the body of the host instead of protecting it. Both humoral (antibody-mediated) as well as cell-mediated immune responses have the potential to lead to hypersensitivity by stimulating other cells—such as the macrophages—or processes, e.g., complement, which have the potential to cause either localized or generalized damage. Various external symptoms include hives, respiratory distress (asthma), vasoconstriction, and even shock. Depending on the type of response elicited, hypersensitivity reactions are classified into four main categories. Types I, II, and III are generally termed the immediate hypersensitivity type responses and are mediated by antibodies. Type I, for instance, involves the participation of IgE antibodies, which are fixed to cells such as the mast cells. This response results in the release of cellular chemicals such as histamines, which stimulate further harm. Anaphylaxis and the formation of urticaria (hives) are examples of type I reactions. In type II responses, the antigens eliciting the response are either cells themselves or associated with cells, and there is the involvement of complement. Thus, cytolysis is a common outcome. Type III hypersensitivity involves the formation of antigen-antibody complexes, which may be deposited at various locations such as the joints, where they attract the attention of various tissue-destroying mechanisms, e.g., in rheumatic fevers. Finally, the type IV response is mediated by antigen-primed lymphocytes and macrophages and does not involve the antibodies at all. Various hypersensitivity reactions are not microbe-specific per se, and nonliving antigens such as pollen, animal hairs, and various foods are often the culprits that set them off. However, these reactions are relevant in the context of this book because the destructive mechanisms that take place usually work to exaggerate the pathogenicity of a microbe and to protect it from the host. *See also* ANTI-BODY; IMMUNE RESPONSE; INFLAMMATION.

hypha

A single filament of a multicellular fungus. Typical vegetative hyphae are horizontal and may be either septate with separate cells or a continuous multinucleate mass. Aerial hyphae usually represent spore-bearing or reproductive portions of a fungus. *See also* FUNGI; MYCELIUM.

I

immune response

When an infectious agent such as a bacterium or virus gets past the defense mechanisms of the body and enters either the bloodstream or tissues, the body responds in specific manner to attempt to rid itself of the offending agent. The system responsible for this activity is the immune system, which attempts to get rid of offending agents through a number of different mechanisms mediated by different types of lymphocytes and a few other blood and tissue cells, most importantly, the macrophages.

A typical immune response occurs in two phases or stages. The first time the body is exposed to a particular antigen, it reacts in what is termed a primary immune response, in which a small population of the appropriate immune cells—either B or T cells—are summoned to the site, where they recognize the "foreignness" of the agent and begin to gather the necessary forces to fight it off. This primary response is a relatively low-level response, but it primes a subset of the immune cells against the specific antigen. Upon a subsequent exposure, these primed cells immediately respond in a far more rapid and vigorous manner because they have already "learned" about the nature of this enemy. This phase is termed the secondary immune response. During this secondary response, the immune system also summons a number of cells such as macrophages as well mechanisms such as complement to help the B cells and T cells neutralize or destroy the threatening agent. A discussion of the details of the immune response is somewhat beyond the scope of this book and readers interested in learning more should consult immunology text books. *See also* IMMUNE SYSTEM; HYPERSENSITIVITY; INFLAMMATION.

immune system

Diffuse system consisting mostly of different types of blood cells, found primarily in bodies of vertebrate animals, whose primary purpose is to mount a specific response against different types of challenges—such as pathogenic microbes and chemicals—to either neutralize their harmful effects or to remove them from the body. The primary players in this system are the lymphocytes, which mediate either a humoral or cellular reaction against the agent, whose components they recognize as foreign antigens. The humoral arm of this system is represented by the B cells, which produce antibodies against the antigens, while a mixed population of T cells serve as the cellular arm. The antibodies or cells summoned by the immune system proceed to eliminate or attempt to eliminate the offending agent from the body, or at least render it harmless. For example, an antibody can combine with a harmful substance to neutralize its toxicity, while T cells may exert a similar action against a virus or virus-infected cell. In addition to the lymphocytes, other cells such as macrophages and monocytes, as well as chemical agents such as complement, also play a role, particularly in performing the actual killing and neutralization of the infectious organisms.

In contrast to other organ systems in the body, all of which are physically identifiable entities with organs that are connected with one another, the immune system is a conceptual system whose physical boundaries spill over into several others. While a system of lymphatic vessels, regional lymph nodes, and glands does exist, a large part of the body's immunological activity takes place in the bloodstream. Like other blood cells, the primary actors, i.e., the lymphocytes, originate in the bone marrow and spleen. In addition, different tissues such as the subcutaneous layers and the connective tissue of various organs also contain immunologically active cells.

The immune response relies on a few fundamental properties that distinguish it from other innate defense mechanisms in the body. First, as mentioned earlier, it mounts a specific response against a specific agent. The antibodies or cells that are summoned in the case of one type of pathogen typically have little or no effect on other pathogens, unless these share significant similarities with the first agent. The high degree of specificity also implies a high level of adaptiveness on the part of the immune system—it is able to respond to threats it has never encountered before. Yet another important property of this system is its ability to distinguish between self and non-self or foreign bodies. Finally, the immune system also possesses "memory," which enables it to respond to second and subsequent exposures to an antigen more readily and vigorously. *See also* B CELL; COMPLEMENT; IMMUNE RESPONSE; T CELL.

immunity

A term used to refer to all the mechanisms used by the body to protect itself against potentially harmful foreign agents that it may encounter in its environment. Microorganisms pose one of the largest groups of environmental threats against which the body needs such protection. Immunity may be conferred by certain innate characteristics of a body, such as the skin, which presents a physical barrier to various microbes, chemical substances in the mucus surfaces that deter the growth of harmful organisms, and physiological reflexes such as coughing and tearing, which also prevent microbial entry. Within the body, phagocytic cells and events such the development of fever, help fight off various threats that have made it past the first line of defense. In addition to these innate mechanisms, which are present in most organisms, various higher animals, notably the vertebrates, have evolved more specialized systems of acquired immunity against specific invaders. These mechanisms, which come into play relatively late, supplement the protection provided by the innate defenses. The immune system, consisting of lymphocytes and antibodies, is the major player in mediating acquired immunity against specific pathogens. *See also* IMMUNE RESPONSE.

immunization

The practice of inducing an individual's immunity to a specific infectious disease prior to exposure to the causative agent, so that the person is protected against the disease. Protection may be induced via either active or passive means. Examples of substances used for active immunization include killed pathogenic organisms (e.g., whole bacterial or protozoan cells), non-virulent variants of the pathogens, inactivated or attenuated viruses, and denatured toxins. The underlying property of these vaccines is that they are all antigenic in nature. Thus, when they are introduced into the body, they stimulate the immune system to mount a primary response that is specific for the pathogen, so that upon a later exposure to the disease agent in nature, the individual is able to ward off infection by responding to it with a stronger, secondary response. In contrast, passive immunization involves the introduction of immunoglobulins (antibodies) or lymphocytes

Table 6. Recommended Childhood Immunization Schedule, United States, January–December 2000

Vaccines[1] are listed under routinely recommended ages. [Bars] indicate range of recommended ages for immunization. Any dose not given at the recommended age should be given as a "catch-up" immunization at any subsequent visit when indicated and feasible. (Ovals) indicate vaccines to be given if previously recommended doses were missed or given earlier than the recommended minimum age.

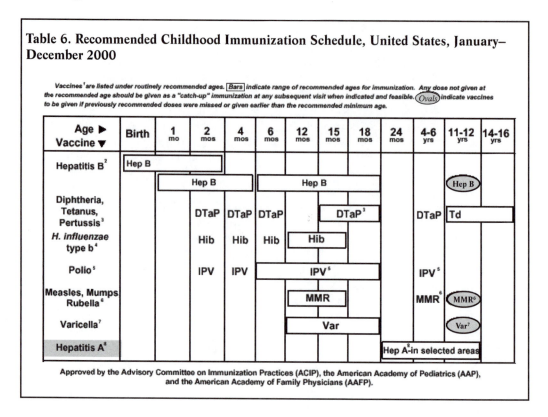

Age ▶ Vaccine ▼	Birth	1 mo	2 mos	4 mos	6 mos	12 mos	15 mos	18 mos	24 mos	4-6 yrs	11-12 yrs	14-16 yrs
Hepatitis B[2]	Hep B	Hep B			Hep B						(Hep B)	
Diphtheria, Tetanus, Pertussis[3]			DTaP	DTaP	DTaP		DTaP[3]			DTaP	Td	
H. influenzae type b[4]			Hib	Hib	Hib	Hib						
Polio[5]			IPV	IPV	IPV[5]					IPV[5]		
Measles, Mumps, Rubella[6]						MMR				MMR[6]	(MMR[6])	
Varicella[7]						Var					(Var[7])	
Hepatitis A[8]									Hep A-in selected areas			

Approved by the Advisory Committee on Immunization Practices (ACIP), the American Academy of Pediatrics (AAP), and the American Academy of Family Physicians (AAFP).

directly into the bloodstream of the individual being immunized. The transfer of immune cells is also called adoptive transfer. This type of immunization typically confers immunity for a shorter time period, albeit with greater immediacy, corresponding to the lifetime of the antibody or cells administered. Because the body itself had no prior exposure to the antigen, it will not be able to mount a secondary immune response. Immunoglobulins may also be given passively as a means of bolstering general immunity rather than against a specific disease. In instances of severe immunodeficiency (e.g., after radiation therapy), passive immunization may be achieved by bone marrow or thymus transplants to supply the individual with the tissue that synthesizes immune cells.

Immunization is often used synonymously with vaccination, although the latter term is sometimes used in a more restrictive sense to refer only to active immunization, in the manner first developed by Edward Jenner.

Nowadays, vaccines against several viral, bacterial, and parasitic diseases are administered during childhood as a routine practice, especially in the developed countries. Still other vaccines are given when the risk of exposure increases for some reason, such as imminent travel to an endemic area or in case of a disease outbreak. See Table 6 for the childhood immunization schedule recommended by the CDC in 1998. *See also* IMMUNE RESPONSE; IMUNE SYSTEM; JENNER, EDWARD; VACCINATION.

immunoglobulin. *See* ANTIBODY

immunology

Branch of science or medicine that studies the structure and functioning of different components of the immune system. Although it is now a full-fledged discipline concerned with a system present predominantly—if not exclusively—in higher animals, the subject is mentioned here because of its close historical ties to microbiology. Indeed, the roots of immunology may be said to lie in microbiology because the first investigations into the body's defense systems against pathogens began with the study of antimicrobial immune responses. Even today, some of the most effective means of disease prevention and treatment come from the field of immunology.

impetigo

Also known as streptococcal pyoderma, impetigo is a superficial skin infection with *Streptococcus pyogenes*, characterized by crusted lesions. This may be followed by inflammation of the kidneys (glomerulonephritis) some three weeks after the skin infections, but other common sequelae of streptococcal infections, such as rheumatic fevers, are not seen. *See also* STREPTOCOCCUS PYOGENES.

IMViC tests

Panel of biochemical tests used to distinguish different bacteria of the Enterobacteriaceae family. The acronym derives from *i*ndole test, *m*ethyl red (MR) test, *V*oges-Proskauer (VP) test, and *c*itrate test.

The indole test detects the ability of an organism to produce indole from the amino acid tryptophan. The specimen is cultured for 48 hours in a medium containing tryptophan and then tested with Kovac's reagent, which turns a pink or red color in the presence of indole.

The MR test looks for the production of acid by an organism growing in a buffered, glucose-containing medium for 48 hours. When added to the culture, a neutral solution of methyl red, which is yellow at pH values above 6.2, will turn a pink color, indicating an acidic (or lower) pH.

The VP test detects the production of acetoin, an intermediate metabolite in the fermentation of glucose to butanediol, which occurs under anaerobic conditions. Two- to five-day-old cultures in a glucose-containing medium are tested with a reagent, which reacts with acetoin to give a red color.

The citrate test identifies the ability of different coliforms to use citrate as the sole carbon source. Test tubes with special citrate media are inoculated with saline suspensions of the test organism and examined for growth (in terms of increasing turbidity) after 1–2 days of incubation at 37°C.

inclusion body

Any physically delineated mass of material in a cell, which has no discernible role insofar as the normal functioning of the cell is concerned. The inclusion bodies may be present in either the cytoplasm or the nucleus. They are often composed of foreign material such as viral proteins or metal granules, but they may also be composed of inert cellular matter such as lipids or proteins. Viral inclusions usually demonstrate altered staining behavior from the rest of the cell; they consist of accumulated viral proteins or nucleic acids, due to the active synthesis of these viral components in the cell. They may be seen both in tissue culture as well as *in vivo*, e.g., the negri bodies of rabies viruses. *See also* NEGRI BODY.

incubation period

The interval between the infection of an individual with a pathogen and the time when the disease first manifests itself. During this period, the infected individual appears asymptomatic but nevertheless supports the multiplication and possibly also the spread of the pathogen in the body. When dealing with vector-borne diseases, the incubation period in the vector refers to the time between the entry of the organism into the vector and the time when the infection may be transmitted to a new host. In this case, too, the incubation period represents the period of growth and multiplication of the organisms.

indigenous microflora

Population of different microbes whose natural habitat is a specific organ or location in a higher animal. For instance, organisms such as *Escherichia coli* and *Lactobacillus acidophilus* are part of the indigenous microflora of the human intestine, and *Staphylococcus epidermidis* and *Micrococcus* species are nor-

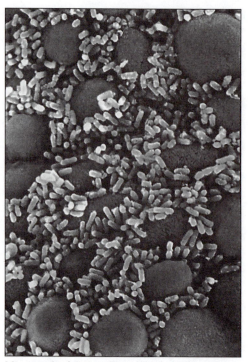

Micrograph of *Escherichia coli* bacteria inhabiting the intestinal villi. © *Science Source/Photo Researchers.*

infection/communicable disease control

Epidemiological programs of surveillance of a disease, generally within a healthcare facility, designed to investigate, prevent, and control the spread of various infections and their causative agents.

infectious mononucleosis

Acute viral disease caused by the Epstein-Barr virus (EBV), with clinical symptoms of fever, tonsilar and pharyngeal inflammation, lymphatic involvement (especially near the cervical lymph nodes), and spleen enlargement. The characteristic feature of this disease that distinguishes it from other infections and diseases with similar clinical symptoms is the specific proliferation—usually benign—of lymphocytes (B cells) and the presence of anti-EBV antibodies in the bloodstream. This disease tends to affect adults more severely than children and is widespread throughout the world. Contrary to its name, this is not a terribly infectious or contagious disease, and requires close contact, usually with saliva for transmission. Its mode of infection earned it the name of "kissing disease," although the term no longer seems to be in wide usage. For the most part, infectious mononucleosis is self-limiting and has low fatality rates, with most patients showing complete recovery. There appears to be a loose association with the development of chronic fatigue syndrome, although the evidence for this link is not definite. Treatment is supportive and no vaccine has been developed. *See also* EPSTEIN-BARR VIRUS.

inflammation

One of the body's main lines of defenses against various sources of harm, such as pathogenic microbes and trauma, inflammation is a complex protective process involving a number of cells and cellular events at the site of injury or challenge. The summoning of various immunological cells, such as macrophages and T cells, is part of this process. An inflammatory response is character-

mal residents of the skin. Indigenous microflora may either be parasites, saprophytes, or commensals and are not pathogenic to their host under normal circumstances. Indeed, many of the indigenous microbes fulfill useful, even essential, roles for their hosts, such as degrading cellulose (in the stomach and intestines of ruminants) or producing important vitamins (*E. coli* in the human gut) and maintaining the health of such populations is to the host's advantage. Because of their locale, however, these organisms are also prime candidates for acting as opportunistic pathogens when the host's defenses are down or when the microbes are transferred to sites in the host other than their regular niche. For example, *Clostridium perfringens* is a normal resident of the intestinal lining, where its presence does not affect the host in any significant manner. However, if these bacteria gain entry into the bloodstream or the peritoneal cavity—sites that are typically free of living organisms—they may multiply and damage various organs and tissues, causing such conditions as gangrene or septicemia. *See also* OPPORTUNISTIC PATHOGEN; PROBIOTIC.

ized by symptoms of swelling, redness (if surface tissues are affected), pain, sometimes tenderness, heat, and a temporary loss of function of the tissue in the inflamed area. The overall aim of the inflammatory response is to restore the injured tissue to its normal state, but if the response is prolonged due to the persistence of the source of harm, it can lead to harmful consequences, as is often the case with hypersensitivity reactions. Chronic inflammation may be treated symptomatically with anti-inflammatory drugs, but, unless the root cause is removed, such drugs will not cure the problem. *See also* IMMUNE RESPONSE.

influenza

An acute respiratory tract disease in humans caused by influenza viruses. Compared to infections with other respiratory viruses such as the rhinoviruses and coronaviruses, influenza is more severe. Typical symptoms include high fevers, chills, protracted and severe coughing, congestion, runny nose, aches and pains, and extreme weakness. In addition, influenza infections may also involve the gastrointestinal tract and induce symptoms of diarrhea, nausea, and vomiting, especially in young children. Although the influenza infection itself is self-limiting and disappears in 2–7 days, the illness may last much longer due to complications (e.g., pneumonia arising from secondary infections by bacteria or other microorganisms). While complications can and do occur in almost every age group, the elderly and infirm are more likely to develop the most serious problems. According to the CDC, influenza is linked with an estimated 20,000 deaths and even more hospitalizations per year in the United States.

Influenza is an important health concern all over the world because of the epidemic or pandemic scale of the outbreaks, the rapidity with which these epidemics develop, and the seriousness of the complications. Of the three types of viruses, influenza virus A has been found associated with pandemics or large epidemics, type B with regional or relatively contained epidemics, and type C with localized outbreaks and sporadic cases. The respiratory symptoms result in the shedding of large amounts of virus into the atmosphere in aerosols that are subsequently inhaled by other people. This results in the rapid spread of the virus over large areas. Although vaccines against the influenza viruses have been developed and are somewhat useful in protecting against infections, their large-scale production and use is not advisable because the influenza viruses types A and B are notorious for undergoing periodic changes in their antigenic structure, which alters their susceptibility to the vaccines. Thus, a virus that initiated a particular epidemic may have disappeared and reappeared in a significantly altered guise by the time the infection has spread to another continent, even though the essential features of the disease itself remain unchanged. Further compounding the antigenic complexity of influenza viruses is the property of antigenic shift (demonstrated by the type A viruses only), which is the abrupt and drastic change in antigenic character. When this occurs, large numbers of people, and sometimes entire populations, have no antibody protection against the virus. In the past 100 years, the world has witnessed major pandemics in 1918, 1957, and 1968, each of which resulted in large numbers of deaths. The development of vaccines and a better understanding of the nature of the virus and how to deal with it has resulted in improved management of the more recent influenza epidemics, although the cure for this disease still remains beyond our reach. *See also* EPIDEMIC; INFLUENZA VIRUS.

influenza virus

Enveloped animal virus containing a single-stranded RNA genome, widely associated with epidemics of upper respiratory infections (influenza or "the flu") in humans. The virus has a complex structure. The RNA is present in the form of eight unequal fragments in close association with a specific protein to form a helical nucleocapsid. Different structural genes and enzymes are encoded in these different segments. The outer lipid membrane contains two glycoprotein antigens embedded in its outer surface, which give the influenza virus its "spiky" appearance under an electron microscope. Both proteins also appear to play a major role in virulence. One of the proteins is a hemagglutinin, which is chiefly responsible for the virus' ability to attach itself to host cells. Because the surface antigens are capable of undergoing structural changes (antigenic drift and shift) rapidly and frequently, any antibodies against the influenza viruses are effective only over limited periods of time.

The replication cycle is complex, with a multistep uncoating process that occurs after the virion is adsorbed on the host cells, usually in the respiratory lining. Different phases of genome replication and nucleocapsid assembly take place in the cytoplasm as well as the nucleus. Finally, the virus particle acquires its envelope when it emerges from the cell by a process of budding from specific locations in the cell membrane where the glycoproteins were inserted during the intracellular phase of its life cycle.

The influenza virus is primarily a respiratory pathogen that usually causes epidemics of influenza. There are three main types of virus, depending on the host range and fundamental antigenic differences (i.e., other than those generated by the drift and shift

mechanisms). Type A viruses infect a wide variety of mammals, as well as birds. Pigs and birds are important reservoirs. This strain causes the most severe human infections and is most commonly associated with epidemics and pandemics. Type A viruses are characterized by their ability to undergo antigenic shift. The type B virus has a similar host range but causes a less severe version of the disease. Type C, which is the least studied, appears to cause nonsymptomatic infections of humans only and has not been found in other organisms. The interchange of genetic material between the different types of viruses is possible due to genetic reassortment, when the virus is transferred back and forth from humans and other hosts. *See also* ANTIGENIC DRIFT; ANTIGENIC SHIFT; INFLUENZA.

inoculating loop

An instrument, used almost exclusively in microbiology laboratories, for transferring small amounts (microliters or drops) of culture from one medium to another, particularly when attempting to isolate single colonies from mixed sources. It is simply a

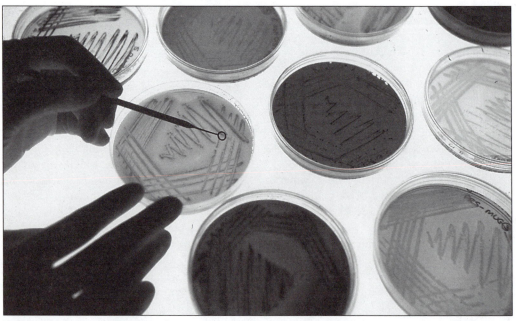

Technician uses a sterile inoculating loop to pick a single colony from a petri dish containing bacterial cultures. © *Science Source/Photo Researchers.*

wire made of some inert, flame-resistant metal such as platinum, which is attached to a handle for easier use. The tip of the wire is twisted into a small loop, which can pick up drops of liquid by capillary action and also glide smoothly over solid agar media. To maintain sterility and avoid contamination from the air, the loop is regularly heated to white heat over an open Bunsen flame before and after transferring the cultures. *See also* INOCULATION.

inoculation

The introduction of living microbes into a living host or into a culture medium. The former type of inoculation—usually carried out for purposes of immunization—may be administered orally, by injection, or by inducing inhalation of aerosols of the specific microbes. Culture media are inoculated with suspensions of the bacteria, with the amount of inoculum and method of transfer (e.g., with loops and droppers) depending on the state of the medium, the nature of the culture material, and the purpose of the culture—i.e., identification or production. *See also* IMMUNIZATION; INOCULATING LOOP.

interferon

A class of chemical substances produced naturally by white blood cells, fibroblasts, and various virus-infected cells, which have some antiviral activity and enhance the body's immune response against viruses and some other agents, such as cancer cells.

J

Jenner, Edward (1749–1823)

Jenner was the first person to experimentally verify and report the success of the procedure that came to be known as vaccination in preventing disease. An English country physician in the late eighteenth and early nineteenth centuries, Jenner had observed the apparent resistance towards smallpox in milkmaids who had previously contracted a similar but milder disease called cowpox. He then proceeded to verify this observation by experiment. He injected the blister fluid from a cowpox patient into a boy who had no prior exposure to either disease, and followed this up with a shot of the smallpox blister fluid after the boy recovered from cowpox. The boy did not contract smallpox, and thus Jenner was able to verify his hypothesis that cowpox protected against smallpox. The procedure came to be called vaccination due to the source of the original virus (*vaca* is cow in Latin) and the disease (vaccinia) that it caused. *See also* COWPOX; SMALLPOX; VACCINATION.

Drawing of a sculpture by Monteverde showing Edward Jenner vaccinating his son. © *Science Source/Photo Researchers.*

K

kala-azar

Chronic, systemic disease caused by various species of the protozoan *Leishmania*. Also called Dumdum fever, kala-azar often occurs as a zoonotic infection that is transmitted from dogs to humans via fleas or ticks. The predominant species involved in this visceral leishmaniasis is *L. donovani*, although *L. tropica*, *L. chagasi*, and *L. infantum* may also cause similar disease. Incidence has been shown to be highest in tropical and subtropical regions of the world such as North and East Africa, Asia, around the Mediterranean, South America, and the Indian subcontinent. In most of these places, children are the disease's main victims, except for India and Bangladesh, where adults are most affected.

Kala-azar has a long incubation period and may occur up to two years after the original infection. How the disease presents is as much a feature of the host's immune status as of the organism's virulence. Small, short-term papules may develop at the site of the insect bite, although these may be overlooked. The onset of the systemic disease may be acute, with a sudden onset of typhoid or malaria-like fever, or very insidious with vague feelings of discomfort and ill health. The actual disease is characterized by a spiking fever (with two daily peaks), liver and spleen enlargement, diarrhea, dysentery, and weight loss. Lymphatic involvement leads to accumulation of fluid in the lymph nodes, especially near the site of the initial lesion. If left untreated, visceral leishmaniasis is fatal, usually due to secondary complications. Diagnosis is by the observation of parasite-containing phagocytes in the circulating blood or serologic tests to detect antibodies against the parasite. Antimony or pentamidine isethionate may be used to treat the infection. *See also* LEISHMANIA.

Kaposi's sarcoma

Probably best known because of its association with HIV and AIDS, Kaposi's sarcoma may be best described as a cancer of the blood vessels—usually apparent as lesions in the skin, but also in the lungs, digestive tract, and lymph nodes. The cancer appears to be induced by an infection with the human herpesvirus 8 (HHV8), which is also known as the Kaposi's sarcoma–associated herpesvirus (KSHV). This tumor was initially found during the late nineteenth century among Mediterranean men and, subsequently, as an endemic tumor in certain parts of Africa, but it became widespread in the West only after the rampant spread of HIV. In fact, the preponderance of this cancer among male homosexuals in the early 1980s earned this disease the nickname of "gay cancer" in the early years of AIDS. The association of Kaposi's sarcoma with a specific herpesvirus was discovered only in 1994.

Kaposi's sarcoma differs from most other cancers in that it does not originate in a single location and spread (metastasize) to other locations, but rather originates at several sites at once. Furthermore, unlike most other cancers, the origins of this tumor cannot be traced to the transformation of any one type of cell. Instead, a number of different cells, including new red blood corpuscles, lymphocytes, and inflammatory cells, are involved. The primary skin lesions consist of patches of painless, reddish-brown to purple discoloration, due to localized proliferation of the cells. Over time, the patches might coalesce to form larger patches. When these patches or lesions develop in the lungs or intestines, patients might exhibit corresponding symptoms of distress, such as breathlessness or nausea and vomiting. In the case of lymphatic involvement, the sarcoma cells crowd around the lymph nodes and restrict the circulation of lymph, which can result in painful fluid accumulation and swelling in the area. Kaposi's sarcoma is rarely fatal in and of itself, but because the virus tends to affect severely immuno-compromised individuals, it is often associated with other fatal conditions. *See also* AIDS; HERPESVIRUS.

kinetoplast

An accessory body found in the cells of certain protozoan species, e.g., *Trypanosoma* and *Leishmania*, consisting of a large mitochondrion, with mitochondrial DNA that lies near the basal body of the parasite's flagellum or undulating membrane. The location of the trypanosome kinetoplast with respect to the nucleus differs in different stages of development of the organism. *See ALSO LEISHMANIA*; MITOCHONDRIA; *TRYPANOSOMA*.

Klebsiella pneumoniae

Also known as Friedlander's bacillus (after the scientist who first isolated it), this bacterium is a normal resident of the intestinal tracts of various animals, along with related members of the Enterobacteriaceae family, such as *Escherichia coli* and *Enterobacter*. Unlike many other members of this family,

however, *Klebsiella* is found nearly as frequently as a free-living organism in soil. The bacteria are frequently found associated with human disease, most commonly pneumonia, but also septicemia and shock if the organisms invade the bloodstream. In rare instances, *K. pneumoniae* may cause peritonitis and meningitis.

Infections are diagnosed by the isolation and culturing of the bacteria from clinical samples. *Klebsiella* species are gram-negative rods with no flagella (unlike most other coliforms) and a very thick capsule. Many strains produce a distinctive yeasty odor in culture. Organisms are microaerophilic, requiring only low levels of oxygen, and they may survive for long periods under anaerobic conditions. The capsule is responsible for the viscous or mucoid appearance of *K. pneumoniae* cultures and is an important contributor to the pathogenicity of these organisms. Scientists have identified more than 70 types based on capsular antigen alone. In addition, some microbiologists have attempted to differentiate these bacteria into three separate species on the basis of the specific nature of the respiratory disorders they induce. While *K. pneumoniae* is associated with classic pneumonia, *K. rhinoscleromatis* has been found in nodules in the nose, and *K. ozenae* is known to hamper smelling abilities. These species are distinguishable only by their capsular antigens.

Koch, Robert (1843–1910)

If Leeuwenhoek is regarded as the founder of microbiology, then the German scientist Robert Koch, along with Louis Pasteur, may be viewed as one of the field's main architects and builders who strengthened its foundations and gave the field its modern shape. His contributions to the understanding of microbial—particularly bacterial—diseases were numerous. Before he conducted his Nobel Prize–winning research (1905) on the cause and pathogenesis of tuberculosis, he had provided scientific proof for the etiology of several diseases, including cholera and anthrax. In doing so, he also helped establish

Bacteriologist Robert Koch. © *Science Source/Photo Researchers.*

1. The proposed organism must be observed in every case of the disease.
2. The proposed organism must be isolated from a diseased animal and grown as a pure culture in the laboratory.
3. A preparation of the pure culture of the proposed organism must cause the same disease when injected into a healthy animal.
4. The proposed organism must be isolated from the newly infected animal.

One of the primary drawbacks in these postulates is immediately apparent; steps 3 and 4 would be unethical and potentially dangerous to prove for human diseases. In these cases, scientists attempt to find an animal model for the disease. Only rarely has this approach failed—a rather notorious example being that of the connection between ulcers and *Helicobacter pylori*. (In that case, the scientist in question actually consumed the bacteria himself to prove his point, as animal models were lacking.) A second obstacle in following these postulates arose in cases where disease agents were hard to isolate or visualize. For instance, until the development of electron microscopes, viruses were impossible to see. In such cases, scientists have looked to indirect means, such as the production of antibodies, as indicators of infection. Regardless of the problems, however, Koch's postulates—while not scientific "laws," in the sense of Newton's laws of physics—are used to this day as a framework for establishing the etiology of infectious diseases. *See also* KOCH, ROBERT.

the basic principles of infection and disease causation that we now refer to as Koch's postulates. Koch was also responsible for developing methods of growing pure cultures of bacteria by cultivation on solid media. This simple but necessary technique facilitated a vast number of bacteriological experiments that were impossible until then. He also conducted important investigations on a wide variety of human and animal diseases, among them, malaria, tropical dysentery, trachoma, typhus, trypanosomiasis, and rinderprest. *See also* KOCH'S POSTULATES.

Koch's postulates

Set of rules or criteria by which a specific microorganism is identified as the cause of a specific disease. According to some sources, these or similar rules were laid down as early as 1840 by a German pathologist named Jacob Henle. However, Robert Koch first used and demonstrated the truth of these criteria in experiments some 36 years later. Koch's postulates may be expressed as a set of four basic principles.

Krebs, Hans A. (1900–1981)

British biochemist who experimentally demonstrated the individual reactions by which living cells oxidize carbon compounds for obtaining energy. The cyclic process or network of reactions was subsequently named in his honor. Krebs received the Nobel Prize in 1953 for this work. He was also responsible for similarly elucidating the urea cycle in living cells. *See also* KREBS CYCLE.

Krebs cycle

The metabolic reactions whereby different kinds of living organisms conduct respiration. Through this cycle, the pyruvic acid generated by glycolysis is converted to carbon dioxide and water, with a net release of energy in the form of ATP molecules. Prior to its entry into the Krebs cycle, the pyruvate is converted to a highly energetic form called acetyl-Co A by virtue of its attachment to Coenzyme A. Acetyl Co-A then reacts with a cellular acid called oxaloacetic acid, resulting in the formation of citric acid. This molecule then goes through a series of enzymatic changes and gradually releases carbon dioxide and ATP. The Krebs cycle is also known as the citric acid cycle or the tricarboxylic acid cycle (TCA cycle) to reflect the nature of the participating molecules. The energy molecules are produced by the coupling of this cycle to the electron transport chain. Some of the intermediates of the Krebs cycle, such as succinic acid and malic acid, are diverted for use in other metabolic and biosynthetic pathways (e.g., amino acids, purines, and pyrimidines). An interesting fact to note about this cycle is that the conversion of pyruvate to CO_2 is only the overall result of the reaction. If the path of the pyruvate were tracked (by radio-labeling, for example), it would be seen that the carbon atoms remain in the cell for two successive cycles and are released as carbon dioxide only during the third repeat of the cycle. In many instances, they show up in other molecules. In many instances, the radiolabeled carbon atoms show up in other molecules instead of carbon dioxide, reflecting the incorporation of the Krebs cycle intermediates into other metabolic pathways. *See also* KREBS, HANS; METABOLISM; RESPIRATION.

Kkurthia

Genus consisting of gram-positive, aerobic, rod-shaped or filamentous bacteria associated with the spoilage of meat and meat products. Organisms are motile with peritrichous flagella. They are metabolically versatile and able to use a number of molecules, such as amino acids and fatty acids, as their carbon sources.

kuru

A neuro-degenerative disease in the spongiform encephalopathy group, seen exclusively among certain tribes in the region called the Fore Highlands in Papua New Guinea. Also known in those regions as the "laughing death," this obscure and hitherto unknown disease, initially described in detail during the 1950s, was the first disease to have brought the human prion diseases into the public eye. Characterized by a gradual loss of muscular coordination followed by dementia and death, kuru is believed to have been transmitted among members of the tribe due to certain religious rituals in which people honored their dead relatives by consuming their brains. With the cessation of this practice, the disease declined rapidly and is now virtually wiped out. *See also* CREUTSFELD-JACOB DISEASE; GAJDUSEK, D. CARLETON; MAD COW DISEASE; PRION; PRUSINER, STANLEY; SPONGIFORM ENCEPHALOPATHY.

L

L forms

Variants of bacterial cells formed in certain cultures that, for some reason, are unable to synthesize all the components of their cell walls. They appear in cultures as granules called "L bodies" that are capable of uniting and forming amorphous masses. The multiplication of these masses gives rise to bacterial cells identical to the parent strain.

β-lactam antibiotics

Group of natural and synthetic antibiotics whose structure is characterized by the presence of a four-membered, nitrogen-containing ring called the β-lactam ring. These compounds act by binding to specific proteins that are involved in synthesizing the peptidoglycan component of the cell wall. This disruption of the cell wall causes the organism to lose control over the passage of materials in and out of the cell, which eventually kills the organism. The β-lactams thus may only kill bacteria that are actively growing; they are of no use against resting cells. Common examples include penicillin and cephalosporin. *See also* ANTIBIOTIC; CELL WALL; CEPHALOSPORIN; β-LACTAMASE; PENICILLIN; PEPTIDOGLYCAN.

β-lactamase

Enzyme produced by certain species or strains of bacteria (e.g., *Haemophilus influenzae)*, which have the ability to degrade the β-lactam ring structure and thus inactivate antibiotics such as penicillins. The production of this enzyme by an organism will thus confer resistance against the β-lactam antibiotics. This gene for the enzyme is often carried on a plasmid and may therefore be transferred easily from one species to another. *See also* ANTIBIOTIC; ANTIBIOTIC RESISTANCE; ENZYME; β-LACTAM ANTIBIOTICS.

lactic acid bacteria

Term given to a heterogenous group of gram-positive, non-spore-forming, non-respiring bacteria that produce lactic acid as one of the principal end products of carbohydrate fermentation. Historically, this name was first used in the late 1900s to denote bacteria implicated in turning milk sour. Examples of the important genera that fall under this group are *Lactobacillus, Leuconostoc, Streptococcus,* and *Pediococcus*, and contrary to what was originally believed, they are not closely related to each other in a phylogenetic sense. For instance, the metabolic pathways and enzymes used to produce lactic acid are different in the various species. However, because of their common utility in the food processing/preservation industry, these organisms are grouped together by many scientists, and the term holds up as a practical method of categorization in the study of food microbiology. *See also* LACTIC ACID FERMENTATION.

lactic acid fermentation

Perhaps the most widespread fermentation pathway in cells from prokaryotes to human muscle, this type of fermentation converts pyruvic acid generated during glycolysis to lactic acid in the presence of the enzyme lactic dehydrogenase. In a number of bacteria, called the homolactic fermenters, this is the only end product, while organisms such as *Leuconostoc* are heterolactic fermenters, which produce ethanol or acetic acid and carbon dioxide in addition to lactic acid. *See also* FERMENTATION; GLYCOLYSIS; LACTIC ACID BACTERIA.

Lactobacillus

This genus comprises a rather heterogneous group of over 100 gram-positive bacterial species whose most shared characteristic is the ability to produce lactic acid as a fermentation product. They are known to occupy a diverse variety of habitats and are often found amidst vegetation or as part of the normal intestinal microflora of humans and animals. Usually found as pairs or short chains of small rods, the lactobacilli may be either homolactic or heterolactic fermenters due to a large repertoire of metabolic enzymes. Various species have thus found wide applications in the food industry and are routinely used in the manufacture of dairy products such as cheeses, yogurt, and butter, as well as in other fermented foods. Common examples include *L. acidophilus*, a normal intestinal resident; *L. bulgaricus*, often used in the production of buttermilk; and *L. delbruckei;* considered as the representative species. While some organisms such as *L. casei* are capable of growth in a wide variety of habitats, others may be found only in specific niches, e.g., *L. sanfrancisco*, an organism used for sourdough fermentation. Species may be differentiated on the basis of biochemical and growth properties, such as carbohydrate fermentation patterns, hydrolysis of arginine, or temperature requirements, although the wide range in GC content (32–53 moles percent) indicates that in the future, different subgroups of these bacteria may be reclassified into new genera. *See also* BREAD; BUTTER; LACTIC ACID FERMENTATION; YOGURT.

Lactococcus

This organism, specifically *Lactotococcus lactis* subsp. *cremoris,* is widely used in the dairy industry for the preparation of butter, yogurt, and buttermilk. Until recently, it was classified under the genus *Streptococcus* as *S. lactis* but was renamed when ribosomal sequence analysis revealed the two genera to be significantly different. The same techniques also revealed the similarities between isolates that were originally thought to be different species. Currently, many of these strains are classified as strains/serotypes of *L. lactis*. The genus consists of gram-positive cocci that exist in pairs or short chains. They are mainly heterolactic fermenters that produce a variety of end products due to the fermentation of milk sugars. Some strains are heat sensitive, with the ability to grow at 10°C but not 45°C. Thus, while these bacteria may occasionally be incriminated in the spoilage of certain types of refrigerated foods, the threat is easily removed by pasteurization or even simple boiling. *See also* BUTTER; BUTTERMILK; LACTIC ACID FERMENTATION; YOGURT.

Lancefield, Rebecca (1895–1981)

One of the first women to achieve renown as a microbiologist, Lancefield, an American, was responsible for developing the serological methods to differentiate among different strains of streptococci. She demonstrated that most of the common streptococcal infections of humans, such as scarlet fever, erysipelas, and strep throat, were caused by different serotypes or strains of a single species (namely *S. pyogenes*), which she called the Group A streptococci. She also pinpointed the determinant of these differences—an antigen called Protein M, which plays a role in bacterial virulence. *See also* PROTEIN M; *STREPTOCOCCUS*; *STREPTOCOCCUS PYOGENES*.

Laveran, C. L. A. (1845–1922)

During the last quarter of the nineteenth century—a time when medical bacteriology was burgeoning with information on new links between bacteria and specific diseases—the Frenchman Charles Louis Alphonse Laveran distinguished himself by suggesting that living organisms *other than* bacteria, in particular the protozoa, were also capable of causing human diseases. In a career spanning nearly three decades, Laveran made fundamental contributions to the field of medical parasitology, including the discovery of the malarial parasite, suggesting (although not proving) the mosquito as a possible vector, and the discovery and elucidation of several trypanosomes and their diseases. These contributions earned him the 1907 Nobel Prize in Physiology/Medicine. *See also* MALARIA; *PLASMODIUM*; *TRYPANOSOMA*; VECTOR.

Lederberg, Joshua (1925–)

Justly regarded as the founder of bacterial genetics, American Scientist Joshua Lederberg was the first person to demonstrate bacterial conjugation as well as transduction, thus providing the experimental evidence that processes such as genetic recombination were possible in bacteria. In 1958, he won the Nobel Prize in Physiology/Medicine for this work. *See also* CONJUGATION; RECOMBINATION; TRANSDUCTION.

Leeuwenhoek, Antoni van (1632–1723)

The first person to publicly report his microscopic observations of tiny living organisms, Leeuwenhoek, a Dutchman, is regarded as the founding father of modern microbiology. Before this, scientists had only speculated on the existence of invisible germs but had no evidence to back up their theories. With his exhaustive descriptions and meticulous drawings, some of which have been preserved to this day, Leeuwenhoek showed conclusively that microbes did exist. He conducted exhaustive microscopic researches on a variety of materials and wrote nearly 200 letters de-

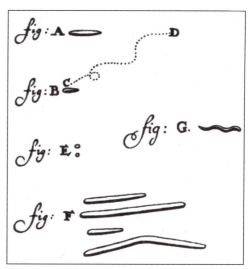

Drawings of "animalcules" from Leeuwenhoek's September 1683 letter to the Royal Society in London. © *Science Source/Photo Researchers.*

scribing his findings to the Royal Society in England. Most of these were translated from Dutch into English and published in the society's *Philosophical Transactions* between 1673 and 1724.

Legionella pneumophila

These gram-negative bacteria was discovered to be the culprit behind an outbreak of a new and atypical pneumonia among the attendees of an American Legion conference in Philadelphia in 1976. The organism, which was believed to have spread via the air-conditioning system, was a hitherto unidentified species, peculiar in its inability to ferment any sugars. Morphologically, *L. pneumophila* are small non-spore-forming rods, which are motile with one to three polar flagella. They are chemo-organotrophs, capable of using certain amino acids as primary carbon and energy sources. The cell walls of these organisms are unique among the gram-negative bacteria in that they contains large amounts of branched fatty acids. This chemical composition requires the organisms to be counterstained with carbol fuchsin for proper visualization. Some species display autofluorescence.

Legionella appears ubiquitously in nature, as long it is well supplied with fresh (nonmarine) water. Outbreaks such as the episode at the Legionnaire's convention have been found associated with air conditioning cooling towers and drinking water systems. Oddly enough, the organism is fastidious in the laboratory, requiring both iron and cysteine for growth, and it is believed that it survives in nature with the help of protozoa such as the *Harmanella, Naegleria,* and *Tetrahymena* species.

L. pneumophila is an intracellular human pathogen that uses the body's own defenses—opsonin antibodies and complement—to enter the cells. Once internalized, these bacteria reside and multiply within a membrane-bound compartment, which keeps them sequestered away from the cellular enzymes. The lungs and respiratory tract are the predominant sites of infection, although wound infections, pericarditis, and other complications may also arise in some cases. See also LEGIONNAIRE'S DISEASE.

legionnaire's disease

An atypical lung infection or pneumonia, caused by *Legionella pneumophila,* named for the circumstances of the first reported outbreak of the disease—a convention of the American Legion held in Philadelphia in 1976. At the time, the infection was believed to have been traced to contaminated air conditioning systems, although more recent evidence suggests that drinking water was the more likely culprit. Primary infection may occur via inhaling aerosols, or perhaps even more significantly, by aspiration—namely, by choking on fluids and sending the contents of the mouth into the wrong passages. Heavy smokers, sufferers of chronic lung conditions, and immuno-suppressed individuals are in the high-risk category for legionnaire's disease. Disease symptoms appear 2–10 days after infection, with nonspecific symptoms of weakness, fatigue, and high fever being the first to appear. Lung infection is characterized by bouts of coughing, often with mucus involved. The pneumonia may also be accompanied by gastrointestinal symptoms of diarrhea, vomiting, and nausea. Other nonspecific

symptoms associated with the flu, such as headaches and muscle pains, are also common. *Legionella* infections need to be diagnosed with special tests, including growth on special media (containing iron and cysteine), a blood test to check for antibodies, and a test for urinary antigens. Infections may be treated easily, especially in the early stages of infection, with antibiotics such as the macrolides and the quinolones. *See also* LEGIONELLA PNEUMOPHILA.

Leishmania

This organism is an obligate intracellular parasitic protozoan responsible for a number of tropical diseases of humans, including such cutaneous lesions known in different places as the Baghdad or Delhi boil, or the Oriental sore, as well as a visceral condition called kala-azar. The genus contains four species that are pathogenic in humans: *L. donovani, L. tropica, L. braziliensis,* and *L. mexicana.* Individual species differ with respect to their geographic distribution and the epidemiology and pathology of the diseases they cause, but are all essentially identical insofar as their life cycles and morphology.

The life cycle of *Leishmania* alternates between mammalian and arthropod (specifically the sand fly) hosts without any sexual phase, but with several developmental stages in either host. A particular characteristic of this group of parasites is the formation of flagellated forms while in the insect gut. These promastigotes infect the mammalian bloodstream via the bite of a sand fly, which regurgitates the contents of its stomach when it feeds on the mammalian blood. Soon thereafter the parasites are taken up by phagocytic cells, which, however, are unable to digest them. Once inside these cells, the flagellate promastigotes are transformed to nonflagellated forms called the amastigotes, which multiply by fission and proceed to invade other cells in the body. Depending on whether these cycles of invasion and release occur locally in the vicinity of the insect bite, or spread to other parts of the body, the leishmaniasis will be either cutaneous or visceral.

The ingestion of amastigote-containing cells by the sand fly initiates the arthropod phase of the parasite's life cycle. The release of these forms in the gut of the insect stimulates their transformation to the flagellated promasti–gotes, which proceed to multiply in the gut where they can reside until transferred back into the mammal. Dogs and rodents are common reservoirs for different species because of which various leishmaniases are often zoonotic in origin, but flies may also transmit organisms from one human to another, especially in the case of *L. tropica*. See also CUTANEOUS LEISHMANIASIS; KALA AZAR.

leprosy

Chronic granulomatous, often mutilating disease in humans infected with *Mycobacterium leprae*. Well recognized since ancient times in the civilizations of Babylon, China, Egypt, and India, leprosy still runs rampant in many parts of the globe. An estimated 10–20 million people have this disease, and, according to the World Health Organization (WHO), about 1 million people become infected every year. The disease seems to be concentrated in the developing nations, with especially high rates of incidence in India, South America, and Africa. Incidence is much lower in the United States, except for certain pockets of endemic infection in the southern states.

Leprosy primarily affects the skin, mucus membranes, and peripheral nerves. Exactly how infection occurs is not clear, although the epidemiology of leprosy suggests that prolonged contact with infectious material is necessary. The disease has a long incubation period that ranges from 2 to 4 years. The main symptoms are due to subversion of the cellular immune response, and the clinical manifestations depend on the immune status of the infected individual. People with the ability to mount a sufficient response will not experience any symptoms at all because they are able to clear the bacteria from their body. Those unable to clear the bacteria will exhibit symptoms ranging from a localized lesion—tuberculoid leprosy—to widespread involvement of the tissues, namely lepromatous leprosy, as well as various intermediate stages. An early, indeterminate form of the disease may be seen in which there is a small decolorized macule which may pass virtually unnoticed into one of the later stages. The lesions in tuberculoid leprosy consist of a few well-demarcated skin lesions that appear flat, somewhat discolored (blanched or whitened), and are anesthetic—i.e., lacking in feeling. Histologically, the lesions appear to contain only a few bacilli but there is active, often severe involvement of the peripheral nerve endings in the area. At the other end of the spectrum is the severely disfiguring lepromatous form of the disease. This form, too, begins in the skin with either diffuse or nodular lesions (the latter are called lepromas), usually in the cooler regions of the body such as the nasal mucosa, anterior region of the eyes, elbows, wrists, and ankles. Microscopically, these lesions are characterized by the accumulation of macrophages that contain massive numbers of the mycobacteria, which they are unable to destroy properly. As the disease progresses, the peripheral nerve endings become involved, leading to a gradual loss of sensation in the area. This loss is related to the disfigurement one typically associates with leprosy; because patients are unable to feel any trauma to the affected area (usually a limb) and so continue to use it; further damage and, eventually, mutation results.

A diagnosis of leprosy is indicated by the presence of the acid-fast organisms in various clinical samples such as skin scrapings, nasal secretions, and lymphatic fluids. These materials should also be cultured in the foot pads of mice or in armadillos to confirm the identity of the infecting organisms. Different drugs such as dapsone and rifampin are used to treat leprosy, and the WHO recommends using a multidrug regimen to deal with the disease properly. Because many of the harmful mechanisms of this disease are related to the workings of the immune system and indeed often use the immune cells against the rest of the body, no effective vaccines have been developed. Patient management includes minimizing disfigurement of affected individuals using antibacterial drugs in conjunction with local corticosteroid therapy. People living in the same place as patients should be careful with respect to their contact with the patients' food, clothes, and objects that the patient might be handling frequently. See also MYCOBACTERIUM; MYCOBACTERIUM LEPRAE.

Leptospira

Genus of spiral-shaped bacteria normally found in such habitats as water, soil, and mud, but which are also important as animal and human pathogens. Pathogenic *Leptospira* species cause a potentially fatal febrile disease in animals and humans. Leptospirosis, as the disease group is known, typically occurs as a water-borne disease, with the organisms gaining entry into the host animals by penetrating the skin or mucus membranes under moist conditions. The disease is seldom fatal in animals, such as rodents, which harbor the bacteria in their kidneys and frequently shed them during urination. Humans frequently contract infections from swimming in contaminated water. Symptoms of fever, chills, and muscular pain become evident 8–12 days after infection. The organism may spread to various parts of the body and cause the inflammation of different organs, including the meninges, lungs, and liver. The most serious form of the disease is an infectious jaundice, accompanied by bleeding of the skin and subcutaneous tissues, which occurs in some 50 percent of the cases.

A diagnosis of leptospirosis is confirmed by the observation of the leptospires in the blood via dark-field microscopy, or by the cultivation of the organisms from clinical samples in enriched media. The Gram stain, that faithful standby of bacteriologists, is not useful for visualizing *Leptospira* because these organisms are too slender. The bacteria are stained more easily using methods such as silver staining, flagellar staining, and immunological staining, which use mordents or antibodies to first coat and thicken the cell walls before the organism takes up the dye. Viewed in a wet smear under a dark-field microscope, the organisms reveal a spinning, cork-screw type of motion, which is due to flagella that originate at either end of the coiled helical bacterial cell. Organisms are chemo-organotrophs that may be either aerobic or microaerophilic and cannot tolerate anaerobic conditions. At least 10 different species have been identified, based on DNA homology and pathogenic abilities. Examples include the pathogenic species *L. interrogans* and *L. noguchii,* and the nonpathogenic *L. biflexa* and *L. hollandia.*

Although the Gram stain is not a practical method for viewing the leptospires, it is nevertheless worth noting that the structure of the cell walls is like other gram-negative bacteria, and thus the bacteria are sensitive to antibiotics such as penicillin and streptomycin. These drugs are useful in treating leptospirosis in the early febrile or pre-jaundiced stages of the disease. Thus far, no vaccines have been developed against the leptospires, and the best method of prevention is to avoid contact with contaminated water. See also WATER-BORNE INFECTION.

Leuconostoc

Often found in milk and other natural food products, *Leuconostoc* species are widely used in the food industry as sources of various flavoring agents, such as diacetyl—which gives butter, yogurt, and other dairy products their characteristic aromas and flavors—and edible acids found in fermented food products such as sauerkraut. It differs from most other lactic acid bacteria because of its heterolactic mode of fermentation. The bacteria are gram-positive cocci that typically occur in pairs or short chains. They are nonmotile, and facultatively anaerobic, although they carry out respiration. The diverse metabolic capacities of these bacteria enable them to adapt to, and survive in, changing environments. *L. mesentroides* is a common species. This species is of particular importance because it has the ability the convert sucrose into dextran, which is used as a substitute for blood plasma. *See also* LACTIC ACID FERMENTATION.

leukocyte

Found circulating in the blood, as well as in organs such as the spleen, lymph nodes, thymus, and bone marrow, these cells are also known as the white blood cells, in contrast to the other major component of blood, the hemoglobin-containing red blood corpuscles or erythrocytes. The main function of the different leukocytes is to help the body defend itself against various pathogens and harmful substances, i.e., in the body's immune system. Based on structure and function, there are five main types of leukocytes: lymphocytes (B and T Cells) and monocytes, which do not contain any granules, and three types of granulocytes, which contain cytoplasmic granules with differing staining characteristics—the basophils, eosinophils, and neutrophils (also the polymorphonuclear lymphocytes or PMNs). *See also* B CELL; IMMUNE SYSTEM; T CELL.

lichens

Composite life forms consisting of fungi in symbiotic associations with either algae or cyanobacteria. This relationship is extremely beneficial to the saprophytic fungi, as the autotrophic abilities of the photosynthetic partner provides a continuous source of organic material on which the fungi sustain themselves. This enables fungi to thrive in barren environments—such as rock faces—where they would not otherwise survive. The benefit of this partnership to the algae or cyanobacteria remains unclear. *See also* ALGAE; CYANOBACTERIA; FUNGI.

Linneaus, Carl (1707–1778)

Carl Linnaeus—von Linnè in his native Sweden—introduced the concept of binomial nomenclature to the field of taxonomy and defined the concept of species in differentiating among organisms within a single genus. Linnaeus first introduced his system in a 1749 dissertation and subsequently spelled out the rules of the system in *Species Plantarum* in 1753. To this day, scientists follow this scheme in naming different living organisms. *See also* BINOMIAL NOMENCLATURE.

Lipmann, Fritz A. (1899–1986)

German-born scientist responsible for discovering Coenzyme A and explaining its integral role in various metabolic and biosynthetic pathways in living cells. Together with Hans Krebs' elucidation of the individual steps of the citric acid cycle (Krebs cycle), the discovery of Coenzyme A enabled scientists to understand in detail the precise mechanisms by which a large portion of the living world derives and uses energy. The two scientists shared the Nobel Prize in Physiology in 1953. *See also* KREBS CYCLE.

lipopolysaccharide (LPS)

Chemical constituent of the outer membrane of gram-negative bacteria, which, as its name indicates, is made of complexes of lipid and polysaccharide (carbohydrate) molecules. The main function of these macromolecules is to provide the organisms with an additional layer of impermeability to harmful substances such as antibiotics. LPS is also a major constituent of bacterial endotoxins, which give rise to symptoms of fever and shock when released into the bloodstream of host animals. *See also* OUTER MEMBRANE.

Lister, Joseph (1827–1912)

A British surgeon widely regarded as the founder of antiseptic surgery, Lister introduced the practice of washing and disinfecting surgical wounds in diluted carbolic acid, which is toxic to various germs. *See also* ANTISEPSIS.

Listeria

Although described as far back as 1926, this bacterial genus came into the public eye during the 1980s when there was a sudden rise in the incidence of both human and animal diseases of this organism as food-borne infections. Small, gram-positive rods, *Listeria* species are found ubiquitously in nature and have been isolated from a variety of sources, including fresh and waste water, mud, soil, and decaying fruits and vegetables. *Listeria* species are typically motile due to the presence of peritrichous flagella and may be either aerobic or microaerophilic.

The main species of the genus is *L. monocytogenes,* so named because it was first isolated from laboratory animals—rabbits and guinea pigs—suffering from marked symptoms of mononuclear leucocytosis, i.e., an infection of these white blood cells. A similar condition has also been described in humans, although this is a relatively rare clinical manifestation of *Listeria* infections (listeriosis). More often, human listeriosis takes the form of meningitis, encephalitis, or septicemia, and if the individual is pregnant, spontaneous abortions. The severity and outcome of disease is directly linked to the immune status of the affected individual. Farm animals, such as cows and pigs, are also similarly affected. One of the main routes of infections in both humans and animals is the consumption of contaminated food or feed. However, humans may also be infected via simple contact, presumably by the entry of organisms through scratches and wounds. Timely diagnosis—through identification of the infectious agent and through serological tests—is important because listeriosis has a mortality rate of 20–40 percent. All strains are susceptible to ampicillin, which is administered in conjunction with aminoglycoside antibiotics. Properly treated, the prognosis is much improved. Control measures are directed at minimizing exposure to the organisms—i.e., adopting safe food-handling practices, washing hands, and treating (washing) raw fruits and vegetables.

lithotroph

An organism that derives its energy from inorganic chemical sources. *See also* AUTOTROPH.

lockjaw

A common name for the disease tetanus, derived from one of the most noticeable symptoms of this disease, namely, a prolonged spasm of the jaw muscle resulting in an inability to move the mouth. *See also* CLOSTRIDIUM TETANI; TETANUS.

lumpy jaw

A chronic infection of cattle caused by *Actinomyces bovis* and, occasionally, *A. israelii.* The condition is characterized by the formation of a hard immovable protuberance from the region of the jaw, which develops sinus tracts and discharges pus. Usually the organisms do not spread to the lymph nodes and the infection remains localized. *See also* ACTINOMYCES.

Luria, Salvador E. (1912–1991)

Winner of the 1969 Nobel Prize in Physiology/Medicine along with colleagues Max Delbrück and Alfred Hershey. Luria was recognized for his contributions to the development of the bacteriophage system as a quantitative model for genetic and virology research. The Italian-born Luria was also responsible for investigating and elucidating the phenomenon of phage conversion.

lux genes

Cluster of genes found in bacteria such as *Photobacterium*; these genes are responsible for the ability of these bacteria to emit light. The cluster consists of the genes required for the biosynthesis of luciferin, a long-chain fatty aldehyde that emits radiation when oxidized in the presence of the enzyme luciferase, which is also encoded in the same gene cluster or operon. These genes are both necessary and sufficient for the property of bioluminescence, which is to say that it is possible to induce other bacteria to produce light merely by introducing these genes via a plasmid or bacteriophage. The genes have found broad applications in a number of innovative techniques, e.g., in biosensors for the detection of minute quantities of metals and toxins. They are also often used as genetic markers in gene transfer experiments. *See also* BIOLUMINESCENCE.

Lyme disease

Tick-borne, zoonotic disease of humans believed to be caused by an infection with the spirochete bacterium, *Borrelia burgforferi*. Named for the Connecticut county of Old Lyme, where it was first discovered, Lyme disease was initially called Lyme arthritis, to reflect one of the predominant signs of the disease. However, joint involvement is only one of the clinical pictures presented by an infection with this organism, and the disease was renamed to encompass the much broader spectrum of possible outcomes.

The main reservoirs of *B. burgdorferi* are mammals such as rodents and deer, from which the bacteria are transmitted to humans, mainly via the nymph stages of the *Ixodes* ticks. About a week after exposure to the ticks, patients may develop one to several reddish lesions, called "erythema migrans" (EM), at the site of the tick bites. It is important to note that while the EM lesion is typical of Lyme disease, it is not a constant feature, and the absence of these lesions should not rule out the possibility of Lyme disease without other tests. Concurrent symptoms include fever, headaches, malaise, fatigue, stiff neck, and muscular pains, as well as lymphatic symptoms—swelling, etc. that may last for up to several weeks if left untreated. The second phase of Lyme disease may occur after several weeks and or even months following the primary bout of symptoms. Possible complications during this phase include early nervous system damage due to inflammation (i.e., aseptic meningitis, neuritis, myelitis) and cardiac abnormalities (such as the blockage of major vessels as well as enlargement of the muscles). In addition, patients may suffer from intermittent swelling and pain in the joints and, eventually, develop chronic arthritis.

Because many of the complications are serious and irreversible, it is important to make an early diagnosis and treat the disease while still in the early stages. Diagnosis is based on clinical findings as well as serological tests for *Borrelia*. The latter should be used with caution because they have not been standardized properly and there is a high margin of error in interpreting findings. The observation of blood smears, while a great method for confirming infection, is not enough in and of itself, because the detection of the bacteria is not always guaranteed. In recent times, the use of PCR (polymerase chain reaction) to detect and amplify *Borrelia* sequences has proven a valuable addition to the battery of Lyme disease diagnostics. The material used for this genetic test may be blood, cerebrospinal fluid, skin, connective tissues, or urine. Due to the extreme sensitivity of this technique, special care must be taken to avoid false positives. Lyme disease is treated with various broad-spectrum antibiotics. Early phases of the disease respond well to oral antibiotics such as amoxicillin and doxycyline, although the latter is contraindicated in children and pregnant women. Neurological and systemic stages of Lyme disease require intravenous antibiotic therapy. See also BOR-RELIA.

lymphocyte

The principal components of the immune system of animals and birds, these cells act to fight off various infectious agents and harmful substances that enter the system. All lymphocytes look alike but can be differentiated into two main types—B cells and T cells—on the basis of their specific functions and the structure of specific receptor molecules on the cell surface. A small population of lymphocytes fall in neither category and are called null cells. *See also* B CELL; IMMUNE SYSTEM; T CELL.

lymphogranuloma venereum

A sexually transmitted disease characterized by lymphatic involvement, caused by *Chlamydia trachomatis*. Infections begin with a small, often unnoticeable lesion at the site of infection—the penis or vulva—followed by inflammation of the regional lymph nodes when the organisms travel to the site. Signs of lymphatic infection typically manifest themselves several days, or even months, after the initial infection. This stage is characterized by painful swelling and suppuration—i.e., pus formation—and often the extension of inflammation to neighboring tissues. Lymph nodes heal spontaneously, albeit slowly. Complications arising from the infection include the formation of abscesses around the region of the anus, as well as fistulas. Diagnosis is achieved by the demonstration of the Chlamydiae in the lesions, and effective treatments include broad-spectrum antibiotics such as tetracycline, erythromycin, or sulfonamides. Draining the infected lymph nodes by aspiration is helpful in speeding the healing process, but the lesions should not be incised. The most common mode of transmission of lymphogramuloma venercum is via sexual intercourse, especially with the exposure of one partner to an open lesion. The disease is endemic in Asia and Africa where it is seen to occur with higher frequency among the lower socioeconomic classes. Fewer than 400 cases are reported per year in the United States, predominantly in the southern tropical regions. *See also* CHLAMYDIA TRACHOMATIS.

lyophilization

Laboratory technique, also known as freeze-drying, used for the preservation of bacterial strains. Bacterial cells are suspended in a solution of sugar/peptones, dried under a vacuum at a temperature of about −40°C, and sealed in vials, also in a vacuum. The sudden freezing and desiccation brings the metabolic activities of the bacteria to a halt, while leaving them still viable. The viscosity of the suspending solution prevents formation of ice crystals that would damage or kill the bacterial cells. Lyophilized cultures can survive for several years without losing their viability or undergoing mutation. This is the form in which cultures are shipped from suppliers such as ATCC to various laboratories.

lysozyme

Enzyme that catalyzes the cleavage of chemical bonds between adjacent sugar units in the backbone of the peptidoglycan molecules of bacterial cell walls, thus causing the disruption of the cell wall and eventual lysis of the cells. The presence of lysozyme in natural substances such as tears prevents bacteria from thriving in those environments. When suspensions of gram-negative bacteria are treated with lysozyme, they become thick and viscous, indicating the leakage of nucleic acid due to lysis. Gram-positive cells, which have a much thicker peptiglycan wall than the gram-negative bacteria, show a higher degree of resistance because the enzyme does not penetrate to the deeper layers. Hence, suspensions of these organisms do not lyse and become viscous. A related enzyme called lysostaphin specifically cleaves a similar but not identical bond found almost exclusively in the peptidoglycan of the gram-positive organism *Staphylococcus aureus*.

M

MacLeod, Colin (1909–1972)

Member of the scientific team that discovered DNA as the material of genes through the analysis of bacterial transformation, and co-author of the first 1944 paper that describes these findings. *See also* AVERY, OSWALD; MCCARTY, MACLYN; TRANSFORMATION (BACTERIAL).

mad cow disease

Also known as bovine spongiform encephalopathy, this disease, which made international headlines during the mid- to late 1990s, is believed to be a prion disease of cattle. The symptoms and histological findings resemble those of scrapie in sheep, and cows have been known to contract the disease from protein material isolated from the brains of diseased sheep. *See also* PRION; PRUSINER, STANLEY; SPONGIFORM ENCEPHALOPATHY.

magnifying power

Numerical measure of the ability of a microscope to magnify or enlarge an object. The magnifying power of a microscope is the size of the image in relation to the original; for example, a power of 100 means that the image we observe through the microscope is 100 times larger than the original object. The magnifying power of a microscope is dependent on several factors, including the types of lenses used, the manner in which lenses are assembled, and perhaps most important, the wavelength of the enlarging medium, which may be light or X-rays or electrons. *See also* MICROSCOPE; RESOLUTION LIMIT.

malaria

A parasitic disease caused by members of the protozoan *Plasmodium*, malaria is arguably the infectious disease that has had the greatest impact on humankind in history, even in the face of the overwhelming attention—be it public or scientific—that AIDS has garnered in the past few decades. Most common in the tropical and subtropical areas of the world, where it is endemic, malaria affects some 250 million individuals worldwide and claims about 1 million lives annually. A significant proportion of the deaths occur in children, particularly in the African countries,

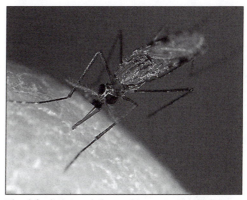

Blood-feeding *Anopheles gambiae* mosquito. *CDC/James D. Gathany.*

which account for nearly 90 percent of the malaria-related deaths. In the United States, the Centers for Disease Control receives about 1,000 case reports per year, although actual incidence is believed to be considerably higher.

About 95 percent of malaria cases are caused by infections with *P. vivax* and *P. falciparum*. The disease is endemic in several parts of the world, such as Africa and India, and outbreaks in North America are often attributable to the patient's exposure to the parasite while visiting these countries. The primary mode of infection is via the bite of the female *Anopheles* mosquito, which is the definitive host for the malarial parasites. The disease may also be transmitted between humans via contaminated blood transfusions, shared hypodermic needles, and—possibly—congenitally via the placenta. The incubation period between the infectious mosquito bite and the first appearance of symptoms is at least one week, corresponding to the pre-erythrocytic phase of the parasite's life cycle. However, infections may surface several months after exposure, particularly if the disease has been kept in check by inhibition with antimalarial drugs, which do not kill, but merely stall parasite propagation.

One of the characteristics of malaria is the periodicity of the symptoms, believed to synchronize with the parasite's life cycle. Clinical symptoms include anemia, spleen enlargement, and the classical paroxysmal appearance of flu-like symptoms. A typical paroxysm begins with a cold stage and rigors that last for 1–2 hours, after which the patient develops a high fever and feels hot and dry. This is followed by several hours of sweating, which lowers the body temperature to normal or lower-than-normal levels. Another typical feature of malaria is the anemia that may arise due to a number of different mechanisms. In addition to the massive destruction of red blood corpuscles (RBCs) due to parasitic invasion and propagation, there is also an increased fragility of the RBCs, because the formation and maturation of new cells is unable to keep pace with destruction. Also, the spleen, which responds to infection

by attempting to clear infected red cells, clears a large number of normal blood cells. The parasites also inhibit the incorporation of iron into the heme groups, which further reduces RBC function. Finally, autoimmune mechanisms may become activated and destroy additional RBCs. In addition to these general features, the different species of *Plasmodium* have certain specific effects. Examples include causing relapses several years after initial infections, due to the establishment of a latent cycle in the liver, and settling in the blood in a low-grade chronic infection, with occasional flaring to cause what is known as recrudescence.

Malaria is one of the few parasitic infections that is considered a medical emergency, particularly in the case of the potentially fatal *P. falciparum* infections. Consequently, an early diagnosis is of paramount importance. An important element in achieving a correct and timely diagnosis is taking a careful medical history, especially asking specific questions about the patient's place of residence, recent and past travel, previous malaria attacks, possible transfusion history, the symptoms (with particular reference to periodicity), and medications. In addition, blood smears need to be examined and interpreted carefully so as to properly identify the correct species and administer appropriate treatment. At least two drawings of blood in a 36-hour interval are necessary, and both thick and thin smears are recommended. The standard stain used in parasitic blood work is the Giemsa stain, which should be performed as soon as possible upon drawing blood. Counterstaining to visualize reticulocytes is also recommended, especially for the identification of *P. vivax*, which is the most common malarial species.

Treatment, in the early stages, is simple and relies on the administration of appropriate drugs to which the specific strain of parasite is not yet resistant. Quinine, derived from the bark of the tropical tree cinchona, was the first material found to cure malaria naturally, and chemical derivatives such as chloroquine have been used effectively for many years. However, the usefulness of these drugs has decreased because the parasites have begun developing drug resistance. This problem has become especially serious with regards to new *P. falciparum* infections. These drugs also serve as means of chemoprophylaxis when administered to potential targets of the disease, such as unexposed travelers to endemic regions. Malarial control is targeted mainly at controlling the mosquito populations because satisfactory vaccines are yet to be developed. Control measures include improving drainage and sewage facilities, especially in Third World countries. The use of mosquito repellents and nets while sleeping are recommended for travelers to endemic areas. Insecticides such as DDT, once useful in controlling these vectors, are now losing efficacy because mosquitoes are developing resistance. The complex nature of the body's immune response has posed several problems in developing effective vaccines against malaria, which remains one of the major goals of the parasitology/public health community. *See also* PLASMODIUM.

Marburg virus

Structurally identical but antigenically distinct from the Ebola virus, the Marburg virus is a member of the filovirus family that appears to be endemic in parts of Central and East Africa. Like others in this family, the virus causes a hemorrhagic fever. The first documented outbreak occurred in 1967 in Germany (hence the name Marburg) and in Yugoslavia among laboratory personnel working with African green monkeys that had been imported from Uganda. *See also* EBOLA VIRUS; FILOVIRUS; HEMORRHAGIC FEVER.

McCarty, Maclyn (1911–)

Member of scientific team that discovered DNA as the material of genes through the analysis of bacterial transformation. McCarty was a co-author on all three major papers that offered the first evidence and explanations of these findings. *See also* AVERY, OSWALD; MACLEOD, COLIN; TRANSFORMATION (BACTERIAL).

measles

Best recognized by the presence of a characteristic rash all over the body, measles is an infection with a member of the paramyxovirus group called the measles or rubeola virus. Like most other viral infections caused by this group, measles originates in the respiratory tract. Following this primary infection, the viruses enter the local lymphatic glands—either as free particles or in association with macrophages from the host's bloodstream—and spread to other lymphatic organs. After about 6 days, there is a generalized viremia as a result of which the virus is seeded in several sites all over the body. Clinical manifestations of measles correspond to a necrotic phase following viremia—9–10 days after the initial infection, there is a sudden onset of symptoms, including cough, runny nose, and conjunctivitis. During this phase, the patients also suffer from fever and malaise, which worsen until the appearance of a rash, usually day 14 or 15 of the infection. This rash is, at least in part, due to the cell-mediated immune response by the host to the virus, and its appearance usually signifies the end of the infective period of the patient (the period of the highest infectivity is between days 8 and 14). The measles rash appears first on the head and spreads progressively to the chest, trunk, and finally the limbs.

The infection typically resolves itself after about two weeks, due to clearance of the infected cells by the immune system. However, because viral infection lowers the body's overall resistance to disease, measles may often be complicated by secondary infections, particularly in the respiratory tract. Measles has also been associated with a number of central nervous syndromes, including encephalitis and a fatal condition known as subacute sclerosing panencephalitis (SSPE). Clinical diagnosis of measles is straightforward and thus seldom requires corroboration in the laboratory. A few decades ago, this disease, along with mumps and chicken pox, numbered among the common childhood diseases that every parent expected to see This is no longer the case in the developed world, where

MMR vaccines are available and administered to children as a matter of course. However, this highly preventable disease still poses a major problem in the underdeveloped world, where poverty and malnutrition leave children highly susceptible to measles in its worst form, and where help in the form of vaccines is still not affordable to all. To date, measles has not been eradicated, although the World Health Organization is taking steps to ensure that the most susceptible populations do receive vaccinations. In addition, passive immunization with normal immunoglobulin is also used as a treatment option (post infection but before onset of symptoms) in unvaccinated children. Antibiotics are required to treat secondary bacterial infections that may complicate the initial disease. *See also* MMR VACCINE; PARAMYXOVIRUS; SUBACUTE SCLEROSING PANENCEPHALITIS.

medium

A solid or liquid preparation of nutrients used in the laboratory for the cultivation of microbes (bacteria and fungi) and cell lines. Together, the different ingredients of a medium provide the necessary compounds—carbon sources, sugars, minerals, salts, vitamins, and growth factor—that an organism requires for growing, with the exception of the gases, either oxygen or carbon dioxide, which are obtained from the atmosphere. Liquid media (broth) usually consist of different ingredients dissolved in water. In solid media, agar is used to transform the solution into a solid gel, containing the nutrients in a uniform suspension.

By far the most common medium used for growing bacteria in the lab is nutrient medium, which contains meat extract, peptone, salt, and water. The meat extract and peptones provide most of the important nutrients, including amino acids, vitamins, and minerals, in addition to serving as the primary carbon source. Salt is used to provide the osmotic balance. It is routinely used for the simple isolation and maintenance of a large number of microbes from the environment. Because the exact compositions of the ingredients are unknown, nutrient medium is a complex medium. When scientists want to study the nutritional requirements of specific organisms, they need to know the precise amount of each substance in the growth medium so that they can set up adequate controls. Media that are made from highly purified and specific chemical compounds, rather than nutrient-rich extracts from natural sources, are called defined media.

The basic nutrient medium may be enriched with high-protein materials such as blood, serum, or milk proteins, which allow the medium to support the growth of a larger spectrum of microbes. Examples include blood agar and chocolate agar—basic nutrient agar enriched with defibrinated blood from sheep or other animals. Such media are especially useful for cultivating pathogenic organisms because they provide the bacteria with materials akin to their natural milieu. Blood agar is prepared by adding blood to sterile medium after it has cooled to temperatures below 56°C, which retains the integrity of different proteins and cells in the blood. This medium is particularly useful because it allows the detection and identification of bacterial toxins like hemolysins, which lyse red blood cells. When grown on blood agar plates, the hemolysin-producing bacteria, such as *Streptococcus,* produce characteristic clear or discolored zones around the colony. Chocolate agar differs from blood agar in that the blood is added to the medium at a temperature of about 80°C. This causes the lysis of cells and the partial breakdown of blood proteins which, in turn, renders these nutrients more readily available to bacteria.

In addition to nutrients, media may contain specific reagents that selectively inhibit or enhance the growth of individual species—selective media—or allow the differentiation of different organisms on the basis of their chemical action on the reagents—differential media. For example, blood agar containing sheep's blood will selectively inhibit the growth of *Hemophilus influenzae*, while different types of coliforms, such as *E. coli* and *Salmonella,* may be selected for by adding bile salts to the medium, which inhibit all other types of organisms. The addition of an indicator dye for acid would enable a scientist to tell an acid-producing fermentative organism from a nonfermentative type. Often a reagent that renders a medium selective will also enable the differentiation of one species from another. *See also* CULTURE.

Mendel, Gregor (1822–1884)

This Austrian monk conducted detailed and rigorous experiments on hereditary characters in pea plants between 1857 and 1865. His studies led him to predict patterns of inheritance of various characters and demonstrate the existence of dominant and recessive behaviors. Mendel's work, published in the early 1860s, was largely ignored within the scientific community during his lifetime and only gained widespread renown after 1900.

metabolism

Chemical reactions that occur in living cells for various purposes, including harvesting energy from surroundings and the synthesis of various chemical components. Most metabolic reactions involve the participation of specific enzymes that help the organisms convert different substrates into the required end products. Many biochemical reactions do not occur in a single step but through a sequence of reactions known as a metabolic pathway. An example of a common metabolic pathway is glycolysis, the pathway through which glucose is broken down to smaller molecules with a net yield of energy for the organism. Many of the metabolic reactions and pathways seen in bacteria and other prokaryotes do not occur in eukaryotes, and vice versa. Specific reactions and pathways are described under the pertinent headings. *See also* ATP; GLYCOLYSIS; RESPIRATION.

metal decorating

Technique used in the preparation of samples for electron microscopy to increase the level of contrast (in this case, electron-reflecting abilities) between the specimen and its surrounding medium. A very thin layer of electron-opaque material such as tungsten is deposited on the sample—already sliced into an ultrathin section and mounted on a grid—under a high vacuum at an angle to the supporting medium. The tungsten atoms accumulate on the surface of the sample (object)—like a plaster cast—and produce a high level of topographic contrast, which can be detected by the reflection of the electrons bombarding the sample for the formation of the magnified image. *See also* ELECTRON MICROSCOPE; SCANNING ELECTRON MICROSCOPE.

Metchnikoff, Elie (1845–1916)

The Russian scientist Elie Metchnikoff discovered the phenomenon of phagocytosis by certain cells of the immune system. His first observations of this phenomenon were in the invertebrate system, where he found that an injection of thorns from a tangerine tree into starfish larvae attracted mobile cells to the site. Extending the basic principles of his findings to vertebrate systems, Metchnikoff reasoned that leukocytes performed a similar task in the bloodstream of animals. His studies of the immune reaction to anthrax bacillus, as well as the immune system of crustaceans, further confirmed his idea that phagocytosis was a mechanism used by hosts to help rid themselves of pathogens. In 1908, he shared the Nobel Prize in Medicine with Paul Ehrlich, also a pioneer in immunology. *See also* PHAGOCYTOSIS.

Methanobacterium thermoautotrophicum

Autotrophic, methanogenic, anaerobic, rod-shaped archaebacterial species that exhibits a preference for growth temperatures in the range of 65–70°C. The organism is capable of using various materials, including elemental sulfur, sulfites, thiosulfates, and ammonium sulfide, as a sulfur source, and urea and glutamine as its nitrogen sources. The genetics of this organism have been studied in some detail, and it is known to possess a 1.7-Mbase chromosome and a plasmid of about 4.4 kilobases. This organism is interesting because of its potential industrial uses, both as a source of various metabolic enzymes that function at high temperatures without denaturation, as well the methanogenic genes, which scientists hope to clone into other (more easily cultivatable) organisms and harness for the production of natural gas.

Methanococcus jannaschii

A member of the Archaebacteria, this spherical organism was discovered in 1983, when a submarine off the coast of Baja California collected some material from a hydrothermal vent. As the generic name suggests, the organism is a methanogen, a methane-producing chemolithotroph that derives energy and nutrition from hydrogen and carbon dioxide. *M. jannaschii* is also thermophilic, living at temperatures between 48 and 94°C. It is strictly anaerobic, intolerant of oxygen, and is capable of nitrogen fixation. Morphologically, it is a motile spherical cell (again suggested in its generic name) distinguished by the presence of two bundles of flagella in opposite orientation emerging from the same pole.

methanogen

Obligately anaerobic, chemolithotrophic archaebacteria that obtain energy by a complex series of reactions in which both water and carbon dioxide are used and methane is produced as a metabolic waste product. Many of these species are also capable of synthesizing methane from methanol or acetate. Methanogenic bacteria may be found in muddy environments (e.g., swamps), where they contribute to swamp gas, as well as in the rumen of cows and other animals. Examples include *Methanobacterium* and *Methanobrevibacter*.

Methylophilus methylotrophus

This organism was grown commercially for several years as a source of protein for animal fodder (commercially known as Pruteen), but it is no longer used because of the drop in prices of more easily available protein sources. *M. methylotrophus* is marked by its requirement for methane as a carbon and energy source, i.e., it is an obligate methylotroph.

methylotroph

Organism that use simple carbon compounds, such as carbon monoxide (CO), methane (CH_4), formaldehyde (CH_2O), or methanol (CH_3OH), as the sole source of carbon and energy. The common properties of these compounds are that they are more reduced than carbon dioxide and lack carbon-carbon bonds. The methylotrophy of different species may be either obligate or facultative. Due to their peculiar nutritional requirements, the methylotrophs live in environments that typically sustain few or no life forms other than themselves.

microbiology

The scientific study of minute life forms including bacteria, microscopic fungi, algae, and protozoa, as well as entities such as viruses and prions, which are not considered as true living organisms.

microscope

Laboratory instrument to visualize structures too small to be seen with the naked eye. It functions by producing an enlarged image of an object, e.g., a cell or an organism, under scrutiny. The most common instruments, called optical microscopes, use light to form these magnified images. Specialized types of microscopes, such as X-ray microscopes and electron microscopes, use other sources such as X-rays or electrons and produce images that are enlarged several orders of magnitude greater than ordinary light microscopes. In very simplified terms, a microscope works by bending light or other rays from the object and focusing them in another plane to produce an enlarged image. This bending is achieved by a system of lenses, made up of different materials depending on the enlarging medium. Ultimately, the functionality of a microscope is measured by two main properties: its magnifying power and its resolution limit, which are determined by the medium used to form the images and the type of lens system it contains. *See also* CONTRAST; ELECTRON MICROSCOPE; MAGNIFYING POWER; OPTICAL MICROSCOPE; RESOLUTION LIMIT.

microtome

A device that can slice various materials into sections less than $0.1\mu m$ thickness, usually by use of a very sharp diamond-edged knife. Such ultra-thin sections are used for preparing specimens for viewing under an electron microscope. *See also* ELECTRON MICROSCOPE.

Miescher, Friedrich (1844–1895)

In 1869, this Swiss physician was the first person to isolate a new type of chemical distinct from proteins from the nuclei of white blood (pus) cells. The importance of this material, which Meischer named nuclein for its source, was not realized until several decades later, but this discovery gave the chemical definition for the substance that would later be found to be the carrier of genetic information. *See also* DNA.

mitochondrion

Often referred to as the "powerhouses" of cells, the mitochondria are semiautonomous organelles that serve as the sites of respiration/energy generation in most eukaryotes. These organelles are suspected to be the modern-day descendants of prokaryotic endosymbionts in eularyotic cells. Partial evidence for this lies in the fact that these organelles contain their own DNA—under autonomous control—as well as ribosomes that resemble prokaryotic ribosomes rather than the eukaryotic types. Mitochondria vary considerably in shape from small rod-like packages to large branched structures. They consist of a double membrane, with the inside layer folded into several cristae, which contain the components of the electron transport chain. The fluid enclosed within the mitochondrial sac contains respiratory (Krebs cycle) enzymes, as well as enzymes of the urea cycle. *See also* CYTOPLASM; ENDOSYMBIONT HYPOTHESIS; EUKARYOTE; RESPIRATION.

mixed acid fermentation

Fermentative reactions carried out by enteric bacteria, in which the pyruvate produced by glycolysis is converted to a number of different end products besides ethanol and lactic acid. Organisms such as *Escherichia coli* and *Salmonella,* for instance, contain a separate set of enzymes called the formate hydrogen lyase system, which is activated under acidic conditions, to convert pyruvic acid into formic acid and further break this down into carbon dioxide and hydrogen. This type of gas-producing fermentation is said to be aerogenic. On the other hand, there are also other enteric mixed-acid fermenters, such as *Shigella*, which are anaerogenic and stop at producing formic acid. The relative amount of different end products varies according to the growth conditions. *See also* FERMENTATION.

MMR vaccine

A combined vaccine for protection against three common childhood viral infections: measles, mumps,and rubella (also known as German measles). All three vaccines consist of live attenuated viruses, which are grown in chick embryos (measles and mumps) or, in the case of the rubella virus, in duck embryos or human-derived cell lines. Nowadays, the recommended vaccination schedule is a first dose at 12–15 months of age, followed by a second dose at 4–6 years. In addition, the U.S. Centers for Disease Control (CDC) recommends that all unvaccinated adults born after 1956 receive a dose of the vaccine unless they are immune to the diseases (having contracted one or more of them during childhood) or have contraindications for a particular vaccine. Women who are pregnant or suspect pregnancy should *not* receive this vaccination. Adults with a special need for the vaccine include international travelers and immigrants—especially from countries where the vaccination is either not available or where the CDC-recommended guidelines for immunization schedules are not followed—and people who routinely encounter large crowds, such as college students, hospital workers, and cruise ship travelers/workers. Once familiar in nearly every household as common childhood diseases, the incidence of these potentially dangerous diseases in the United States has dropped by about 99 percent since the introduction and licensing of the MMR vaccines in the 1960s. One of the objectives of the U.S. Public Health Service Year 2000 project is to eliminate measles, rubella, and congenital rubella syndrome from the country, and reduce the incidence of mumps to fewer than 500 reported cases per year. *See also* ATTENUATED VACCINE; GERMAN MEASLES; IMMUNIZATION; MEASLES; MUMPS; RUBELLA VIRUS; VACCINATION.

monkeypox virus

Member of poxvirus family, which naturally infects monkeys, but is capable of causing zoonotic infections in humans as well. The disease is mostly similar to classical smallpox, although it is typically milder and there is a higher degree of lymphatic involvement and swelling of the glands in the lower jaw. Most of the reported cases of this disease have been from Africa, particularly in the Congo and neighboring regions. According to recent studies, this virus has been shown to occur naturally in wild squirrels, in addition to their primate hosts. Recent reports indicate that there is a higher incidence of monkeypox in human populations than ever before. Although scientists are not sure whether this is due to earlier masking of this cause by the smallpox virus, there is considerable concern in the medical/scientific community about the appearance of a new emerging disease. *See also* POXVIRUS; SMALLPOX.

Montagnier, Luc (1932–)

French scientist credited with the discovery of the retrovirus later designated as HIV and believed to be the cause of AIDS. *See also* AIDS; GALLO, ROBERT; HIV.

Morbidity and Mortality Weekly Report

Publicly available, free reports of national public health data in the United States, published weekly by the Centers for Disease Control and Prevention (CDC). Information is presented in the form of morbidity tables for nationally notifiable diseases, as well as summaries of mortality figures due to influenza and pneumonia. Much of the information in these reports is based on the compilation and analysis of data supplied to the CDC on a weekly basis by state health departments. In addition to data on specific disease outbreaks and reports of incidence and case fatality rates, the reports also contain information on infectious and chronic diseases, environmental hazards, natural and human-created disasters, occupation-related health concerns, and injuries. Some information on topics of international relevance, as well as events of interest to the public health/epidemiology community, is also included. To learn more about the reports and their data—or to subscribe—visit the CDC website at <www.cdc.gov/subscribe.html>. *See also* CENTERS FOR DISEASE CONTROL AND PREVENTION.

morbidity rate

A numerical measure of the disease incidence, morbidity is expressed in terms of the proportion of patients per unit of population who have a particular disease in a specified period of time (usually one year).

mortality rate

In mathematical terms, the mortality rate of a population may be calculated by dividing the number of deaths that occurred in the population over a particular period of time (e.g., one year or one decade) by the total number of people who were at risk of dying during that period (e.g., the entire population or a specified subset). Most epidemiological estimates of mortality rates are specific for a disease. Surveillance data on case fatality rates from different epidemics may be used to calculate mortality rates for various diseases.

motility

Ability of a living cell or organism to move through space from one point to another by using some motive force of its own. Thus, motility typically involves the use of energy. The floating of microbes in a flowing stream of water is not representative of motility. In Leeuwenhoek's time, motility was the single property that formed the basis for deciding that the minute beings he observed under a microscope were indeed living. Mechanisms used by motile microbes include structures such as flagella and cilia, which help propel them along in liquid and semi-solid media or on surfaces, or pseudopodia, which enable the organisms to "walk" on surfaces by extending their protoplasm. *See also* CILIA; FLAGELLA; PSEUDOPOD.

mucormycosis

This label is given to the spectrum of conditions induced by an infection with fungi such as *Mucor* and *Rhizopus*—better known as pin or bread molds. While most people are resistant to these saprophytic organisms, which are commonly present in the environment, these fungi can cause serious problems in immuno-compromised individuals or those weakened by other diseases such as diabetes. Inhalation of the fungal spores and use of hypodermic needles contaminated with spores have been observed as the most common routes of infection. Once in the body, these fungi have a special affinity for the blood vessels, where they cause thrombosis and infarction. Frequently affected sites include the lungs and pulmonary blood vessels, blood vessels of the craniofacial region, and the gastrointestinal tract. Antifungal drugs such as amphotericin B may be used for treating severe cases.

Muller, Hermann (1890–1967)

Winner of the 1946 Nobel Prize in Physiology and Medicine for his 1926 discovery that X-ray radiations could induce heritable mutations in fruit flies (*Drosophila*). This fundamental finding opened up the field of genetics to lines of experimentation that had not been possible, or even considered, until then, and thus, led the way to a molecular understanding of the subject. *See also* MUTAGEN; MUTATION.

mumps

Viral disease most commonly associated with the inflammation of various glands in the body. The causative agent, known as the mumps virus, is a member of the paramyxovirus family, which also includes the viruses causing other common childhood diseases such as measles. Mumps originates with an infection of the respiratory tract, from where the virus quickly spreads via the lymphatic system to various glands. The most commonly affected are the parotid and salivary glands, presumably because of their proximity to the initial site of infection. This gives rise to the classic symptoms of mumps—a painful swollen face due to edema or fluid retention in the inflamed glands, which makes it difficult to swallow and talk—some 14-18 days after initial infection. The infective period for a patient begins a few days before clinical symptoms appear and usually lasts until about a week after. Other susceptible glands include the pancreas, ovaries or testicles, thyroid, and occasionally breast tissue. The multiplication of the virus at any of these sites induces inflammation and swelling. Mumps is one of the most common causes of viral meningitis and, rarely, a form of encephalitis. Another serious (though uncommon) by-product of this disease is a long-term deafness in one ear. Small children may contract mumps virus infections without exhibiting glandular symptoms of any kind.

The most common form of mumps, i.e., with parotid and salivary gland involvement, is easily diagnosed by the clinical symptoms. However the atypical infections present some difficulties. The virus may be isolated and cultured from a number of specimens including saliva, throat swabs, urine, and—in the case of meningitis—from the cerebrospinal fluid. The cultured virus is identified by means of either an hemadsorption test or enzyme immune assays (EIA). Like measles and chicken pox, mumps was a familiar childhood disease until a few decades ago. Nowadays, however, children are routinely dosed at 15 months with a vaccine of live-attenuated virus in combination with the measles and rubella vaccines (MMR). Experts also recommend a booster shot during the late teens or early twenties. *See also* MMR VACCINE; PARAMYXOVIRUS VIRUS.

mutagen

Any agent that can induce mutation in a DNA sequence. The most common mutagens are chemicals, but high-energy radiation and viruses may also induce mutations under appropriate conditions. *See also* MUTATION.

mutant

Any cell or organism that contains a mutation in its genome. The organism from which the mutant was derived is called the "wild type." In a certain sense, *all* organisms are mutants because they were invariably derived by the mutation of genes in some parent cell. However, for practical purposes, mutants may be considered as organisms in which the mutations were introduced *deliberately* in the laboratory by means of chemical mutagens, radiation, or viruses. Wild type organisms and their mutants typically differ from one another in just one or a few observable characteristics. *See also* AUXOTROPH; MUTATION.

mutation

A heritable change in the DNA sequence of a cell or organism, i.e., an alteration that is passed on to successive generations. A mutation may take place in either resting or dividing cells, and it can occur due to the elimination, insertion, or substitution of one or more bases (nucleotides) in the DNA strand. Depending on the specific location of the mutation, a change may or may not have an outward manifestation—i.e., change

the phenotype—of the organism. For instance, a single base change might occur in a non-coding region of DNA, or, when it does occur in coding sequence, it may either be compensated for by the redundancy of the genetic code or it may alter the amino acid in a protein sequence. This latter change may, in turn, affect the structure or function of the protein. The addition or removal of bases, particularly in coding sequences, will have more visible effects. The term reversion is specifically used to describe a mutation in one generation that appears to reverse or correct a mutation that occurred in a previous cycle of gene replication.

A number of different agents and mechanisms may cause mutations. High-energy radiation and chemicals are examples of agents that can act on resting cells by physically removing one or more bases from the DNA strand so that the newly generated sequence gets passed on to progeny. Chemicals might also act by binding with nucleotides on one of the two strands and thus interfering with proper duplication. In these instances, the mutation will be inherited asymmetrically because the complementary strand will be replicated faithfully. Mutations in the nature of single-base substitutions may also occur due to reading errors on the part of the enzymes involved in synthesizing new DNA. *See also* DNA; MUTAGEN.

mutualism

Symbiotic association between two living organisms, in which both participants derive benefits from one another. Mutualism can be seen, for example, in the lichens, a life form in which photosynthetic algae procure energy from light and heterotrophic fungi derive nutrition from organic material in the soil. It also can be seen in the nodules of leguminous plants, where bacteria (*Rhizobium* species) fix nitrogen, which allows the plants to thrive in nitrogen-poor soils, in return for nutrients and protection. *See also* LICHENS; *RHIZOBIUM LEGUMINOSARUM*; SYMBIOSIS.

mycetoma

A clinical syndrome characterized by swelling and suppuration of subcutaneous tissues, usually in a foot or the lower legs and occasionally at other sites. A common local name for this condition in India is "Madura foot." The lesions may be induced by infection with a number of different species of actinomycetes and fungi. A characteristic feature is the presence of visible granules—called sulfur granules, although they are not composed of this substance—in the pus of mycetoma lesions. Symptoms are similar to osteomyelitis caused by staphylococci, and specific diagnosis is made by the presence of granules and the microscopic and cultural identification of the causative agents.

Mycobacterium

This bacterial genus contains a vast number of species, of which the best known are *M. leprae* and *M. tuberculosis*, the causative agents of leprosy and tuberculosis, respectively. According to *Bergey's Manual*, there are about 50 different species of *Mycobacterium*, which are distinguishable from most other bacteria by the presence of special fatty molecules called mycolic acids in their cell walls. These lipid complexes make the cell walls less permeable, so that even though the mycobacteria are gram-positive (i.e., with a thick layer of peptidoglycan and no outer membrane), they are able to resist decolorization with alcohol in the gram-staining procedure. Indeed, the mycoloc acids render these cells acid-fast and resistant even to treatment with heated acid-alcohol solutions.

As a group, the mycobacteria are aerobic, non-motile rods that have very slow generation times of up to 20 hours. Thus, the isolation and identification of *Mycobacterium* species in the laboratory may take several weeks. This slow growth also is the reason that these bacteria tend to cause chronic rather than acute infections. Different animal and human bacteria are often found to share biochemical, serologic, and even pathogenic properties, for which reason they are grouped into mycobacterial complexes. For example, *M. tuberculosis* is grouped together with other species such as *M. bovis*, *M. africanus,* and *M. microti* into a single complex of organisms that may cause tuberculosis. The *M. avium* complex consists of the eponymous organism and a second species called *M. intracellulare*. With the exception of *M. leprae*, all nontubercular species are known by the acronym MOTT for "mycobacteria other than tubercle" bacilli. A common feature of the pathogenic species seems to be the induction of the cellular, rather than humoral, immune response. In both tuberculosis and leprosy, the organisms appear to be able to turn this reaction to their advantage so that they are protected, rather than destroyed, within the body.

Mycobacterium leprae

Also known as Hansen's bacillus (named for the scientist who discovered it), this organism is the cause of the infamous disease leprosy, which has plagued humankind for centuries. It can be seen in large numbers from the lesions of lepromatous leprosy, usually in regular masses grouped together like bundles of cigars, but never in chains. Electron microscopic images of tissue from leprosy lesions reveal large masses of these bacteria both inside and outside the cells, usually embedded within capsular material. Only viable bacteria seem to be able to give the acid-fast reaction during staining, and this distinction is important in terms of choosing the right therapeutic approaches in leprosy patients.

More so than any other species, *M. leprae* appears very fastidious. Laboratory cultivation of this organism is achieved by injecting suspensions of the bacilli into the footpads of mice and the nine-banded armadillo. This latter animal model is now the most common source for these bacteria in medical and research laboratories. *See also* LEPROSY.

Mycobacterium tuberculosis

Like its leprosy-causing cousin, *M. tuberculosis* is the causative agent of a chronic disease that was recognized for centuries before scientists made the definitive link between germs and disease. It is the primary representative of the *M. tuberculosis* complex of mycobacteria, which are capable of infecting various animals (notably humans and cows, but also other mammals) and causing tuberculosis. The organisms possess the ability to elicit a delayed-type hypersensitivity response and then establish themselves in the host organism. Like other members of its genus, *M. tuberculosis* is acid-fast in nature. Unlike the leprosy agent, the bacteria that cause tuberculosis may be cultured *in vitro*, but they grow extremely slowly, with a generation time of 18–24 hours. Bacterial colonies begin to appear only after two week's incubation. *M. tuberculosis* produces very distinctive colonies on solid media, with serpentine cords due to the presence of mycolic acids in the cell wall. *See also* TUBERCULOSIS.

mycoplasma

The term mycoplasma is used to represent an entire group of organisms, as well as being the name of a single genus within the group. This section deals mainly with the group as a whole.

The mycoplasmas constitute a taxonomically separate category of living organisms—grouped in the class Mollicutes—that are the smallest known type of life form capable of independent existence outside other living cells. The single most distinctive characteristic of this group, which differentiates it from the true bacteria, is the lack of an extra-protoplasmic cell wall. This makes these organisms extremely pleomorphic and sensitive to osmotic changes in their environment, but also resistant to β-lactam antibiotics such as penicillin. The mycoplasmal cell membrane is also considerably different from a bacterial membrane and contains complex compounds such as sterols and glycolipids in addition to the phospholipid bilayer and membrane proteins.

Although the mycoplasmas might appear to be more primitive than true bacteria at first glance, their genetic sequences indicate that they probably evolved from gram-positive ancestors that also gave rise to modern clostridia and *Lactobacillus* species. In this case, evolution took the form of the loss of genetic material resulting in an extremely tiny genome, which has severely limited the biosynthetic capabilities of the mycoplasmas. For this reason, most species are found coexisting in commensal or parasitic relationships with other organisms such as arthropods, mammals, and plants. Different mycoplasmas are commonly found as contaminants in eukaryotic cell cultures, but paradoxically, their isolation and cultivation is rather difficult, requiring complex media enriched with nutrients and co-factors. These organisms form extremely tiny colonies that are barely visible to the naked eye and require the use of a microscope for observation and identification. A typical characteristic of many mycoplasmas, notably *M. hominis,* is the "fried egg" morphology of their colonies on agar due to the penetration of the medium at the center where there is a higher rate of multiplication.

Currently, the mycoplasmas are classified into six separate genera (in four families), which differ from one another with respect to their nutritional requirements and host organisms. Different organisms have been implicated in various human and veterinary diseases. The main genera to colonize humans are *Mycoplasma* and *Ureaplasma*, although two species of *Acholeplasma* have also been isolated in certain rare instances. Perhaps the best-characterized human pathogen is *M. pneumoniae*, the causative agent of a number of respiratory conditions, including tracheo-bronchitis in children, pharyngitis in adults, and pneumonia in all age groups. *M. genitalium*, originally isolated from cases of urethritis in 1981, has since been found to cause respiratory infections as well as joint diseases. *M. hominis* and *U. urealyticum* have been isolated from clinical specimens in a number of urino-genital infections of adults and are also known to cause congenital or neonatal pneumonia and respiratory disease. These and other species of mycoplasma have been isolated from AIDS patients, and while the role of these organisms is not yet clear, some scientists suspect that they might be a co-factor in AIDS pathogenesis. Other reports suggest that different mycoplasmas may contribute to the development of chronic fatigue syndrome. *M. hyopneumoniae* is an important veterinary pathogen associated with serious economic implications for farmers due to its impact on animal health and productivity.

One of the most important factors in the pathogenicity of these organisms is their ability to adhere to host cells. The proteins involved in adhesion—called adhesins—are the major candidates of vaccines against mycoplasmal infections. Thus far, the most potent antimicrobial agents against *M. pneumoniae* have been the macrolide antibiotics. Certain broad-spectrum antibiotics may be used, but any antibiotic whose major target is the cell wall—e.g., the β-lactamases—are ineffective against these organisms.

mycosis

A disease caused by infection with a fungus.

N

nalidixic acid

Synthetic antibiotic against gram-negative bacteria that acts by inhibiting the enzyme gyrase, which is involved in DNA replication. *See also* ANTIBIOTIC.

nanobe

The recent reports concerning these miniscule creatures is living testimony to the fact that no matter how much we have learned about the world of microbes since Leeuwenhoek first reported his animalcules, there is always something more lurking just around the corner. The term "nanobe" is the name given by geologist Philippa Uwins of the University of Queensland in St. Lucia, Australia, to a novel discovery made around 1996 of extremely tiny, possibly living organisms found in sandstone fossils. The name is derived from the scale of the size of these creatures; where most bacteria fall in the micrometer (.001 of a millimeter) range, the nanobes are, at most, one-tenth as small.

Uwins and her colleagues stumbled upon the nanobes while studying samples of sandstone that they had retrieved from 3–5 kilometers below sea level. When they broke open these rocks, they discovered the presence of an unusual organic substance, which appeared able to colonize and grow on the surfaces of laboratory containers and equipment.

The scientists proceeded to conduct a detailed analysis of this substance and found the material to be composed of filaments with spores and fruiting bodies very akin to the fungi and actinomycetes in appearance, but several magnitudes of order smaller. Although individual units of this structure appeared to be too tiny to sustain life independently, Uwins and others noted that the nanobes were capable of propagation without living hosts (unlike viruses, for example) and it was possible to clearly discern between what appeared to correspond to the cytoplasmic and nuclear regions of a larger cell. Furthermore, their experimental evidence strongly suggested the presence of DNA in all nanobe samples. Metabolically, the organisms appear to be strict aerobes that require adequate supplies of water and grow at ambient temperatures and atmospheric pressure.

Whether nanobes are true living creatures is still a matter of controversy in the scientific community, and it should be noted that Uwin is not the first person to make a claim for the existence of such tiny living organisms. The confirmation of the existence of such creatures may provide the impetus for us to re-examine our current stage of knowledge about the possibility of extraterrestrial life, due to its similarity in size—but not morphology—to the purported fossilized microbes found on the Martian meteorite ALH84001.

necrosis

Cellular death characterized by the breakdown and destruction of cellular components, usually due to some external factor such as trauma, infection, or enzymatic action. Death due to outside influences is distinct from apoptosis, which is death programmed within the cells' own genome. In medicine, the term necrosis is often used to describe localized lesions in soft tissues, e.g., necrotic lesions of gas gangrene.

negative staining

Staining technique used in microscopy to aid in the visualization of different objects—e.g., microorganisms or cellular components—in which the background, rather than the specimen, is stained. In optical microscopy, negative stains are particularly useful for observing organisms that have thick capsules, which are impervious to most dyes and therefore will not allow the bacterial cells to take on any color. India ink is an example of a negative stain. Bacterial cultures mixed with the dye and are spread out in a thin layer on a slide and observed under a microscope with a beam of light illuminating the slide from underneath. The bacteria will appear as bright, unstained objects against a black background. Negative staining techniques also have wide applications in electron microscopy. Here, chemicals that contain heavy atoms such as phosphotungstate or molybdenum are spread in a layer in the background, leaving the specimen itself unaltered. Electrons will then reflect off the background only and produce light-colored images of the outlines of the specimens. *See also* CONTRAST; ELECTRON MICROSCOPE; MICROSCOPE; OPTICAL MICROSCOPE; STAINING.

negri bodies

Inclusion bodies formed in cells of a host animal infected with the rabies virus. They are typically found in the brain cells of infected animals. *See also* INCLUSION BODIES; RABIES VIRUS.

Neisseria

Considering the havoc wreaked by these bacteria on the human race—in the guise of such dreaded diseases as gonorrhea and spinal meningitis—it is something of a shock to discover that *Neisseria* is one of the most delicate and fastidious organisms of the bacterial kingdom. These bacteria are strictly aerobic, can tolerate only a narrow range of temperatures (22–42°C), and most species, particularly the pathogens, cannot survive for more than 24 hours outside the moist mucus membranes that constitute their normal habitat.

The bacteria were named for the German physician who is credited with their discovery as the causative agent of gonorrhea in 1879. They are gram-negative, non-motile cocci that are typically found in pairs with adjacent sides flattened against each other. Two species, *N. gonorrhoeae* and *N. meningitidis,* are confirmed pathogens (respectively responsible for gonorrhea and meningitis), while there are about 14 other strains and species that appear to be non-pathogenic or occasional opportunistic pathogens. Many of these latter species, e.g. *N. mucosa, N. animalis, N. ovis,* and *N. pharyngis,* may be found inhabiting the upper respiratory tract, as is *N. meningitidis,* while *N. gonorrhoeae* is typically found associated with the genital tract (and hence also known as the gonococci). The two pathogenic species are virtually indistinguishable in all respects except for their habitat and the occasional presence of a capsule in some strains of *N. meningitidis.* Growth in the laboratory requires complex media such as chocolate agar, which contains lysed blood cells. When incubated for 24-48 hours on such media, the neisseriae form smooth, elevated, bluish gram colonies of about 1–2 mm in diameter. The development of these colonies is a useful visual clue for diagnosing *Niesseria* infections. *See also* GONORRHEA; SPINAL MENINGITIS.

Nicolle, C. J. H. (1866–1936)

The winner of the 1928 Nobel Prize in Medicine for his work on understanding the centuries-old disease of typhus, Charles Jules Henry Nicolle was a prominent contributor to medical pathogenesis. He was the first person to demonstrate the role of the body louse in transmitting typhus. He made fundamental contributions to the understanding of such diverse diseases as Malta fever, rinderprest, tick fevers, tuberculosis, trachoma, and measles. He was also one of the first scientists to successfully cultivate the human parasite *Leishmania* in the laboratory.

Nif genes

Cluster of genes found in nitrogen-fixing bacteria, e.g., *Azotobacter, Azomonas, Rhizobium*, and *Klebsiella*, which code for the protein components of the nitrogenase enzyme complex, which is responsible for catalyzing biological nitrogen fixation. *See also* NITROGENASE; NITROGEN FIXATION.

nitrate assimilation

The assimilation of nitrate is one of the most important routes through which inorganic nitrogen is incorporated into living matter, and thus is a key step in the nitrogen cycle. It is used widely in the microbial kingdom, and even organisms that possess nitrogenase prefer to reduce inorganic nitrates whenever possible because the process requires far less energy than reactions such as nitrogen fixation. In chemical terms, nitrates are assimilated by a process of serial reduction of nitrate—first to nitrite (NO_2^-) with the help of an enzyme called nitrate reductase, and then from nitrite to ammonia in the presence of nitrite reductases. In contrast to nitrogenase, whose structure is highly preserved across different bacteria, the reductases vary widely with respect to structure and character in different organisms. *See also* ASSIMILATION; NITROGEN CYCLE.

nitrate respiration

Anaerobic respiration reaction in which chemotrophic organisms use inorganic nitrate as the oxidizing agent. Different species convert nitrates to nitrites and then to ammonia,

nitrogen, or nitrous acid, thereby participating in various steps of the nitrogen cycle. The main difference between nitrate respiration and the reductive nitrate assimilation reactions is that the former processes yield energy while the latter do not. Examples of bacteria that produce nitrites and ammonia include *E. coli* and species of *Clostridium*. *See also* ANAEROBIC RESPIRATION; DENITRIFICATION; NITRATE ASSIMILATION; NITROGEN CYCLE.

nitrification

Process by which different types of living organisms convert organic nitrogen—previously converted to ammonium compounds—to inorganic form. It is the key step in the nitrogen cycle for restoring the nitrogen assimilated into living cells back to the environment. Nitrification is a two-step oxidation process, performed by two different groups of nitrifying bacteria, in which ammonium ions (NH_4^+) are first converted to nitrite (NO_2^-) and then nitrate (NO_3^-). The first conversion takes place as follows:

1. $NH_3 + \frac{1}{2} O_2 \rightarrow NH_2OH$ (hydroxylamine)
2. $NH_4OH \rightarrow HNO$ (nitroxyl)
3. $HNO + \frac{1}{2} O_2 \rightarrow HNO_2$ (nitrous acid)

None of the metabolites in the reaction series above is ever found free in the cells. The first two seem to exist only in an enzyme-bound manner, while nitrous acid ionizes almost instantaneously to release free nitrite ions. The oxidation of ammonia to nitrite also results in a small amount of nitrous oxide (N_2O) as a by-product. The exact mechanism of this reaction is not clear, but it most likely occurs during step 2.

The second reaction, which is performed by the "nitrobacteria," is a single step oxidation from nitrite to nitrate, catalyzed by an enzyme called nitrite oxidase. In addition to the autotrophic nitrifying bacteria, some heterotrophic organisms such as *Arthrobacter* and *Aspergillus* (a fungus) are capable of nitrification. *See also* NITRIFYING BACTERIA; NITROGEN CYCLE.

nitrifying bacteria

Obligately aerobic, autotrophic microorganisms, living in soil and aquatic environments, which obtain energy via the process of nitrification and thus form a key link in the nitrogen cycle. There are two distinct groups of nitrifiers: bacteria that oxidize ammonium ions to nitrites (e.g., *Nitrosococcus* and *Nitrosomonas)* and nitrobacteria, which oxidize nitrites to nitrates (e.g., *Nitrobacter* and *Nitrococcus)*. *See also* NITRIFICATION; NITROGEN CYCLE.

Nitrobacter

These bacteria are among the most abundant nitrifying bacteria in nature, deriving their energy by oxidizing nitrites to nitrates. They are widespread in soil and water and are usually found along with ammonia oxidizers, which supply them with nitrites. *Nitrobacter* species are gram-negative, autolithotrophic, non-spore-forming, and acid-fast. Although not photosynthetic, these organisms are capable of using CO_2 directly from the atmosphere rather than relying on organic sources. Strains of different species may be motile. Three main species that have been identified are *N. winogradskyi, N. hamburgensis,* and *N. vulgaris*. Species are differentiated on the basis of GC content (range is 59–62 moles percent), DNA homology, antigenicity, and the arrangement of their heme proteins. *See also* NITRIFICATION; NITRIFYING BACTERIA; NITROGEN CYCLE.

Nitrococcus mobilis

Gram-negative, autolithotrophic, marine bacterial species that derives energy by converting nitrites to nitrates. Organisms are motile cocci with one to two flagella. *N. mobilis* consists of a tubular membrane system whose inner surface is covered with particles, which are presumably the site for nitrification. Various granules such as carboxysomes, PHB, and polysaccharides are found in the cytoplasm. GC content is 61 moles percent. *See also* NITRIFICATION; NITRIFYING BACTERIA; NITROGEN CYCLE.

nitrogen cycle

A series of interlinked processes though which nitrogen is passed into and out of the living world—the biosphere. The assimilation of nitrogen into living cells is of paramount importance because nitrogen is a key constituent of proteins and nucleic acids. But although it is the most abundant material in the earth's atmosphere, nitrogen remains the limiting nutrient in a majority of ecosystems because its uptake into living cells is somewhat more complicated than that of carbon and oxygen (two other essential components of living molecules). With the exception of certain nitrogen-fixing bacteria (e.g., *Azotobacter, Azomonas, Rhizobium,* a few species of *Acetobacter* and *Klebsiella*, and the cyanobacteria), most organisms are incapable of using elemental nitrogen directly. Rather, the nutritional demand for nitrogen is met by different nitrogen-containing organic or inorganic compounds—most animals, for instance, can get their nitrogen only by consuming proteins in plants or other animals. Various chemical processes that constitute the nitrogen cycle include nitrogen fixation, the assimilation of nitrogenous compounds (such as ammonia, nitrites, and nitrates), nitrification, and denitrification. The nitrogen cycle does involve a few non-biotic processes—such as the conversion of atmospheric nitrogen to ammonium ions by lightning—but most reactions can be performed by at least some living organisms. *See also* AMMONIFICATION; AMMONIA ASSIMILATION; DENITRIFICATION; NITRATE ASSIMILATION; NITRATE RESPIRATION; NITRIFICATION; NITRIFYING BACTERIA; NITROGEN FIXATION.

nitrogen fixation

Process of converting atmospheric nitrogen (N_2) into ammonia and ammonium compounds. As indicated by the name, one may visualize this process as a means of catching some of the nitrogen floating about in the air and "fixing" it to the ground, as one would a kite. Nitrogen fixation is an important step in the earth's nitrogen cycle because it is the only means of incorporating atmospheric nitrogen into the earth's crust and the biosphere. It is the first step in a long chain of reactions that is ultimately responsible for providing all living creatures (including humans) with nitrogen, which is an integral component of our vital macromolecules, such as proteins and nucleic acids. Only a small fraction—about 15 percent—of the earth's total nitrogen fixation can be accounted for by non-biological processes such as lightning and combustion; the remainder is through biological means. Within the living world, nitrogen fixation is confined to simple prokaryotic organisms and does not occur in higher plants or animals. Examples of nitrogen-fixing organisms include photosynthetic cyanobacteria, symbiotic bacteria such as *Rhizobium* that reside in the roots of leguminous plants, and free-living bacteria that may be either anaerobic (e.g., *Clostridium* and *Klebsiella* species) or aerobic (*Azomonas* and *Azotobacter*).

In chemical terms, nitrogen fixation involves the conversion of atmospheric nitrogen to ammonia or ammonium compounds. These molecules are almost instantaneously converted to less-toxic compounds, usually through ammonia assimilation. All nitrogen-fixing bacteria contain a complex of enzymes known as nitrogenase to catalyze this conversion. The enzymes are encoded in special genes called the *Nif* genes. *See also* AMMONIA ASSIMILATION; *NIF* GENES; NITROGEN CYCLE; NITROGENASE.

nitrogenase

Enzyme that catalyzes the biological conversion of atmospheric nitrogen (N_2) into a form such as ammonia (NH_3) that can be easily assimilated into living cells. Actually, the enzyme is a complex of metal ions and proteins, consisting of at least two catalytic components that successively catalyze the breakdown of the triple bond holding the two nitrogen atoms together and the conversion of the gas to ammonia.

$$N_2 + (6\ H^+ + 6\ \text{electrons}) \rightarrow 2\ NH_3$$

Most nitrogenases contain the metal ions iron and molybdenum. Some bacteria con-

tain a backup nitrogenase with vanadium rather than molybdenum. Nitrogenases are rapidly inactivated by oxygen, and consequently must be protected. Many nitrogen fixers—such as species of *Clostridium* and *Klebsiella*—are anaerobic, while cyanobacteria sequester the enzyme within a special cell called the heterocyst. Aerobic organisms such as *Azotobacter* appear to protect their enzyme by maintaining a high rate of respiration, which keeps the intracellular oxygen at very low levels. The nitrogenase proteins are encoded in a special cluster of genes called the *Nif* genes. *See also NIF* GENES; NITROGEN CYCLE; NITROGEN FIXATION.

Nobel, Alfred, (1833–1896)

The name of this Swede has been immortalized in what are arguably the most prestigious of all scientific awards created in the twentieth century. Nobel discovered dynamite, which made him extremely wealthy, although its effect on war was the opposite of what he had hoped—it turned out to be one of the most destructive weapons of its time, rather than helping end war quickly. He channeled his wealth into establishing the Nobel Prizes in various fields of scientific and humanitarian endeavors.

Nocardia

Described in the fourth volume of *Bergey's Manual of Systematic Bacteriology, Nocardia* is a bacterial genus associated with a constellation of cutaneous and systemic human diseases known as nocardiosis. Morphologically, these bacteria are similar to the genus *Actinomyces* because both share a filamentous cellular structure that causes them to be often mistaken for fungi, even though both groups of organisms are true bacteria with prokaryotic organization and gram-positive (peptidoglycan) cell walls. Despite these similarities, *Nocardia* is considerably different from other actinomycetes and displays markedly different cell wall structure and metabolism. The differences between the two groups of bacteria are important considerations during the diagnosis and treatment of various diseases. Unlike the parasitic *Actinomyces*, *Nocardia* species are capable of living independently and are frequently found in soil rather than in association with live animals. In addition, the cell walls of *Nocardia* contain a number of uncommon compounds that make these organisms partially acid-fast, as well as more resistant than other gram-positive organisms to various environmental threats, including enzymes such as lysozyme. Yet another notable feature of these organisms is their ability to use unusual substrates, such as fatty acids and hydrocarbons, as their sole carbon and energy source, which makes them very adaptable to a diversity of environments. When grown on solid media, different species grow in a variety of textures and colors (buff, white, orange, red, or brown). Some species produce white vertical or aerial hyphae. Organisms can multiply by fission or fragmentation of filaments and occasionally form spores.

Three distinct species of *Nocardia* have been isolated and identified in various forms of human nocardioses. These species differ with respect to their antigenic structure and, to a lesser extent, the nature of the disease that they induce. *N. asteroides* causes a systemic nocardiosis, which usually begins in the lungs due to the inhalation of spores or viable cells of the bacteria. Depending on the immune status of the individual, this initial infection can be transitory, acute, or chronic, and it may either remain localized or spread to other locations in the body. *N. asteroides* species have been shown to have some affinity for brain tissue; in fact, the organism was first isolated from a brain abscess of a meningitis patient. Multiplication of the organisms at any site results in local tissue destruction and abscess formation. Nocardioses with cerebral involvement are associated with high mortality rates (>85 percent). The other two species, *N. brasiliensis* and *N. caviae,* cause the development of mycetoma, a granulomatous condition induced by a delayed-type hypersensitivity response to subcutaneous infections. Organisms most likely gain entry into the host tissue via scratches or abrasions. Usually the mycetoma lesions are found in the extremities—the legs and, less frequently, the arms—where they induce the development of abscesses and draining sinuses. Typically, these infections remain localized and the organisms do not enter the bloodstream, although they may spread to contiguous tissues such as bones and joints. In addition to subcutaneous lesions, *N. brasiliensis* has also been known to induce a lymphocutaneous syndrome resembling sporotrichosis.

Specific diagnosis of nocardiosis depends on demonstration of organisms in clinical samples—pus or exudates from lesions. Organisms may be differentiated from *Actinomyces* and other related organisms on the basis of colony morphology and specific biochemical and antigenic reactions. Nocardioses are typically treated with sulfonamides, often in combination with other antibiotics such as streptomycin or erythromycin. *See also* ACTINOMYCES, MYCETOMA, *SPOROTHRIX SCHENCKII.*

Norwalk virus

This single-stranded RNA virus, classified in the calicivirus family, was first discovered in 1972 as the etiologic agent of an outbreak of epidemic diarrhea in the town of Norwalk, Ohio. The virus has since been found in nearly all parts of the world, commonly associated with outbreaks of diarrhea and less frequently in sporadic cases. Infection with the Norwalk virus is typically a self-limited disease with mild to moderate symptoms of diarrhea and abdominal pain—typically lasting 24–48 hours—accompanied by nonspecific symptoms of malaise, muscle aches, and low-grade fever. The exact mechanism of viral spread is not known, although the fecal-oral route is suspected. Undercooked shellfish appears to be one of the common sources in the United States. Because the disease is so short lived and causes relatively few complications, it is typically allowed to run its course without any drug or antibiotic therapy. In the event of acute symptoms, which are accompanied by heavy fluid loss, fluid and electrolyte replacement may be necessary. Proper domestic hygiene and vigilant monitoring of the shellfish are among the best means to prevent the spread of the disease. Although sporadic cases of Norwalk virus infection do not qualify as "notifiable" diseases, reporting epidemic outbreaks to the CDC or a public health agency is mandatory. *See also* CALICIVIRUS; DIARRHEA.

nosocomial infection

Public health terminology for diseases acquired by exposure to infectious agents at a hospital or medical clinic. For example, during surgery, indigenous microbes from one location (e.g., the intestinal lumen) may be transferred to tissues or organs such as the peritoneum, where they are not normally present and where they may induce disease. A physician or nurse, making rounds through different wards, may serve as a carrier for certain other infections, particularly if pre-

cautions are not taken. In addition, contact among patients may also lead to the transmission of infections. Examples of infections that are contracted at hospitals include *Pseudomonas* and *E. coli*.

notifiable disease

Any human or animal disease that has the potential to erupt into an epidemic, the occurrence of which needs to be reported to some public health authority so that appropriate measures can be instituted for the containing, controlling, and treating of the disease. Human diseases are usually reported first to a local authority, such as a county or state health department. For the purposes of efficient health administration, different authorities have official lists of the reportable diseases within given geographical areas. See

Table 7. Infectious Diseases Designated as Notifiable at the National Level, United States, 1997

Acquired immunodeficiency syndrome (AIDS)	Malaria
Anthrax	Measles (Rubeola)
Botulism	Meningococcal disease
Brucellosis	Mumps
Chancroid	Pertussis
Chlamydia trachomatis, genital infections	Plague
Cholera	Poliomyelitis, paralytic
Coccidioidomycosis	Psittacosis
Cryptosporidiosis	Rabies, animal
Diphtheria	Rabies, human
Encephalitis, California serogroup	Rocky Mountain spotted fever
Encephalitis, eastern equine	Rubella
Encephalitis, St. Louis	Rubella, congenital syndrome
Encephalitis, western equine	Salmonellosis
Escherichia coli 0157:H7	Shigellosis
Gonorrhea	Streptococcal disease, invasive, Group A
Haemophilus influenzae (invasive disease)	Streptococcus pneumoniae, drug-resistant
Hansen disease (leprosy)	Streptococcal toxic-shock syndrome
Hantavirus pulmonary syndrome	Syphilis
Hemolytic uremic syndrome, post-diarrheal	Syphilis, congenital
Hepatitis A	Tetanus
Hepatitis B	Toxic-shock syndrome
Hepatitis C/non-A, non-B	Trichinosis
HIV infection, pediatric	Tuberculosis
Legionellosis	Typhoid fever
Lyme disease	Yellow fever

Source: "Summary of Notifiable Diseases, United States, 1997," *CDC MMWR Weekly* 46(54): Table A. See: <http://www.cdc.gov/epo/mmwr/preview/mmwrhtml/00056071.htm#00003630.htm>.

Table 7 for the list of diseases designated as notifiable by the Centers for Disease Control (CDC).

novobiocin

Natural antibiotic, produced by *Streptomyces niveus*, that acts by inhibiting gyrase and, hence, DNA replication. For reasons that are not entirely clear, novobiocin appears to preferentially target gram-positive over gram-negative organisms. *See also* ANTIBIOTIC; *STREPTOMYCES*.

nucleus

Membrane-bound compartment found exclusively in eukaryotic cells, whose function is to house all the genetic information an organism requires for living. The principal component of the nucleus is the cellular DNA, packaged into one or more chromosomes, depending on the organism. In addition, the nucleus may contain special DNA-binding proteins such as histones and various nucleic-acid-polymerizing enzymes. This compartment is the site for DNA replication as well as for the transcription of genes into RNA during gene expression and protein synthesis. The nuclear membrane is a phospholipid bilayer with two distinct faces—nuclear and cytoplasmic—containing embedded proteins as well as pores that form the route of communication between the nucleus and the rest of the cell (cytoplasm). The nuclei of many cells may be visualized easily under an ordinary light microscope with the use of selective stains. *See also* CELL; CHROMOSOME; CYTOPLASM; EUKARYOTE.

O

oncogenic virus

A virus that is capable of inducing the development of cancer in its host by disrupting one or more of the processes that control the normal growth cycle of the host cell it infects. With the exception of the human T cell lymphotropic virus (HTLV-1), all known oncogenic viruses of humans are DNA viruses. However, many RNA viruses are known to induce cancer in birds and other animals. *See also* CANCER.

Oparin, Alexander (1894–1980)

Russian biochemist who first suggested the idea of chemical evolution as the process underlying the origins of life on earth. His ideas are presented in his book, *The Origin of Life*, which was published in 1924. *See also* ORIGIN OF LIFE.

operon

A cluster of genes—usually for a system of proteins with related or sequential functions—whose sequential expression is regulated by a common mechanism. For example, the *lac* operon encodes three enzymes required for breaking down lactose and is under the control of a single promote/repressor system. *See also* GENE.

opportunistic pathogen

A microbe that does not usually behave as a pathogen—i.e., cause disease—in a host organism, but is capable of doing so under special conditions. An example of such a circumstance is a weakened immune system in the host. For instance, organ-transplant recipients who receive immunosuppressive drugs to prevent transplant rejection are unable to clear infecting organisms from their bloodstream or tissues as well as they normally would be. The continued presence of bacteria in their blood and tissues can give rise to a variety of adverse effects due to the depletion of nutrients or the production of harmful substances by the organism.

Another situation in which an organism may function as an opportunistic pathogen is its accidental transfer to a site in the body where its growth may cause damage, e.g., from the skin or intestinal lining (where it may be found under natural circumstances) to an internal organ. Such transfers have been known to occur inadvertently during surgery or when the bacteria enter the bloodstream via a wound and subsequently lodge themselves in a remote location. See Table 8 for examples of microbes that constitute the indigenous microflora of various parts of the body but have been implicated as opportunistic disease agents. *See also* INDIGENOUS MICROFLORA.

Table 8. Opportunistic Infections by Indigenous Microflora

Organism	Normal location	Disease Association
Bacteria		
Acinetobacter [1]	Skin and gastrointestinal tract	Post-operative wound infections
	Respiratory tract	Pneumonia, meningitis.
	Urinogenital tract	Urethritis
Actinomyces	Mouth	Dental plaque and caries
Aeromonas	Gastrointestinal tract	Postoperative infections and osteomyelitis
Bacillus species	Skin, ears, eyes and gastrointestinal tract	Wound infections, meningitis.
Bacteriodes	Respiratory tract	Lung abscess
	Gastrointestinal tract	Peritonitis, intestinal abscesses.
Campylobacter	Gastrointestinal tract	Diarrhea
Corynebacterium [2]	Skin, respiratory tract	Bacterial endocarditis
Enterococcus	Respiratory tract	Meningitis, pneumonia and bacterial endocarditis
	Urinogenital tract	Pyelonephritis and cystitis [3]
Haemophilus aegyptius	Eyes	Eye infections
Mycoplasma	Respiratory tract	Atypical pneumonia
	Urinogenital tract	Urethritis
Neisseria	Respiratory tract	Meningitis
Peptococcus	Gastrointetinal tract	Peritonitis and abscesses
Propionibacterium	Skin and gastrointestinal tract	Pimples and acne; bacterial endocarditis
Rothia dentocariosa	Mouth and respiratory tract	Dental caries and abscesses
Sarcina	Urinogenital tract	Postoperative complications
Staphylococcus epidermidis	Respiratory tract	Endocarditis
	Skin	Pimples, boils, carbuncles;
S. saprophyticus	Urinogenital tract	Urinary tract infections in women
Streptococcus agalactiae	Urinogenital tract	Neonatal meningitis, endocarditis, osteomyelitis and myocarditis.
Fungi		
Candida albicans	Skin	Cutaneous candidiasis
	Mouth	Thrush
	Urinogenital tract	Vaginal yeast infections
	Gastrointestinal tract	Postoperative wound infections
Trichophyton	Skin	Dermatophycoses
Pityrosporium	Skin	Tinea versicolor and folliculitis
Epidermophton floccosum	Skin	Athlete's foot and other cutaneous lesions
Protozoa		
Trichomonas	Urinogenital tract	Vaginitis

Notes:
[1] Specifically *A. calcoaceticus* on the skin and miscellaneous species elsewhere.
[2] This includes non-pathogenic species, i.e. the diphtheroids.
[3] Kidney and bladder infections respectively.
*Information culled from Boyd and Hoerl's *Basic Medical Microbiology*, [Little, Brown and Co., Fourth Edition, 1986]

opsonin

A serum protein that enhances the susceptibility of an antigen (e.g., a protein on the surface of a bacterium) to phagocytosis by macrophages and other cells of the immune system. The protein coats the antigen (i.e., opsonizes the antigen) so that it can be recognized and bound by specific receptor sites on the surface of the phagocyte. Examples of some well-characterized opsonins include the antibody IgG and the protein fragment C3b of the complement system. *See also* ANTIBODY; ANTIGEN; PHAGOCYTOSIS.

optical microscope

A microscope that uses ordinary visible light to produce magnified images of objects too small to be seen with the naked eye. The simplest type of optical microscope is a simple magnifying glass, consisting of a single convex lens made of transparent glass or plastic. The magnifying power of a lens depends on several factors, such as the refractive index of the lens material, the shape of the lens, and how finely the lens has been ground and polished. Single lenses typically have resolution limits of about $10\,\mu m$ or 0.1 mm, which is not enough magnification power to view microbes.

The most commonly used optical microscope is a compound microscope, which consists of two or more lenses—the magnified image produced by one lens (called the objective lens) is further enlarged by the second lens, known as the ocular or eyepiece. This assembly is housed in a hollow tube such that the distance between the two lenses can be adjusted to bring the image into focus.

The first optical microscopes, invented around 1590, were crude instruments that could enlarge to about 10–20 times the size of the original object. By the middle of the next century, Robert Hooke, Curator of Experiments of the newly established Royal Society of London, began to look at a variety of objects under such a microscope and in 1665 published his classic work, *Micro–phagia*. Soon thereafter, Antoni van Leeuwenhoek opened up the field of microbiology. He ground his own lenses and made microscopes with magnification powers of up to 270´ but was so secretive about his methods that microscopes of matching quality were not manufactured until well over a century after his death. The development of light condensers in the latter half of the nineteenth century improved visualization by illuminating the background of the specimen with a strong, even beam of light. Ultimately, however, the resolution power of a compound light microscope is limited by the wavelength of visible light and the smallest resolvable distance is 0.5 μm. In microbiology, optical microscopes are useful for visualizing organisms such as bacteria, protozoa, and amoebae. Smaller entities such as viruses or subcellular components like ribosomes and mitochondria cannot be seen with these instruments.

Because microscopy requires us to scrutinize objects held under an assembly of lenses, specimens require special handling. Microorganisms need to be "fixed" onto some sort of support, and they require adequate illumination so that the light rays emanating from them are bent appropriately in the formation of the enlarged image. In addition, illumination enhances the contrast between the object and its surrounding medium and helps us to see the object more easily. A typical light microscope consists of a solid stage, which can hold transparent glass slides upon which bacterial and other specimens are fixed. The stage contains a hole directly beneath the lens tube so that the specimen can be lit from below. *See also* CONTRAST; MAGNIFYING POWER; MICROSCOPE; RESOLVING POWER; STAINING.

orbivirus

A genus of arthropod-borne viruses (arbovirus) of the reovirus family of double-stranded RNA viruses, found in both animals and humans. The generic name is derived from the large ring or doughnut-shaped protein units or capsomers that make up the inner capsid of these viruses (*orbi* = ring). The segmented genome contains 10 segments, each of which encodes a specific viral gene. Some examples of orbiviruses include the causative agents of the blue tongue disease of sheep and African horse sickness. The only known human orbivirus is the Kemerovo virus. Normally carried by a species of Siberian ticks, this virus has been isolated from the blood and cerebrospinal fluid of certain patients with meningo-encephalitis. Antibodies to this virus have also been found in Siberian patients suffering from fevers of uncertain origins. *See also* ARBOVIRUS; REOVIRUS.

organoleptic

A term used to describe various qualities of food that are perceived by our senses, such as taste, texture, and aroma. Microbial ac-

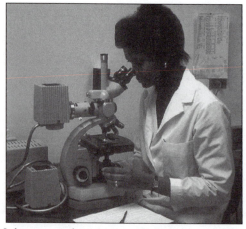

Laboratory worker using an optical microscope. *CDC.*

tion on various food products have direct effects—both desirable and undesirable—on their organoleptic qualities. For example, the "creamy" flavor and consistency of yogurt is due to the action of *Streptococcus thermophilus*; while the soft rots caused by the action of *Erwinia carotovora* gives fruits and vegetables an unpleasant mushy texture.

organotroph

Organism that derives either its energy or carbon (or both) from organic molecules in the environment. Much of the microbial world, including the fungi and many bacterial species, is organotrophic. *See also* HETEROTROPH.

origin of life

Among the earliest recorded beliefs among scientists regarding the origins of life—dating back over 2,000 years, to the time of Aristotle—was that living things arose spontaneously from lifeless matter. Examples used to support this theory of spontaneous generation or "abiogenesis" included the seeming birth of maggots and other insects from decaying food and earthworms from wet earth. It was not until the seventeenth century that the veracity of this belief was tested by controlled experiments. Francisco Redi showed that certain white worms—believed to arise spontaneously from decaying meat or fish—would develop only if the material were left exposed and allowed contact with common houseflies, which deposited eggs on the decaying food. This experiment spurred others to test the theory of spontaneous generation, and careful observations demonstrated that living organisms could arise only from a living parent.

But scarcely had theory of abiogenesis been debunked for higher organisms, than the controversy was stirred up anew when Leeuwenhoek reported his observations of microorganisms in 1673. Unlike maggots, the microbes did seem to arise spontaneously from meat and other material without eggs or other visible starting material. Toward the middle of the eighteenth century, another Ital-

ian, Lazzaro Spallanzani, offered evidence against abiogenesis of microbes, but it was not until nearly a century later that the work of French scientist Louis Pasteur put an end to the debate on spontaneous generation. Under the existing conditions on earth, it is not possible for lifeless matter to generate a living organism by itself. Only life can beget life. Moreover, a living organism can only produce another organism like itself—it is not possible for a dog to give birth to kittens, for instance, or for a bacterial cell to divide and produce fungi! But scientists believe that in the distant past, conditions on earth were sufficiently different so as to enable the formation of the first living organism from nonliving matter.

The most widely accepted theory about the origin of life was first proposed in the 1920s by two scientists working independently: Alexander Oparin, a Russian biochemist, and a British chemist by the name of J. B. S. Haldane. Their idea was that life originated on earth not as a single random event, but rather through a process of evolution of chemical reactions that gradually developed into a self-sustaining and self-perpetuating system—namely, a living cell. Like Darwin's theory of biological evolution, this idea of chemical evolution proposed a gradual transition from simple to more complex processes—from the elements, to simple inorganic compounds, to organic molecules of gradually increasing complexity, and finally to life. The feasibility of at least some of these reactions has since been borne out by calculations and experiments.

According to the ideas put forward by the American chemist Harold Urey (who won the 1934 Nobel Prize in Chemistry), the conditions on primitive earth would have resulted in the combination of various elements such as carbon, hydrogen, nitrogen, oxygen, sulfur, and phosphorus to form simple inorganic compounds such as carbon monoxide (CO), carbon dioxide (CO_2), methane (CH_4), and water (H_2O). Urey also deduced that under the highly energetic conditions prevalent at that time—solar radiation, electrical storms, and high temperatures—the chemical bonds holding these smaller compounds together could be broken down and reforged between more than one species to give rise to larger organic molecules. This was subsequently shown to be possible by a scientist named Stanley Miller, who designed a laboratory apparatus to simulate the conditions of primitive earth. He mixed methane, ammonia, hydrogen, and water vapor and subjected the mixture to an electrical spark. The product resulting from this reaction was condensed into a liquid and analyzed. True to Urey's predictions, the mixture did indeed consist of significant quantities of several different organic molecules, among them simple sugars, amino acids, and nitrogenous bases.

According to Oparin's model, these small organic molecules then settled into the ocean and over millions of years became concentrated enough to form a kind of primordial soup from which life eventually developed. He speculated that the concentration of these materials would have first given them the opportunity to join together to make larger organic molecules or polymers. This, too, has been shown to be possible in the laboratory. The next step in this theory was the formation of aggregates or coacervates of the large organic molecules. This possibility is supported by the natural tendency of organic polymers to cluster together and form discrete packets when mixed together in high concentrations. Oparin and Haldine speculated that, over millions of years, these coecervates developed mechanisms to sustain themselves and also to replicate themselves, and thus a living organism slowly emerged. The ideal molecules for performing any task were probably chosen by a process of natural selection. This last step—representing the transition from lifeless organic matter to living cells—is the least understood among scientists. So far, no one has been able to develop an experimental system to duplicate the conditions of process that may have finally given rise to life. *See also* Leeuwenhoek, Antoni van; Oparin, Alexander; Redi, Francisco; Spallanzani, Lazzaro.

Oroya fever

First identified in 1870 among construction workers who were building a railroad from Oroya to Lima in Peru, Oroya fever is a systemic form of disease caused by an infection with the organism *Bartonella bacilliformis*. Primary infection is due to the injection of the organisms into the body via the bite of the sandfly; the geographical distribution of the disease is limited to the areas where these insects live—namely, Peru, Ecuador, and parts of Colombia. The disease is not spread from human to human except by the transfusion of infected blood. Typical symptoms include irregular fever, malaise, muscle and joint aches, severe anemia, and generalized lymphatic involvement. Symptoms correspond to the multiplication of the bacteria within the erythrocytes and other cells and the subsequent destruction of these cells. If untreated, the case fatality rate of Oroya fever may be as high as 90 percent, but the disease may be treated effectively with a number of different antibiotics including penicillin, tetracycline, and chloramphenicol. Diagnosis is achieved by the demonstration of the bacteria in blood smears—either visualized by Giemsa staining or by culturing clinical samples in enriched media—and by serological tests. Attempts at controlling this disease are aimed at vector control and minimizing exposure to sandflies at night because these insects are known to feed only from dusk till dawn. *See also* Bartonella.

outbreak

An epidemiological term for the sudden, often unexpected, incidence of a disease. In an outbreak, there is a noticeably higher occurrence of the disease—in the population of a given geographical area over a certain period of time—than would be expected to occur randomly. The concept also encompasses epidemics. The types of diseases usually associated with outbreaks are acute infections that are easily spread by contact, droplet infection, or food and water. *See also* EPIDEMIC.

outer membrane

Outer layer of the cell wall in gram-negative bacteria. The basic structure of the outer membrane resembles the cell membrane, containing a lipid bilayer studded with different membrane proteins. However, unlike the cell membrane, which has symmetrical layers of lipid, the external layer of the outer membrane is made of lipopolysaccharides (LPS), whose structure differs from organism to organism. LPS is responsible for eliciting symptoms of septic shock when these organisms invade the bloodstream. Thus, the membrane confers specific properties of pathogenesis and antigenicity to gram-negative bacteria. *See also* LIPOPOLYSACCHARIDE.

oxygen cycle

Term given to the chemical reactions through which the element oxygen is incorporated into living matter and returned to the atmosphere. Of all the elements that constitute organic and living matter—namely, carbon, hydrogen, oxygen, nitrogen, sulfur, and phosphorus—oxygen is probably the easiest to process, with respiration and photosynthesis balancing each other, for the most part, in terms of the amounts of oxygen used and released. Respiration—performed by a variety of microbes, as well as all higher animals and plants—fixes atmospheric oxygen by converting it into water, which is essential for virtually all forms of life. Photosynthesis is the primary biological mechanism for returning the oxygen trapped in water and organic materials back to the atmosphere. This arm of the cycle is carried out by green plants, algae, and—most important—the cyanobacteria. *See also* PHOTOSYNTHESIS; RESPIRATION.

P

pandemic

An epidemic that spreads with such speed that within a short period of time people from most parts of the globe are affected. An example of a disease that is frequently associated with pandemics is viral influenza. The unpredictable behavior of the virus due to antigen drift and antigenic shift has led to difficulties in containing outbreaks within limited geographic areas. In the nineteenth century, the rise in mass movements of people across different continents led to several pandemics of cholera, which, up until then, had been almost completely confined within the Indian subcontinent, where it had been endemic for centuries. *See also* EPIDEMIC.

papillomavirus

Group of oncogenic (cancer-causing) DNA viruses in the papovavirus family that exhibit the properties of extreme host-specificity and marked preference for the epithelial cells of the skin (a property known as tissue tropism). Thus, the tumors caused by this virus originate in the skin and often take the form of warts or papillomas. The first virus in this group to be isolated was the Shope's papillomavirus from rabbits in 1933. The human papillomavirus, which was not discovered until much later, is difficult to study in the lab because no one has yet managed to grow this virus in tissue culture. Much of what we know about this group of viruses is based on studies with the bovine virus, for which there is a laboratory model.

Exactly how initial infection with the papilloma viruses occurs is still not completely understood. The distinctive DNA expression and replication, however, has been studied in some detail. The papilloma viruses employ two mechanisms of genome replication, depending upon the state of differentiation of the host epithelial cell that it has infected. The first mode of replication, which is called plasmid replication, occurs only in the basal cell layer of the skin. A single strand of the viral DNA amplifies to form anywhere from 50 to 400 copies with the help of the host enzymes. After this, the number of copies per host cell remains constant and replication occurs with each cell cycle. The second mode of replication, known as vegetative replication, occurs in the fully differentiated epidermal cells and is characterized by the amplification of the viral gene to several thousand copies per cell, with no control over the number. In neither case do the DNA copies form new virions or get released from the host cell to infect neighboring cells. Thus, warts are not infectious. A single wart formed by this virus represents the clone of a single host cell. *See also* ONCOGENIC VIRUS; PAPOVAVIRUS; WART.

papovavirus

Family of small animal viruses that mainly infect mammals and are oncogenic or tumor-inducing in nature. The family was named for the first three viruses found within the group, namely the *pa*pilloma, *po*lyoma, and simian *va*cuolating agent. As a group, the papova viruses are small (45–55 nm), naked (non-enveloped) particles with icosahedral capsids and circular, double-stranded DNA genomes. Different viruses are classified into two main categories—the papilloma and the polyoma viruses. *See also* ONCOGENIC VIRUS; PAPILLOMA VIRUS; POLYOMA VIRUS; WART.

Paracoccidioides brasiliensis

A dimorphic fungus, endemic to South and Central America, that causes a systemic mycosis in humans. Both the fungus and the mycosis share many characteristics with other fungi such as *Blastomyces* and *Coccidioides;* in fact, the disease (called paracoc–cidioidomycosis) was originally termed South American blastomycosis. The fungus assumes a yeast-like appearance in the body but grows as a mold showing only asexual reproduction in culture. The natural habitat and ecology of *P. brasiliensis* are still somewhat uncertain, although it is believed to reside in soils, especially in rural and undeveloped areas. Similarly, while the route of infection has not been established, infection is believed to occur due to the inhalation of spores, an inference borne out by the observation that the mycosis first manifests itself in the lungs. As is the case in many systemic fungal diseases, the clinical presentation depends heavily on the nature of host-parasite interactions as well as the immune status of the infected individual. Adult paracoccidioidomycosis is a chronic granulomatous condition characterized by the formation of patchy lesions in the lung and ulcerative lesions of the skin and mucus membranes in the oral, nasal, and gastrointestinal passages. There is inflammation and swelling of the regional lymph nodes. The adrenal glands seem especially susceptible. The disease may lead to fatal consequences if not treated properly. A less common version of paracoccidioidomycosis is an acute form seen in juveniles, characterized by the involvement of the blood vessels

and bone marrow problems. Diagnosis is confirmed through histological and serological tests and by culturing infective material. Drugs of choice are ketoconazole and itraconazole, and amphotericin B should be reserved only for the severely ill. *See also* *BLASTOMYCES DERMATITIDIS.*

parainfluenza virus

This virus—actually a group of viruses within the paramyxovirus family—is a common cause upper respiratory tract infection in the general human population, as well as the leading cause of "croup" in young children. There are four individual types of parainfluenza viruses (numbered in the order in which they were first discovered/isolated), of which types 1 and 3 are classified together in the genus paramyxovirus, and types 2, 4a, and 4b are grouped along with the mumps virus, due to antigenic differences. Infection with the parainfluenza viruses follows a fairly typical course. Viruses enter via the respiratory tract and induce primary symptoms of pharyngitis and occasionally bronchitis with a low fever. In adults, the infection is often self-limiting and resolves in a matter of a few days. In young children whose immunity is not well developed, the infection may spread to the larynx and trachea and lead to symptoms of respiratory distress. Occasionally this may progress to an obstruction of the larynx, which may require surgical intervention.

Diagnosis of parainfluenza infections is best achieved through serological methods to detect viral antigens and antibodies in the bloodstream of the patient as these viruses grow slowly and often induce few noticeable changes in tissue culture. Treatment is, for the most part, palliative. No good vaccines have been developed against these viruses. The best means of controlling the spread of the viruses, especially among young children, is to minimize exposure. *See also* PARAMYX-OVIRUS.

paramyxovirus

This family of RNA viruses contains such well-known human pathogens as the measles and mumps viruses, as well the respiratory syncytial virus and parainfluenza viruses. The viruses are fragile particles ranging in size from 150 to 300 nm and consisting of a helical nucleocapsid enclosed within a lipid envelope. The genome of the paramyxoviruses is a single molecule of single-stranded RNA of negative polarity, i.e., it is unable to serve directly as a template for protein synthesis or genome replication. Depending on the viral species, anywhere from 7 to 10 genes are encoded in the genome.

Almost all the known paramyxoviruses have been found associated with respiratory diseases, especially in younger populations. Among the important factors contributing to the pathogenicity of these viruses are two proteins found in the viral envelope. One of these plays a role in ensuring the attachment of the virus particles to cells in the host, due to the specific recognition of molecules or receptors present on the surface of these cells. Often this protein also has the ability to induce the clumping or agglutination of blood cells in the host. The second protein is called the fusion protein because it induces the fusion of membranes. This enables the virus to enter the host cell by fusing the envelope with the membrane and then releasing the nucleocapsid into the cell. In addition, the fusion protein also induces the fusion of membranes of adjacent host cells, i.e., syncytia formation. This property prolongs the duration of survival of the virus in a host because it bypasses the need for cell lysis and reentry on the part of the viruses. For humans, these proteins—which are in chemical terms glycoproteins—serve a useful purpose as the means of identification of the different viruses. *See also* MEASLES; MUMPS; PARAINFLUENZA VIRUS; RESPIRATORY SYNCITIAL VIRUS.

parasitism

Symbiotic relationship between two organisms, in which one organism—the parasite—lives within the second (host) organism at the expense of the host, which may suffer varying degrees of damage or even death. Consequently, many microbial parasites of higher

animals are natural pathogens. Interestingly enough, the death of the host is not the most favorable outcome in a parasitic relationship because it requires the parasite to go searching for a new host. An ectoparasite is established on the exterior of the host, while an endoparasite is established within its host's body. *See also* HOST; SYMBIOSIS.

parvovirus

Group of simple, single-stranded DNA viruses that are the smallest viruses known to infect eukaryotic cells. Virus particles are naked icosahedral particles with a protein capsid that encloses a genome of roughly 5 kilobases in length. There are over 50 viruses within this family, which are classified into six genera in two subfamilies, divided according to whether the host is a vertebrate or invertebrate. In practical terms, there are two main types of parvoviruses: (1) autonomous viruses, which have all the necessary genetic information for replication with the help of the host cell only; and (2) defective viruses, which require the presence of a second complete virus—called the helper virus—for replication in the host cell. Adenoviruses and occasionally herpesviruses are known to function as the helpers.

The only known human parvovirus is the virus B19, which infects the reticulocytes in bone marrow, leading to anemia or a depression in red blood cell production. This is followed by a rash, called erthaema infectiosum (also slapped-cheek disease or fifth disease), which is not dissimilar to the rubella virus. This rash is much more common in children than adults. About 80 percent of patients may also develop temporary joint pains that resemble arthritis, and infection during the early stages of pregnancy may be associated with spontaneous abortion. *See also* EPIDEMIC.

Pasteur, Louis (1822–1895)

This French scientist may be justly regarded, alongside Robert Koch, as one of the primary shapers of the science of microbiology. Trained as a chemist, Pasteur worked on numerous and varied questions in the fields of

Bactereologist and chemist Louis Pasteur. © *Science Source/Photo Researchers.*

fermentation, bacteriology, and medicine, and with every advance added information of fundamental importance to our understanding of how microbes function and how they cause disease. His introduction to the world of microbes came when he was asked by a winemaker and distiller to explore the reasons for the contamination of alcohol with undesirable substances during the fermentation process. In the course of his investigations, Pasteur discovered that fermentation was linked to microbial activity. This line of research also led him to deliver the killing blow to the theory of spontaneous generation for the origins of life. From the fermentation of wine, Pasteur turned to problems of the wine spoilage by microbes and developed the technique, which we now call pasteurization in his honor, to prevent spoilage, not only of alcoholic beverages, but also of milk and other food items. He then turned his attention to diseases—first of silkworms and then of humans—and thus gave shape to the "germ theory of disease." He was responsible for discovering at least three disease agents in humans, the streptococci, staphlyococci, and pneumococci (later re-

named as *Streptococcus pneumoniae*) and also for establishing effective asepsis in hospitals. Last but not least, Pasteur made vital contributions to the area of disease prevention. Working first on the bacterial diseases anthrax and chicken cholera, and then on the rabies virus, Pasteur developed techniques for attenuating disease agents to make them avirulent and using these to vaccinate against disease. In 1886, Pasteur made history when he treated and cured a nine-year-old boy named Joseph Meister—who had been badly bitten by a rabid dog—with the anti-rabies vaccine that he had developed. *See also* FERMENTATION; KOCH, ROBERT; ORIGIN OF LIFE; PASTEURIZATION; VACCINATION.

pasteurization

Method of treating food substances by heating at relatively low temperatures (e.g., 50–60°C for about 20–30 minutes, or 70°C for 10–15 minutes) to prevent spoilage or destroy pathogens. Although this treatment does not achieve sterility, it kills the vegetative forms of most organisms, thus increasing the shelf life of the food without significantly altering properties such as taste and texture. Louis Pasteur developed this method in the late nineteenth century to prevent the spoilage of beer and wine, and today it is the technique of choice for processing milk, and juices. *See also* FOOD PRESERVATION.

pathogen

A microorganism that typically causes a disease when it, or a toxic substance it produces, enters a human or animal host. It should be noted that not every pathogenic organism causes disease every time, but that disease is the most common outcome. The link between a pathogen and the disease symptoms may be highly specific—e.g., tetanus, in which muscular spasms occur in response to the exotoxin produced exclusively by *Clostridium tetani*—or somewhat generalized—e.g., fever (elevated body temperature) or diarrhea (loose stools). In some cases, an organism may just act as a triggering device for harmful processes in the body. The occurrence of any disease in response to a pathogen depends on many factors, including the virulence of the microbe as well as the the immune status of the host organism. *See also* PARASITISM.

patient isolation

The practice of separating a person with some disease from other people for a specified period of time. In most cases, the isolated patients have a communicable disease, and the purpose of isolation (or quarantine) is to prevent the disease from spreading to other people. Depending on the disease type, its severity, and its mode of transmission, the patient's isolation may be either strict or moderate. In some instances, isolation is for the benefit of the patient, e.g., AIDS patients or transplant recipients, who are severely immuno-compromised, may be isolated to avoid exposure to opportunistic pathogens. *See also* QUARANTINE.

penicillin

Penicillin is a β-lactam antibiotic produced by the fungus *Penicillium notatum.* The discovery of this antibiotic heralded a new era in antimicrobial therapy. *See also* ANTIBIOTIC.

Penicillium

A ubiquitously distributed filamentous mold, *Penicillium* is a genus of fungi whose activities have more beneficial than harmful effects for human beings. Perhaps the best known example is that of the antibiotic penicillin—produced by *P. notatum*—the discovery of which opened up a whole new avenue of antimicrobial therapy. But centuries before Sir Alexander Fleming observed the antimicrobial activity of this fungus on his contaminated petri dish, other *Penicillium* species had been providing humankind with a variety of cheeses, among them, soft cheeses like Camembert—ripened by *P. camembertis*—and hard varieties such as Roquefort, Stilton, and Gorgonzola, all of which are products of the activity of *P. roqueforti.*

Penicillium species may be found in a variety of environments, including soil, decomposing organic matter, and in the air (as spores). Asexual spores produced by various species are often colored bluish-green, which gives the hard cheeses their characteristic color. Most species are nonpathogenic, although the fungi may be implicated in certain respiratory and urino-genital infections under weakened circumstances. Mere isolation of these fungi from clinical specimens is not considered an adequate diagnosis, and care should be taken to check the samples for other infectious agents as the primary pathogen.

peptide bond

Specific type of chemical bond found in proteins between two adjacent amino acids, in which the amino group of one acid is linked to the carboxyl group of the next, with the release of a water molecule. *See also* AMINO ACID; PROTEIN.

peptidoglycan

The main chemical constituent of bacterial cell walls. In general, the gram-positive organisms have a much thicker layer of this substance than the gram-negative bacteria. Chemically, peptidoglycans are complex molecules resembling a three-dimensional net around the cell. The net is composed of linear heteropolysaccharide chains (containing g1 and g2 as the monomeric units), which are cross-linked by short peptides. The synthesis of this material is also a complex process, involving the participation of more than one synthetic enzyme. During cell division, new peptidoglycan material is added to the cell wall at the inner surface of the cell, resulting in an inside-to-outside growth pattern. Peptidoglycan, or rather the synthesis of peptidoglycan, is the target for the activity of the β-lactam antibiotics, and the incomplete synthesis results in the generation of very thin-walled cells that eventually lose their ability to retain their shape and thus undergo lysis. *See also* CELL WALL; GRAM STAIN.

Peptococcus niger

Gram-positive, strictly anaerobic bacteria found as part of the normal microflora of the skin, alimentary canal, and genital tract of humans and some animals, and only rarely, if ever, in air and water. The bacteria are small, non-motile, non-spore-forming cocci, typically present in pairs or irregular clumps. The generic name is derived from their ability to metabolize peptones and amino acids. *Peptococcus* metabolizes glucose to form capronate and butyrate, and sometimes a small amount of gas, but not lactic acid. In the laboratory, these organisms require complex, protein-rich media such as blood agar, or chopped meat medium, and must be maintained under completely anaerobic conditions. Not usually pathogenic, *P. niger* can, however, be involved in wound infections—associated with foul-smelling pus due to the breakdown of peptones—and in mixed infections of soft tissues along with other organisms such as *Bacteroides*.

phage. *See* BACTERIOPHAGE

phagocytosis

Cellular process of engulfing solid particles and making then a part of the cytoplasm, employed by cells (called phagocytes) for a variety of purposes. Derived from the Greek word (*phagos*) for eat, in many instances, the purpose of phagocytosis is nutrition. For instance, an amoebae may engulf its target substance—either inanimate nutrient particles or living cells—and retain it inside a vacuole, where the material is digested by appropriate enzymes and assimilated. In multicellular organisms like humans, phagocytic cells such as macrophages and other leukocytes serve a protective function by removing potentially harmful organisms such as bacteria from the body. These cells often contain special receptor molecules on their surface that recognize and bind to the targets—an important first step in the phagocytic process. In other instances, special antibodies called opsonins aid in phagocytic recognition. After attach-

ment, the target material is ingested into a vacuole, where it is killed by the action of various intracellular enzymes. *See also* IMMUNE REACTION; IMMUNE SYSTEM; METCHNIKOFF, ELIE; OPSONIN.

phase-contrast microscope

A specialized type of optical microscope that makes use of the differing optical properties of various transparent and translucent objects to enhance the contrast between them without the use of stains or dyes. This instrument takes advantage of the fact that when light rays are reflected from a non-opaque object, they are scattered or diffracted according to the depth of penetration of the incident light. The microscope is fitted with a special device called a phase plate, which separates diffracted and nondiffracted light rays so as to form an enlarged image with adequate contrast. *See also* CONTRAST; MICROSCOPE; OPTICAL MICROSCOPE.

phenotype

External behavior or appearance displayed by a living organism. The phenotype of an organism or cell bears an unequal relationship with the genotype because the presence of a gene does not automatically entail its expression. In fact, an organism never expresses all of its genes at any given instant of its life cycle. However, the converse is always true; for a protein or mRNA to be produced, the specifying gene must be present. The case of adaptive enzymes for the fermentation of certain uncommon sugars is a good example of the phenotype not matching the genotype. *See also* GENE EXPRESSION; GENOTYPE.

phosphorylation

In chemical terms, phosphorylation is simply the creation of a chemical bond between any molecule and a phosphate (PO_4) group. In the biological context, however, this term is almost always used to refer to the formation of high-energy phosphate bonds, in molecules such as ATP, which make up the energy currency of living organisms. Biological phosphorylation usually takes place in the presence of a membrane-bound enzyme called ATPase, which is coupled to the membrane-bound electron transfer chains that generate the energy by moving electrons across a gradient. When the ATPase is coupled to a photosynthetic chain, the phosphorylation reaction is called photophosphorylation. Oxidative phosphorylation reactions involve the oxidation of some external agent such as oxygen. Various energy-generating steps of the Krebs cycle result in ATP formation through oxidative phosphorylation. When the energy from a chemical reaction is used directly in ATP synthesis from ADP without an electron transport chain, the reactions is called substrate-level phosphorylation, e.g., in glycolysis. *See also* ATP; ELECTRON TRANSFER CHAIN.

Photobacterium

Characterized by their property of bioluminiscence—the ability to convert chemical energy into light—these bacteria constitute the normal microflora of many deep-sea animals and are responsible for emitting the flashes of light than may be observed in the deep. They are gram-negative, facultatively anaerobic, non-spore-forming organisms that are classified in the same family as the cholera-causing *Vibrio* species. Species include *P. phosphoreum* and *P. fisheri* (also called *V. fisheri* in some references). *See also* BIOLUMINISCENCE.

photolithotroph

Photosynthetic organisms that use inorganic compounds such as hydrogen sulfide (H_2S) instead of water to assimilate inorganic carbon—principally carbon dioxide—from the environment into cells. *See also* AUTOTROPH; PHOTOSYNTHESIS; PHOTOTROPH.

photoorganotroph

Photosynthetic organisms that harvest sunlight via a chemical reaction using such organic compounds as methanol and formic acid as the primary reducing agent to convert atmospheric carbon dioxide into living matter. *See also* AUTOTROPH; PHOTOSYNTHESIS; PHOTOTROPH.

photosynthesis

Process by which certain living organisms convert energy from sunlight into chemical energy that is more readily usable by the cells. In simple terms, this process involves the conversion of inorganic carbon from the environment to an organic compound in the presence of light. A wide spectrum of organisms, including prokaryotic cyanobacteria and groups of purple bacteria, as well as eukaryotes such as green plants and algae, rely on photosynthesis to harness the sun's energy.

The most common type of photosynthesis utilizes water molecules to reduce carbon dioxide, which results in the formation of sugar—primarily glucose—which an organism may either use immediately or store for later use. Oxygen, which is vital for the existence of many other organisms (including humans), is a by-product of this reaction. The overall reaction can, be expressed by the following chemical equation:

$$n\ H_2O + n\ CO_2 \rightarrow (CH_2O)_n\ [sugar] + n\ O_2$$

Photosynthesis actually proceeds in a stepwise manner, with a "light" reaction followed by a "dark" reaction. A few photosynthesizers, specifically the purple bacteria, replace water with hydrogen sulfide (H_2S) and produce elemental sulfur instead of oxygen. The reaction takes place in the presence of special membrane-associated pigments called chlorophylls, which impart their green or purple colors to the organisms.

Photosynthesis is one of the most important processes in the living kingdom because it furnishes cells with two of the fundamental requirements of life—energy and carbon. Because the sun is the ultimate energy source for the earth, photosynthesis is the ultimate energy-deriving reaction. Most of the organisms that do not perform the reaction themselves consume the photosynthetic products from organisms that do. For instance, animals—which are incapable of photosynthesis—eat plants. In the microbial world, fungi use organic substances produced by algae or bacteria. Thus, directly or indirectly, all life may be said to hinge upon photosynthesis. *See also* ALGAE; AUTOTROPH; BLUE-GREEN ALGAE; CHLOROPLAST; CHLOROPHYLL; CYANOBACTERIA; METABOLISM; PHOTOLITHOTRAPH; PHOTORGANOTROPH; PHOTOTROPH; THYLAKOID.

phototroph

An organism that derives energy for metabolic activities directly from the sun via photosynthesis. By definition, the phototrophs are also autotrophic because they synthesize their main components themselves without depending on other living organisms for reducing power. Most phototrophic bacteria use water as the reducing substrate, but some species may be lithotrophic or organotrophic. *See also* AUTOTROPH; PHOTOLITHOTROPH; PHOTOORGANOTROPH; PHOTOSYNTHESIS.

phytoplankton

Photosynthetic microorganisms, such as the cyanobacteria and unicellular algae, including diatoms, dinoflagellates, and some green algae found on the surface and upper 100 meters of oceans, lakes, and rivers. These organisms may be present singly, in chains, and in clumps, the latter sometimes visible to the naked eye. Phytoplankton need to be close to the surface of water to harvest energy from the sunlight. They are among the primary participants in the earth's carbon, oxygen, and nitrogen cycles, besides representing the lowest stage of the food web in marine and other aquatic ecosystems. *See also* ALGAE; BLUE-GREEN ALGAE; CYANOBACTERIA; DIATOM.

picornaviruses

These tiny viruses—whose average size is about 25 nm—are the underlying cause of such debilitating diseases as poliomyelitis and hepatitis A. The family is composed of several non-enveloped animal viruses with RNA genomes, as is reflected in their name—*pico* (for small) and *rna* viruses. Various members of the family are classified into five main genera on the basis of host range and pre-ferred sites or tissues of infection. The five groups are: the enteroviruses, which infect mostly humans and have a predilection for the gut (the best known example is the poliovirus); the rhinoviruses, which tend to inhabit the nasal passages and respiratory tract; the hepatovirus (hepatitis A), which has a tendency to infect the cells of the liver; the cardiovirus, which affects the heart; and the aphthovirus, which causes disease in livestock. Individual viruses may be distinguished on the basis of antigenic character, receptor sites on the host cells, cultural characteristics, and pathogenicity in the host organisms.

The antigenic properties of these viruses are provided by four different capsid proteins, all of which are encoded within the viral genome. In addition to the capsid proteinsthe genome it also encodes a polymerase enzyme and another small peptide.

Replication proceeds exclusively in the cytoplasm of the host cell. The virus gains entry into the cell by attaching itself to specific receptors on the cell surface. The m-RNA genome acts directly as the messenger for the synthesis of proteins. This molecule is first translated into a single large molecule containing known protein-cleaving (proteolytic) sequences. These sequences then act on the precursor protein to cut it at specific locations, thereby generating the different viral components. In addition, the genome generates at least one negative-sense RNA molecule, which serves as the template for new copies of the viral genome. Virus particles are assembled by the association of different protein components within the host cell and are released upon lysis of the cell. *See also* APHTHOVIRUS; CARDIOVIRUS; ENTEROVIRUS; FOOT AND MOUTH DISEASE; HEPATITIS A; HEPATITIS A VIRUS; POLIOMYELITIS; POLIOVIRUS; RHINOVIRUS.

piedra

A superficial fungal infection of the hair shafts. Depending on the cause and appearance, there are two main types of this disease. White piedra is caused by *Trichosporon beigelii*, which grows inward into the hair shaft and forms soft, easily detachable, white or tan nodules at irregular intervals. The formation of nodules may cause the hair to break. Often, several nodules clump together at various locations to form a large mass around the hair. White piedra may be seen in different parts of the body where hair growth is abundant, including the scalp, beard, genital, and axillary areas. Shaving and cutting the infected hairs, as well as topical fungicides, may be used to manage the spread of the infection. This disease occurs mainly in tropical and sometimes in temperate climates.

The fungus *Piedraia hortae* causes the second type of piedra, which is called black piedra. Infection in this case starts under the hair cuticle and results in the formation of hard, black, spore-containing nodules on the distal end of the hair shaft. Unlike white piedra, the nodules of black piedra are tightly attached to the hair shaft, which they envelope completely with their hyphae. Consequently, the nodules do not disengage easily and are best removed by cutting off the hair. The disease is seen most commonly in tropical South America, the Pacific Islands, and parts of eastern Asia.

pilus

Extracellular tubular proteinaceous structures present specifically on the surface of gram-negative organisms—e.g., *E. coli*—that has the ability to undergo conjugation. Usually, only one or a few pili (the plural of pilus) are present in an organism. These structures vary in shape according to the environmental conditions under which conjugation takes place; thus, they may be long, thin, and flexible, or short and rigid. They form the physical passageway for the transfer of genes from one organism to another during conjugation. Genes for pili are typically encoded in bacterial plasmids. *See also* CONJUGATION.

pinta

A nonulcerative skin infection caused by the spirochete *Treponema carateum*. The disease is endemic in certain tropical regions, including the Philippines and other Pacific islands, and in Central and South America, where it seems to affect only dark-skinned people. Infections may be transmitted by direct contact or by the *Hippelates* fly. *See also* TREPONEMA.

pityriasis versicolor

Chronic yeast infection of the skin, with the occasional involvement of the hair follicle, caused by *Malassezia furfur* or *Pityrosporon orbiculare*. The infection, also known as tinea versicolor, is seen most frequently in hot and humid regions of the world. It is characterized by the development of scaly, spreading, and often itchy, red to yellow-brown lesions on the smooth skin of the face, arms, and abdomen. If exposed to ultraviolet light, these lesions show a pale yellow fluorescence. A possible reason for the skin discoloration is the toxic effects of the dicarboxylic acids produced by the yeast on the melanocytes, the skin cells responsible for pigmentation. Excessive perspiration, pregnancy, malnutrition, and poor general hygiene are predisposing factors. One of the most effective treatments for this condition appears to be washing the infected areas with compounds such as zinc pyrithione or selenium disulfide—the active ingredients in antidandruff shampoos.

plague

Acute infectious zoonotic disease caused by the bacterial species *Yersinia pestis*. The primary hosts or reservoirs for this organism are rats and rodents, and the principal vector involved in transmitting the disease to humans is the rat flea. In addition to fleas, the bacteria may also be passed along from one human to another via droplet infection.

An incubation period of 1–5 days following the flea bite is normal, during which time the bacteria proliferate at the site of entry and migrate to the regional lymph nodes. During this phase, the patient may experience vomiting and high fever in response to the release of the bacterial endotoxins. The multiplication of the bacilli in the lymph nodes—most often in the area of the groin, but also in the regions of the armpit or neck—

results in the inflammation and enlargement of these nodes to form painful, tender, and often suppurating buboes, which give this form of the disease the name "bubonic plague." Following this, the bacteria migrate from the lymph nodes to the bloodstream, which causes a widespread septicemia that disseminates the organisms to various organs and tissues throughout the body. A flea feeding on humans during this phase is certain to be infected. One of the most serious consequences of septicemia is the invasion of the lungs, which results in the development of the pneumonic plague, in which the multiplication of the bacteria results in massive hemorrhages in the lungs and the release of the toxins into the bloodstream. This form of the plague is also characterized by other symptoms of pneumonia, including coughing and respiratory tract inflammation, leading to the release of a large number of bacteria into the immediate environment. Thus, the human version of plague is at its most contagious for other humans during the pulmonary phase.

The principal means of diagnosis of the plague is the isolation and identification of *Y. pestis* in blood, sputum, or aspirates from the buboes, depending on the nature of the disease. The bacteria may be identified on the basis of their characteristic bipolar staining reaction, or by fluorescent antibody staining techniques. Due to the seriousness of the disease and severity of the symptoms, immediate treatment is imperative. Broad-spectrum antibiotics such as streptomycin, chloramphenicol, and tetracycline are effective against these bacteria. The immunity in patients who have recovered from plague infections is generally high, and the disease does not usually occur more than once in the same individual. One of the main objectives in preventive strategies against the plague is to minimize the opportunity for contact between the bacteria and the humans. In case of a zoonotic, infection this translates into controlling the animal reservoirs and vectors of the disease. Because eradicating rats is not a practical solution in terms of either cost or even feasibility, most measures are aimed at eliminating or reducing the flea population. For instance,

the fumigation of ships that travel over long distances and frequently carry rats has proven effective in limiting the transmission of the disease. Public health officials also advise minimal and careful contact with wild rodents, as well as periodic surveys of rat/rodent populations to nip any potential epidemics in the bud. Strict isolation of patients is important, particularly in the case of pneumonic plague. Despite the long-term immunity induced by *Y. pestis*, vaccines against the organism have met with only partial success and do not rank among the routinely administered childhood vaccines. A formalin-inactivated whole-bacillus preparation is used to immunize people in areas where the disease is endemic or enzootic, or among laboratory workers who handle these organisms regularly.

The plague holds a prominent place in the history of human disease. It should be noted that the term "plague" was used nonspecifically in the past to describe virtually any epidemic/pandemic that killed people in massive numbers, due to which there may be some confusion in making a positive identification of the diseases that devastated populations in bygone "plague" epidemics. Nevertheless, a correlation between the description of symptoms and the spread of the infection with our modern knowledge of the disease and its epidemiological patterns allows us to make educated guesses about the identity of various historical "epidemics." According to medical historian Erwin Ackerknecht, the first unmistakable reference to this disease may be found in Greek texts dated around the third century BC. The infamous Black Death in 1348 and the Plague in London in 1665 are perhaps the best known historical epidemics of the plague to have occurred in Europe. Despite modern advances in medical technology and pest control, the disease continues to pose a threat to humans in areas with high wild rodent populations, including the western half of the United States, South America, eastern and southern Africa, and Southeast Asia, although no pandemics have occurred since the beginning of the twentieth century. *See also* Appendix 1. *See also* YERSINIA PESTIS.

plaque

A physical manifestation of the infection of bacterial cells with lytic or virulent bacteriophages (viruses), analogous to a bacterial colony. When a suspension of bacteria is distributed evenly in a small volume of solidified medium—such as in a petri dish—and allowed to incubate at ambient temperature, the organisms will multiply and form an even, confluent growth of cells throughout the medium rather than individual colonies. When virulent bacteriophages are added to such a confluence, they will infect the bacterial cells—one phage per bacterium—and multiply to form progeny phages, which will then emerge from the host by lysis. The progeny thus formed will proceed to infect neighboring bacterial cells and continue the cycle, so that after a certain amount of time, a clear zone will develop at each of the sites of initial phage infection. These clear zones are called plaques, and the number of plaques formed is more or less equal to the number of phages that were introduced to the bacteria at the outset of the experiment. *See also* BACTERIOPHAGE.

plasma membrane. *See* CELL MEMBRANE

plasmid

Extrachromosomal piece of DNA present in bacteria and capable of independent replication. The most common plasmids are circular, covalently closed molecules although linear plasmids also exist. Usually smaller than the chromosome, a plasmid contains only a few genes, which encode functions that are useful but not essential for the organism. Examples include genes for the production of, or resistance to, specific antibiotics, the ability to metabolize certain substrates, and even structural elements such as gas vacuoles (in strains of *Halobacterium*) and pili. Plasmids may be transferred from one organism to another by the process of conjugation.

Being smaller and hence more manageable, these pieces of DNA are useful tools in genetic engineering and recombinant DNA technology. *See also* BACTERIA; CONJUGATION; DNA; PROKARYOTE.

Plasmodium

Genus of protozoan parasites that are the causative agent of malaria in humans. Four distinct species of this organism are associated with similar yet distinct forms of the disease.

Plasmodium parasites have a complex, two-host life cycle that includes both sexual and asexual phases. The definitive host, which harbors the diploid phases of the parasite, is the female *Anopheles* mosquito. Haploid male and female gametocytes present in the red blood corpuscles—known as the micro- or macro-gametocytes, according to their respective sizes—are taken up by the mosquito's gut when it takes a blood meal. Once in the gut, these forms mature to form gametes, which break out of the ingested blood cells and invade the stomach lining of the mosquito. The maturation of the male cell involves a process of exflagellation, resulting in a motile gamete, which fertilizes the female macrogamete to form a diploid zygote or ookinete. This form then moves outside the stomach wall and forms an oocyst. Within this structure, there is a single cycle of meiosis to form haploid structures called the sporozoites, which undergo further cycles of simple mitosis so that an oocyst, now called a sporocyst, becomes filled with large numbers of sporozoites. The rupture of the oocyst releases these sporozoites into the body of the mosquito, where they spread to various locations, including the salivary glands. A human being is infected by the entry of the sporozoites into the site of the mosquito bite when the insect takes its next blood meal. Fully infective sporozoites typically appear in the mosquito's salivary glands some 2–3 weeks after it first ingests the gametocytes.

The human portion of the *Plasmodium* life cycle is divided into distinct phases. First there is a pre-erythrocytic cycle beginning within

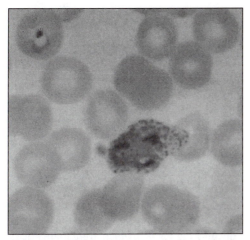

Micrograph of *Plasmodium vivax,* inmature schizont, in blood smear. *CDC.*

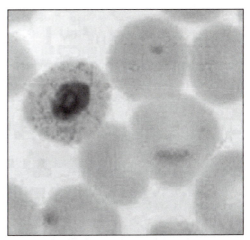

Micrograph of *Plasmodium ovale,* including a growing trophozoite with "ring" nucleus. *CDC/ Dr. Melvin.*

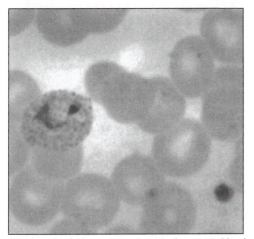

Micrograph of *Plasmodium vivax* trophozoite in blood smear. *CDC.*

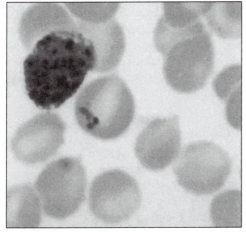

Micrograph of *Plasmodium vivax,* mature schizont, 24 nerozoites. *CDC/ Dr. Melvin.*

an hour of infection, during which the sporozoites infect the liver cells (one sporozoite per cell), acquire rounded shapes, and multiply by a process of schizogony to form cells called merozoites. The second stage in the malarial infection is the erythrocytic cycle, which is initiated by the invasion of the red blood cells as well their precursors (reticulocytes) by the merozoites. Once inside the blood cells, the parasites assume an amoeboid, vacuolated, often ring-shaped form with a single nucleus called the trophozoite. As the trophozoite grows, it feeds on the hemoglobin and the metabolic breakdown products, and excess proteins that are left behind combine to form the characteristic malarial pigment. Eventually, the trophozoite also undergoes schizogony to give rise to a multinucleated intermediate called a schizont, which eventually gives rise to new merozoites—usually a fixed number, depending on the species—which are released into the bloodstream. Once released, the merozoites may infect new blood cells and initiate yet another cycle.

The final stage of *Plasmodium* infection in the human is called gametogony and involves the generation of the gametocytes, the precursors of the sexual forms. It takes place after multiple cycles of trophozoite schizogony, but the exact mechanism governing this switch to gametogony is unknown. Only one of the two sexes is formed in a single erythrocyte. Gametogony is a terminal event in the human. That is to say that sexual reproduction may only take place in the mosquito, and if the gametocytes are not taken up, the life cycle of the parasite comes to an end. *See also* MALARIA; SCHIZOGONY.

Plasmodium falciparum

This is the most deadly of all malarial parasites because of the ability of its schizonts to induce biochemical and physiological changes in the host's erythrocytes, due to which the cells develop a "knobby" surface and become prone to agglutination and lysis. The identification of these knobby erythrocytes is, in fact, a diagnostic indicator of *falciparum* infection, which is also termed malignant malaria. In addition to blood cell damage, falciparum infections are marked by the occurrence of schizogony in the blood vessels of the brain, internal organs, and placenta (in pregnant mothers), which causes the plugging of blood vessels and hence damage to the organ due to reduced oxygen and ruptured blood vessels. Brain involvement leads to cerebral malaria, which in turn may lead to coma and death if not treated in a timely fashion. *P. falciparum* infections exhibit cyclic paroxysms every 36–48 hours. These parasites may be differentiated from the others on the basis of their crescent-shaped gametocytes and small multiple trophozoites in the blood cell. In terms of its epidemiology, *P. falciparum* is generally confined to the tropics and accounts for about 15 percent of the total number of malaria infections worldwide.

Plasmodium malariae

Agent responsible for quartan malaria, a condition named for the occurrence of fever paroxysms at intervals of 72 hours (i.e., on every fourth day). This parasite is relatively rare, shows no particular geographic preference, and seems to occur sporadically all over the world. Of diagnostic value are the observations that *P. malariae* merozoites can invade only mature red blood cells and that their trophozoites may be seen to form a band across the blood cell during the early stages of schizogony. Infection with this species is often associated with one or a series of recrudescent episodes due to a low-grade parasitemia, which may result in the development of symptoms due to a weakening of the immune system.

Plasmodium ovale

The clinical picture of ovale malaria, including its periodicity of 48 hours, is similar to that of the classical *P. vivax* disease, but symptoms are much milder. The parasite appears to be confined to regions of West Africa. Both this and the vivax species are capable of establishing secondary exoerythrocytic cycles of infection due to invasion of the liver cells by merozoites. The parasitic form, called a hypnozoite, may lie quiescent (dormant) for years together and cause relapses some stimulated (by factors unknown) to release the parasites, which then invade the blood cells and cause a relapse of the original symptoms. The drug primaquine is used to kill the liver-infecting phase of these organisms.

Plasmodium vivax

This organism is responsible for the vast majority—nearly 80 percent— of malaria cases worldwide and is found in most parts of the world where malaria is known to occur. Like the *P. ovale* species mentioned above, these organisms have an exoerythrocytic cycle and form dormant hypnozoites that may cause relapses years after the initial infection. A characteristic feature of this species is its preference for reticulocytes rather than mature erythrocytes. Thus, diagnosis is aided by counter-staining blood smears for these cells.

Plesiomonas shigelloides

Gram-negative, rod-shaped bacterium that forms part of a large group of enteric organisms known to cause nonspecific intestinal and extra-intestinal infections. Although this organism is currently classified in the same family as *Vibrio* and *Aeromonas*, there is genetic evidence to suggest it may be more closely related to the genus *Proteus*. It is a facultative anaerobe, fermentative, non-spore-forming, and motile, with tufts of flagella at its poles. Most strains of *Plesiomonas* are susceptible to the antibiotic ampicillin.

Pneumocystis carinii

First described in 1909 as a co-infectious agent of guinea pigs along with trypanosomes, this organism gained prominence as a pathogen after the advent of AIDS, where it figures as the most common complication. Although it has long been considered as a protozoan parasite, recent genetic analysis of its mitochondrial DNA and ribosomal RNA indicates that, *Pneumocystis* is more closely related to the fungi. However, the life cycle of this organism, the morphology of different stages, and its requirement for animal hosts, make it more similar to protozoa.

The life cycle of *P. carinii* consists of two main forms analogous to the cysts and trophozoites of the protozoa. Both forms may be recovered from the alveolar spaces in the lungs of infected patients and are generally found outside the host cells, although some studies suggest that the parasite may also invade the cells. *P. carinii* is seen to cause an interstitial plasma cell pneumonia in premature, immunosuppressed, or malnourished infants. It also causes pneumonia in adults, although the clinical picture presented differs depending on whether or not the individual is immuno-compromised (either due to AIDS or other reasons). The organism causes mild to subclinical infections in people with intact immune systems. The patient shows signs of coughing and shortness of breath but little or no fever, and at the cellular level, barely any change in leukocyte count. Infection has a much more dramatic impact on immunosuppressed individuals, with death as the final outcome in a large proportion of the cases. Patients suffer from violent coughing and breathing difficulties that gradually lead to cyanosis—the development of a bluish color due to the clogging of the interstitial spaces with the lungs. Diagnosis is by the observation of the causative agent in clinical samples (bronchial brushings or lung biopsies); cultivation of these organisms in the laboratory has not been achieved with any measure of consistency. Samples may be stained with a dye called toluidine blue or with Giemsa stains to visualize the active fungal cells. Several drugs are under evaluation for treating *Pneumocystis* infections, but for treatment to have any real benefit, it is important to treat the underlying immunodeficiency.

poliomyelitis

Disease caused by infection with poliovirus. In its mildest form, the infection—initiated by the entry of the virus into the body via the oral-fecal route—may pass off as a gastrointestinal and sometimes respiratory disorder. The most feared and best known guise of this disease, however, is the paralytic form, which results from the spread of the virus from the gastrointestinal tract and upper respiratory organs (e.g., throat and tonsils) to the central nervous system. This spread typically occurs via the blood, i.e., the virus infects the lymph nodes of the intestines, and from there gains entry into the blood stream, causing a general viremia that exposes practically all the tissues of the body to the virus. Because the poliovirus is neurotopic—with a special preference for the spinal cord and brain—and typically does not even infect the peripheral nerves, the infection is often self-limiting at this stage. However, if the virus gains access to the central nervous system, it will multiply and gradually destroy the nervous tissue, which leads to paralysis of various parts of the body. The final outcome of this infection depends on the part of the body that is paralyzed—a permanent paralysis of the muscles of the respiratory organs or heart, for instance, will cause death, while a patient is more likely to survive paralyses of the limbs. The lack of nervous stimulation to these areas, however, will often lead to muscular atrophy, the loss of use, and sometimes even deformities. *See also* POLIOVIRUS; SABIN, ALBERT; SALK, JONAS.

poliovirus

Single-stranded RNA virus that causes poliomyelitis. The poliovirus is classified with the enteroviruses in the family Picornaviridae. Poliovirus is the most virulent member of this group and may be distinguished from related viruses by its special predilection for the spinal cord and brain tissue (i.e., it is neurotopic). *See also* PICORNAVIRUS; POLIOVIRUS.

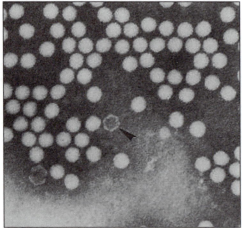

Micrograph of poliovirus type 1. *CDC/Dr. Joseph J. Esposito.*

polymyxin

Antibiotic peptides with detergent action produced by *Bacillus polymyxa*. These peptides act by damaging the outer membrane of gram-negative bacteria and increasing permeability of the cell, which eventually leads to lysis. Their use in humans is somewhat limited because they may cause damage in the kidneys and nerve-muscle junctions. *See also BACILLUS POLYMYXA*; OUTER MEMBRANE.

polyoma virus

These papovaviruses, first isolated from mice in 1953, were so named because of their ability to cause solid-tissue tumors at multiple sites. Their small genomes (in the order of 5 kb), oncogenic abilities, and easy cultivation in tissue cultures have made the polyoma viruses one of the most extensively studied groups of animal viruses, particularly as mod-els for DNA replication. These viruses show a varied host range and infect large numbers of mammals.

The two human polyoma viruses are the BK virus and the JC virus, both discovered in 1971. The BK virus, first isolated from an immunosuppressed patient with a kidney transplant, has been found associated with mild respiratory illnesses in children. Oncogenicity, although suspected, has never been proved. The JC virus was found in a patient with a central nervous system illness and has been implicated in progressive multifocal leukoencephalopathy (PML), particularly in the elderly and immuno-compromised individuals such as AIDS and transplant patients. The simian vacuolating virus, otherwise known as SV40, which was first isolated from primary cell cultures of monkey kidney cells, is also a member of this group. This virus is used as the primary vector for introducing DNA in experimental eukaryotic systems.

The polyoma viruses may exhibit either lytic or abortive cycles of cellular infection, depending on the type of host cell that is infected. Virus replication is far more straightforward than that of the papillomaviruses and takes place entirely in the nucleus. The lytic cycle takes 48–72 hours for completion. *See also* PAPOVAVIRUS; PROGRESSIVE MULTIFOCAL LEUKOENCEPHALOPATHY.

postherpetic neuralgia (PHN)

This condition is one of the major complications of shingles or herpes zoster. Characterized by the persistence of severe pain long after the shingles lesions have crusted and are no longer infectious, postherpetic neuralgia (PHN) is believed to be the outcome of nerve damage by the virus. In addition to pain, the affected areas are also extremely sensitive to heat and cold. The best method of avoiding PHN is the prompt diagnosis and treatment of shingles. Analgesic and steroid treatment may be used to control the chronic pain, but their use is still somewhat controversial. *See also* HERPESVIRUS; SHINGLES.

poxvirus

Named for the characteristic pocks formed as a result of infection (as seen, for example, in smallpox), the poxviruses are a family of large, enveloped, animal viruses with double-stranded DNA genomes. They are the largest and most complex of all known viruses—ovoid or brick-shaped particles that range from 140 to 260 nm in diameter and about 220 to 450 nm in length. These viruses possess a double membrane, both of which consist of lipids—derived from the host's cell membrane—and proteins, which impart antigenicity to these particles. The nucleocapsid core is a dumbbell-shaped structure containing the genome and lateral bodies of unknown function. The genome is a single piece of DNA varying in length from 135 to 350 base pairs, folded over in hairpin loops at both ends such that there are no free ends. Encoded in the genome are genes for various structural components as well as all the enzymes needed for synthesis and replication, including a DNA-dependent RNA polymerase, which most other viruses do not synthesize. Viral replication occurs in the cytoplasm, with gene expression occurring in early and late phases—before and after genome replication—with the formation of large, readily observable inclusion bodies. Because these viruses encode all the enzymes and components required for replication, there is minimal participation by the host cell except in the maturation phase.

A large number of poxviruses exhibit a wide range of hosts, including birds, cows, and monkeys. Only viruses of the genus orthopoxvirus cause human infections. While humans serve as the natural reservoir for smallpox viruses (also called variola virus), the vaccinia virus—which causes a milder human disease—is typically harbored in buffaloes. Other poxviruses of human relevance include the cowpox and monkeypox viruses. *See also* COWPOX; MONKEYPOX; SMALLPOX; VARIOLA; VACCINIA.

prion

Disease agent composed exclusively of protein material without any detectable nucleic acid, implicated in a number of neurological diseases of both humans and other mammals. The credit for the discovery and description of prions—which remain a controversial topic in the scientific community—goes to Stanley Prusiner, who received the 1997 Nobel Prize for his radical idea that an infectious and apparently reproducible entity could contain only protein and no nucleic acid. However, the evidence for Prusiner's claims has steadily accumulated both in his own and other laboratories since he first proposed his ideas during the mid-1980s.

Skepticism with the idea of prions is understandable in light of the fact that their behavior appears to run contrary to every other known disease agent. In addition to being composed solely of protein, prions display a dramatic dual behavior. Various diseases caused by prions—collectively referred to as spongiform encephalopathies due to the sponge-like appearance of the brains of affected victims (both animal and human), arising from the formation of vacuoles in the cortex and cerebellum—have been shown to be either infectious (communicable) or inherited. This picture is further complicated by observations that certain diseases break out sporadically, with no genetic or infectious origins. Well-known examples of prion diseases include the infamous "mad-cow disease," scrapie (a disease of sheep), and Creutzfeld-Jacob disease in humans. Still another unusual characteristic of prions is their apparent mode of multiplication; they appear to increase in numbers within the host cells simply by inducing normal host proteins to change their three-dimensional shape and transform into harmful molecules.

When Prusiner began his research in the 1970s, various spongiform encephalopathies were known to be transmissible in the laboratory by injecting brain tissue extracts of diseased individuals into healthy animals. The culprit was believed to be a slow-acting virus, although no one had been able to isolate a virus from clinical samples. There was, however, a single report from the laboratory of Tikvah Alper in London suggesting that the causative agent of scrapie lacked nucleic acid material because standard techniques for destroying those chemicals failed to inactivate the infectious ability of brain extracts. Following up on these findings, Prusiner's laboratory embarked on an attempt to isolate, purify, and identify the scrapie agent, and concluded that not only did it lack nucleic acids, but that infectivity was greatly reduced when the material was subjected to procedures that denatured proteins. In 1982, Prusiner published a paper describing his results in which he coined the term "prion" for "proteinaceous infectious particle," to describe the new agent. His findings sparked a hot debate that to this day has not been completely resolved within the scientific community.

The exact mechanisms by which the prions wreak their havoc are still being worked out. It is known that the diseases are mediated by a modified version of a normal cellular protein designated prion-protein (PrPc). Diseased individuals contain this same protein, but folded in a different shape—this alternate form is called PrPsc (for prion protein of scrapie). The modification in PrPsc may be introduced by a point mutation in the PrPc gene—thus making the disease inheritable from one generation to the next. However, as mentioned earlier, a single PrPsc, by itself, without help from the gene that generated it, appears to be capable of inducing normal proteins to refold themselves and become the prion. Thus the disease becomes infectious. Prions are typically transmitted from one individual to another by consumption of some infected product. Meanwhile, the sporadic occurrence of a prion disease is attributed to as yet unidentified triggers that induce conformational changes in the PrPc. How the PrPsc itself does damage is also not yet understood.

In addition to mad cow disease and kuru, prions have been implicated in some other rare diseases, including Alper's syndrome, a fatal disease of infants where the prions attack the white matter of the brain, and fatal familial insomnia, a neurodegenerative disease that appears to be heritable rather than infectious. As the idea of prions is gradually coming to be accepted in the scientific community, scientists are also beginning to consider them as a model for attempting to understand the mechanisms underlying even non-prion diseases such as Alzheimer's disease. *See also* CREUTSFELD-JACOB DISEASE; FATAL FAMILIAL INSOMNIA; GAJDUSEK, CARLETON; KURU; MAD COW DISEASE; PRUSINER, STANLEY; SCRAPIE; SPONGIFORM ENCEPHALITIS.

probiotic

Preparation of viable lactic acid bacteria, cultured dairy products, or food supplements containing viable lactic acid bacteria believed to improve the integrity of the intestinal microflora of humans and animals. Ideally, a good probiotic organism should have the following attributes: it should have originated from the natural flora of the host animal in which it is administered, it should be resistant to gastric acid and bile to survive passage through the stomach, and it should have the ability to adhere to and colonize the intestinal mucosa of the host animal. An example of a popular probiotic in humans is *Lactobacillus acidophilus*. This organism has been shown to have beneficial health effects such as reducing gastrointestinal symptoms in lactose-intolerant individuals, improving overall digestion, and discouraging the growth of enteropathogenic organisms such as *Salmonella* and *Helicobacter* species. It may be worth noting that although the term is relatively new, the practice of ending meals with natural foods such as yogurt and cheese (containing many of the constituents of modern probiotics) as digestive aids has been around for centuries among different societies all over the world. *See also* BUTTERMILK; INDIGENOUS MICROFLORA; YOGURT.

prognosis

A forecast (usually made by a physician) regarding the probable outcome of an episode of disease based on the current signs and symptoms, as well as other factors such as the age, immune status, living conditions, and psychological well-being of the affected individual.

progressive multifocal leukoencephalopathy

This is a central nervous system disease caused by infection with a human polyomavirus called the JC virus. The disease is characterized by the demyelination of the nervous tissue in the brain and spinal cord, which leads to progressive loss of various cognitive as well as motor functions. The disease is almost always seen to occur in severely immunosuppressed individuals such as AIDS patients, organ transplant recipients under immunosuppressive therapy, and people with lymphatic disorders. With the rise and spread of AIDS, the incidence of PML, once a rare occurrence, has risen considerably. *See also* PAPOVAVIRUS; POLYOMA VIRUS.

prokaryote

A living organism characterized by the lack of any membrane-bound compartment, such as a nucleus, within its cells. The structure of a prokaryotic cell is thus very simple and consists of different macromolecules required for performing various living processes enclosed within a membrane. The nucleic acid material of these organisms lies free or naked in the cell, usually in the form of a single continuous circular DNA molecule, with no discernible beginning or end. All prokaryotes are single-celled organisms. Examples include bacteria, rickettsiae, and mycolasmas. *See also* EUKARYOTE; PROTISTA.

Propionibacterium

The fermentative action of these bacteria is the reason for the distinctive flavors that we enjoy in Swiss cheeses such as Gruyère and Emmentaler. The flavors are mainly attributable to propionic acid, which is the major end product of sugar metabolism in various species of *Propionibacterium*. Different species of this gram-positive, non spore-forming organism may be found occurring naturally in a variety of dairy products. The most common species is *P. freudenreichii*. A related species called *P. shermanii* is used in the industrial manufacture of vitamin B_{12}. *See also* CHEESE.

protein

Molecule with which living organisms perform the bulk of the tasks required for living. There are many different types of proteins for different roles in a cell. Some proteins function as enzymes that catalyze a variety of metabolic and synthetic reactions; some relay messages between a cell and its surroundings, and others act as vehicles for transport; still others make up various structural elements.

Chemically, proteins are complex macromolecules. They are made up of chains of amino acids in specific sequence linked by peptide bonds, folded into three-dimensional shapes according to specific interactions between different parts of the molecule. The sequence and shape of a protein plays an integral role in its functions and properties. *See also* AMINO ACID.

Protein A

Cell wall component present in most strains of *Staphylococcus aureus*, which plays a dual role in interacting with the body's immune system. On the one hand, Protein A, like many other bacterial proteins, is recognized as foreign and is treated accordingly by the immune system—i.e., it elicits the formation of specific antibodies. But, in addition to this specific binding, certain portions of this protein can recognize and bind to the Fc (nonspecific) portion of antibody molecules of the IgG type. This binding capacity enables Protein A to induce the alternative complement pathway and thus act as a mediator of hypersensitive reactions or allergies. *See also* STAPHYLOCOCCUS AUREUS.

Protein M

Surface antigen found in the cell wall of Group A *Streptococcus pyogenes*, considered to be one of the major virulence factors produced by these organisms. One of its main functions is to protect the bacteria from phagocytosis by the leukocytes, although the mechanisms by which it does so are not clear. A second method by which Protein M appears to enhance bacterial virulence is to promote the colonization of the organisms at certain sites such as the cells of the tonsils. This protein is extremely variable antigenically, and there are several variants, which form the basis for the Lancefield system of classification of the different strains or serotypes of *S. pyogenes*. A specific Protein M serotype is found to correlate strongly with the type of streptococcal disease caused. *See also* LANCEFIELD, REBECCA; *STREPTOCOCCUS PYOGENES*.

Proteus

This bacterial genus was named after Proteus, the shape-changing god of Greek mythology, because of the changing appearance of the bacterial colonies when plated cultures are incubated over a period of days. This property is a reflection of the swarming motility exhibited by these bacteria on solid surfaces, which results in the appearance of confluent growth rather than discrete colonies. *Proteus* belongs to the family Enterobacteriaceae of gram-negative bacteria. Like other members of this family, these bacteria may be found residing in the intestinal tracts of humans and various other mammals, but they are also found almost as frequently in soil and polluted water. Biochemically, they are facultative anaerobes that cannot ferment lactose but produce the gas hydrogen sulfide as a product of metabolism. A distinguishing characteristic is their ability to produce the enzyme urease. *P. mirabilis* is the most important human pathogen and is often associated with wound and urinary tract infections. These infections, once diagnosed, are relatively easy to treat because the organism is susceptible to a wide range of antibiotics. The closely related species *P. vulgaris* (also the representative species of this genus) is an opportunistic pathogen that typically affects only immune-suppressed individuals. These infections are more problematic because *P. vulgaris* is resistant to most antibiotics. The two species are differentiated on the basis of the ability of *P. vulgaris* to produce indole (see IMViC tests). *P. myxofaciens* is yet another member of this genus; it is not a human pathogen but causes a disease in the larvae of gypsy moths.

Protista

Kingdom of living organisms that comprises eukaryotic, unicellular organisms, such as the protozoa, the unicellular algae, and certain groups of fungi. *See also* PROKARYOTE; TAXONOMY.

prototroph

A term for wild-type bacterium capable of growth in ordinary nutrient medium in the laboratory, used in the context of producing specific auxotrophs by mutation. *See also* AUXOTROPH; MUTANT.

protozoa

Group of single-celled, eukaryotic, heterotrophic organisms that require moisture for survival and—like higher animals—are capable of movement from one place to another. If fungi and algae are considered as the progenitors for plants, then the protozoa represent the animal arm of the protist kingdom. Their name is derived from the Greek words for "first animal." Like other protists, they live in a haploid state for most of their life cycle.

Different protozoa are either free-living in nature or exist as parasites in other animals. Parasitism in this case is not limited to intracellular habitats but may take the form of a commensal or mutualistic relationship between host and protozoan. The outcome may be either beneficial or harmful to the host. For instance, the gut of many different animals such as cattle and sheep, and even certain insects, provides a habitat for parasites that return the favor by producing enzymes capable of degrading hard-to-digest materials such as cellulose.

Protozoa share the characteristics of a typical animal cell in that they lack a cell wall and the photosynthetic pigments and machinery. The shape and size of different organisms varies widely depending on habitat, mode of locomotion and nutrition, and methods of reproduction. Some organisms have a very regular shape, while others, such as the amoebae, lack rigidity and therefore have no definite shape. Organisms may be motile due to cilia, flagella, or pseudopodia (false feet). They derive nutrition either by absorbing soluble substances via the cell membrane or by engulfing particles (including other microbes) and subsequently digesting their contents with various enzymes. Many protozoan species have complex life cycles and possesses more than one distinctive morphological form.

Both asexual and sexual modes of replication may occur in these organisms. The former may be a simple mitosis or fission—with the equal distribution of materials occurring before the parent cell splits into two—or a type of multiple fission called schizogony, where a single cell gives rise to several offspring. Sexual reproduction can also occur in more than one way, involving either a simple transfer of genetic material from one organism to another, or the fusion of haploid forms to diploid forms, which then undergo meiosis to produce new haploid cells. Often times, both sexual and asexual reproductive modes may be integrated into a single, complex life cycle.

The protozoa are extremely important from a human perspective because they are responsible for a large number of diseases, especially in the tropical regions of the world. Most of these diseases are caused by parasitic organisms that invade the body and wreak havoc with its normal functions. Transmission to the human host occurs via two main mechanisms—directly, by the ingestion with food or water or the inhalation of aerosols, or indirectly, through an arthropod vector. *See also* CYST; TROPHOZOITE.

Prusiner, Stanley, (1942–)

Scientist who isolated the causative agent of spongiform encephalopathic diseases such as kuru, scrapie and the notorious mad-cow disease. Based on his experiments, Prusiner proposed that these diseases—long thought to be caused by slow acting viruses—were actually caused and transmitted by a new type of agent composed of just protein without any nucleic acid. He called these agents prions. His controversial proposal won him a Nobel Prize in 1997. *See also* KURU; MAD COW DISEASE; PRION; SCRAPEI; SPONGIFORM ENCEPHALOPATHY.

pseudomonad

Term given to a heterogeneous group of gram-negative, aerobic, non-spore-forming bacteria that are widely distributed in nature. The name is derived from the fact that most, if not all members of the group were at one time identified as or classified with the genus *Pseudomonas*. With the advent of molecular techniques, scientists have refined the classification of this group of bacteria and renamed or regrouped various organisms into several different genera and species. *See also* PSEUDOMONAS.

Pseudomonas

Heterogeneous genus comprising gram-negative, aerobic, motile, rod-shaped bacteria. The GC ratio of the genus ranges from 57–70 percent, reflecting the immense heterogeneity within the group. Most species of *Pseudomonas* are strictly aerobic, although a few species may withstand anaerobic conditions in the presence of nitrate. They exhibit motility due to one or more polar flagella and also contain fimbriae for attachment to surfaces and other organisms. Most species consist of an external slime layer or capsule but do not form spores. These organisms are widespread in nature. They thrive best at room temperature but can exist in many different environments because of their large metabolic repertoire. With one or two notable exceptions, *Pseudomonas* is not a serious human pathogen and has only rarely been associated with disease. In the past few decades, however, different species have increasingly come to be implicated in various nosocomial infections. Many more species are known to act as plant pathogens. *See also* NOSOCOMIAL INFECTION; PSEUDOMONAD.

Pseudomonas aeruginosa

This organism is the best known and possibly the most virulent of all the pathogenic pseudomonads. According to some scientists, it is one of the most abundant life forms on the planet. It is the representative species of a subgroup of *Pseudomonas* species that are identified by their ability to produce fluorescent pigments called pyoverdins. *P. aeruginosa* is distinguished from all other members of this group because it also produces a second pigment called pyocyanin, which imparts a greenish color to the organisms. This color is evident not only in bacterial colonies grown on simple nutrient media in the laboratory, but also in individuals infected with this organism. The pigments also impart a characteristic fruity, grape-like odor to infected material. Certain strains of this species may also produce a mucoid extracellular polysaccharide material in addition to their capsules.

Despite its widespread distribution, *P. aeruginosa* infections are relatively rare, a fortunate circumstance because these infections are also extremely difficult to treat due to a high degree and spectrum of antibiotic resistance. In the past, this organism was mostly known and feared as an infective agent of burn wounds and surgical wounds. It is also a frequent cause of bacterial corneal ulcers. In recent times, the mucoid strains of this organism have emerged as the predominant bacterial species isolated from the lungs of cystic fibrosis patients. Diagnosis is confirmed by the isolation of the bacteria from infected wounds or from lung exudates. Telltale signs of *P. aeruginosa* infections include a greenish coloration of the infected areas and the distinctive odor. As mentioned earlier, the development of resistance to a wide range of antibiotics hinders treatment, and the most successful therapies combine antibiotics (after pretesting for susceptibility) with boosters of antibodies to the extracellular products of these bacteria. *See also* ANTIBIOTIC RESISTANCE.

Pseudomonas stutzeri

This *Pseudomonas* species is one of the active denitrifying bacteria found in the soil. It is not typically associated with human disease, although it has been isolated from various clinical samples. Most likely, the organism contributes to, but is not the primary cause of, the lesions or symptoms, if any. *See also* DENITRIFICATION.

pseudopod

Literally translated from Latin as "false foot," a pseudopod is a protoplasmic extension of certain cells that allows them to move on solid or semisolid surfaces and to engulf food and other particles. Pseudopodia are a common feature of the amebae. Phagocytic cells such as the macrophages in higher animals also use these extensions for movement and engulfment of particles. *See also* AMEBA; MOTILITY.

pure culture

A bacterial culture in which all the organisms are of a single species or strain. These types of cultures are almost never found in nature, but the ability to grow them is important for the study of microbial properties in the laboratories. A pure culture is usually prepared by inoculating a small amount of sterile medium with a single bacterial colony and incubating it for some appropriate time. This technique is based on the principle that a single colony usually arises from a single organism. Pure cultures may also be prepared by adding a lyopholized bacterial sample to sterile medium and incubating it, assuming that the sample was itself a pure culture. It should be noted that even a pure culture may contain a mixed population of bacteria due to spontaneous mutations that can occur over several generations. These differences, however are usually minor. *See also* AXENIC; COLONY.

putrefaction

Decomposition of nitrogenous macromolecules that occurs due to the action of various microbes in the environment. The gradual degradation of these compounds—from proteins and nucleic acids to peptides, amino acids, purines, and pyrimidines—leads to the release of ammonia and other gases, which give the process the foul odor typically associated with decaying meat. *See also* AMMONIFICATION; PROTEIN.

Q

Q fever

Zoonotic disease of humans caused by the rickettsial species *Coxiella burnetii*, first described in 1937 following an outbreak in a slaughterhouse. The disease is characterized by a sudden onset of nonspecific symptoms such as high fever, chills, malaise, headache, muscular weakness, and appetite loss. Symptoms, which typically appear 2–4 weeks after infection, are usually short-lived (1–2 weeks) and often are misdiagnosed as the flu. Differentiation may be necessary if the disease persists and treatment is required, because antibiotic therapies are applicable against these bacteria, although not against flu viruses. Mortality is low, although patients with heart disease are at some risk for contracting *Coxiella* endocarditis, which is often fatal.

Q fever is relatively rare, and incidence is highest in regions of southern Europe, northern Africa, and the Middle East, although it may occur anywhere. *Coxiella* is transmitted from animals such as cattle, goats, sheep, and cats, usually by inhalation of organisms from the air. Disease outbreaks frequently originate in such places as slaughterhouses and dairies, where there are large quantities of contaminated material. *See also* Coxiella burnetii.

quality control

Term used in industries of various sorts for procedures to ascertain and maintain a predetermined level of quality in a product or process. Quality control thus encompasses a variety of measures, including the use and maintenance of proper equipment, inspection of the starting materials, and continued monitoring throughout the manufacturing/processing of the product, and corrective action as required. The impact of microbes on various quality control methods is multifaceted. The production of various microbial products such as antibiotics, alcohol, chemical solvents, and enzymes requires such quality control measures as ensuring the use of proper cultures and correct microbial growth conditions. Conversely, in other arenas, notably the food industries, quality control procedures require the control of microbes that may cause spoilage or spread infections.

quarantine

A public health measure to prevent the spread of a communicable disease, in which a per-

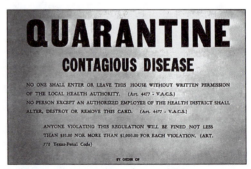

Quarantine sign posted on a house. *CDC.*

son (or animal) is subjected to isolation or to certain restrictions of movement. Often used in an official sense—e.g., for a period of detention of travelers (or their pets) before permitting them to enter a country, or for restriction of the flow of people in buildings known to have patients with highly contagious diseases of epidemic potential—quarantine in some contexts is used synonymously with patient isolation. *See also* PATIENT ISOLATION.

R

rabies

A deadly viral disease of the central nervous system, marked by a progressive and irreversible deterioration of the brain tissue. The most common mode of infection is through the bite of a rabid animal—infected dogs, raccoons, squirrels, and bats—whose salivary glands are laden with the rabies virus. Lab animals have also shown high susceptibility to infections via the aerosol/respiratory route, and there is a strong possibility that humans may be infected this way if the virus concentration is extremely high. There is some evidence suggesting human-to-human transmission via sexual contact, although this is not confirmed. In addition, there have been isolated incidents of rabies infection due to organ transplants—specifically, the cornea. Transmission from mother to fetus has not been reported.

The rabies virus enters the human host, replicates inside the cells at the site of entry, and quickly makes its way to the local peripheral nerve cells. From there it spreads via the nervous tissue to the spinal cord and brain. The time it takes for the typical symptoms of central nervous system damage to manifest depends on the location of the initial infection—6–12 weeks if the person was bitten on the hand or foot, sooner if the bite is closer to the head or spinal cord. Rabies symptoms—which include an initial nonspecific phase lasting from 2–7 days with fever, malaise, nausea, and headaches—only develop after the virus invades the spinal cord and brain. This phase is characterized by behavioral disturbances, in particular, a high degree of intolerance to various sensory stimuli. One symptom commonly associated with rabies is hydrophobia, which is the onset of violent spasms of the throat muscles brought on by attempts to swallow. Death is inevitable and usually occurs within two weeks after the first symptoms appear.

Infection in animals usually follows a similar course to that in humans, although animals may harbor the virus in their salivary glands for several months without visible symptoms. Some animals develop a form of the disease called furious rabies, which is characterized by an initial restlessness followed by extreme agitation. This is the form of the disease that gave rise to the adjective "rabid" to describe irrational and violent behavior. Dogs with furious rabies will growl, bark, and snap, and, as the disease progresses, lose control over their muscles, drip excessive saliva, and exhibit hydrophobia. Death usually occurs 2–3 days after these symptoms appear. Animals with dumb or paralytic rabies are not as excitable and suffer from a paralysis of the throat and jaw muscles. The basis of the difference between furious and dumb rabies is not clear, and scientists no longer use these symptoms as a way to classify the virus and infection, since no discernible differences have been found.

Because the early symptoms of rabies are so nonspecific, early diagnosis, which is essential if the disease is to be stopped, is difficult. If obvious spasms or hydrophobia are seen, then rabies may be confirmed, although by then it is usually too late to control the disease. If history or circumstances suggest rabies, then physicians often recommend isolating and observing the animal (if accessible) for the development of rabid symptoms. One way to confirm diagnosis is to perform a brain biopsy of the animal after its death and look for the presence of negri bodies in the cells.

Rabies is the only known disease for which a post-exposure vaccination is of any use. The reason this method can work is because the disease really manifests itself only after the virus reaches the central nervous system, and it can therefore be stopped before it reaches that location. Once the symptoms of nervous-system invasion become evident, there is no treatment for rabies and the disease is almost always fatal. The vaccine used today is still similar to the version used by Louis Pasteur over a century ago. It consists of a series of gradually increasing doses of virus administered over 7–14 days in the form of intramuscular injections in the abdomen. Vaccines are prepared by growing the virus in the brain tissue of rabbits, sheep, or goats. The success rate of the vaccine is highly variable— varying from 0 to 87 percent. *See also* RABIES VIRUS; RHABDOVIRUS.

rabies virus

Enveloped, single-stranded RNA virus that causes rabies. A member of the rhabdovirus family, it normally resides in mammalian hosts such as dogs, raccoons, and squirrels. The virus has a high affinity for nervous tissue. Recent evidence indicates that this affinity coincides with the presence of certain molecules called acetyl-choline receptors, which are normally abundant on the surface on nerve cell. In addition, it also grows extremely well in the salivary glands of animals and is thus easily spread through bites.

A characteristic feature of rabies virus multiplication in the cytoplasm is the forma-

tion of large cytoplasmic inclusions of up to 27 μm called negri bodies. These inclusions contain the rabies antigen and are highly eosinophilic. Negri bodies are typically seen in undamaged nerve cells of infected animals, and their presence is used as a means of postmortem diagnosis. *See also* NEGRI BODIES; RABIES; RHABDOVIRUS.

rat bite fever

Human disease initiated by the bite of an infected animal—usually by rats but also mice, other rodents, and household pets such as cats. Actually, this name is given to two diseases with similar clinical pictures and epidemiology but different causative agents, namely, either *Spirillum minus* or *Streptobacillus moniliformis*. There is an abrupt onset of symptoms such as chills, fever, muscle pains, and headaches, following the ulceration of the site of the bite. Often the lymph glands are involved and become swollen and painful. The fever lasts for several days in an undulating fashion, during which time patients also develop a maculopapular rash around the extremities. The disease is typically self-limiting unless complicated by some secondary infection, usually also initiated at the site of the bite wound. Antibiotic therapy is only necessary in the latter case. Neither bacterium is transmissible from human to human. *See also* SPIRILLUM MINUS; STREPTOBACILLUS MONILIFORMIS.

recombinant DNA

A piece of DNA that contains genes from more than one source—often from very different organisms such as bacteria and humans or insects—due to the splicing together of pieces from different sources. The splicing is achieved by cutting the pieces of DNA with specific restriction endonucleases and patching in a piece from one source into another by means of enzymes called ligases. An example of recombinant DNA is a human insulin gene spliced into a bacterial genome. Some examples of practical uses of recombinant DNA include modern therapeutic techniques such as gene therapy and the oil-eating "super-bugs" created in the 1960s. *See also* DNA; GENE; GENETIC ENGINEERING; PLASMID; RESTRICTION ENZYME.

recombination

Naturally occurring process of exchange of genetic material between two pieces or molecules of DNA. There are different mechanisms of recombination that take place in various systems. Site-specific recombination, for instance, occurs at specific locations within DNA molecules, usually in the presence of specific enzymes that recognize specific sequences in different molecules. An example of such recombination in nature is the incorporation of prophages and some retroviruses into specific locations in the host genome. Genetic recombination may also occur during DNA replication; pieces of DNA in the same locus (i.e., alleles) are exchanged or physically crossed over between the newly formed chromosomes. Still another form of recombination, called somatic recombination, takes place in the immunoglobulin genes of the B and T lymphocytes during an immune response. This involves the splicing together of different pieces of a fragmented gene and was evolved so as to be able to generate molecules of differing specificities from a single gene. *See also* DNA; DNA REPLICATION; GENE; LEDERBERG, JOSHUA.

red tide

An algal bloom caused specifically by the multiplication of algae called *Gymnodinium breve*, a normal inhabitant of warm saltwater (i.e., oceans) in areas like the Florida Gulf Coast. Unlike most other blooms, red tides are not necessarily related to pollution and eutrophication but seem to be initiated by cysts of the algae which may be carried by ocean currents to locations far from their normal home. In high concentrations, the algae color the waters a ruddy brown hue. Although harmless at lower levels, large populations of these algae can cause significant damage, both as irritants and allergens to swimmers in red tide waters, and by producing potent neurotoxins, which in turn can be ingested either directly or via fish and shellfish. *See also* BLOOM.

Redi, Francisco (1626–1678)

This Italian physician was the first scientist in documented history to attempt to provide experimental evidence against the theory of spontaneous generation by showing that plants and animals can arise only from predecessors of the same kind. His findings did not extend to microbes because these organisms had not yet been discovered. *See also* ORIGIN OF LIFE.

Reed, Walter (1851–1902)

An officer of the Medical Corps of the U.S. Army during the Spanish-American war, Reed was the leader of a medical group that conducted the seminal research on the cause and spread of yellow fever. By 1900, Reed's team was able to report conclusively that the mosquito served as the intermediate host for yellow fever. They also showed that, unlike the malaria parasite, the causative agent of yellow fever did not require a developmental period in the mosquito. The work of Reed's group provided future scientists with the foundations for the discovery of the yellow fever virus. *See also* THEILER, MAX; YELLOW FEVER.

relapsing fever

Systemic bacterial disease caused by a species of *Borrelia*, mainly *B. recurrentis*, which is transmitted by body lice, but also variant species such as *B. hermsii* and *B. parkerii*, transmitted via ticks. The disease is characterized by the onset of general symptoms such as chills and fever 3–4 days after infection; these symptoms are caused by the presence of excessive numbers of the spirochetes in the bloodstream. This may be accompanied by a macular rash, symptoms of jaundice, and tenderness of the liver and spleen, reflecting the propensity of the bacteria for those organs. The first fever lasts for 3–7 days, after which there is an asymptomatic phase followed by a relapse of the symptoms. All together, there may be up to four such relapses, with subsequent episodes having decreasing duration and severity. The recurrences correspond to the immune response of the host, with the symptomatic phases corresponding to a period of the emergence of an antigenic variant of the organism and the body's adaptation to the new serotype. In general, the louse-borne disease predominates in Africa, while the tick-borne variety appears sporadically all over the world. Incidence in the United States is relatively low and appears to favor the western states. *See also* BORRELIA.

reovirus

This family of RNA viruses is unique among the known human viruses because it encodes its genetic information in a double-stranded RNA genome. Members of this family were first discovered as inhabitants of the respiratory and gastrointestinal tracts of humans and animals. They were labeled as "orphan" viruses because they were not seen to be associated with any disease—the name reo is an acronym for *r*espiratory *e*nteric *o*rphan virus. Subsequent investigations have revealed that not all members of this group are nonpathogenic, and viruses of the genus rotavirus are the most important cause of infantile gastroenteritis and diarrhea all over the world.

Reoviruses have a distinctive structure, being composed of two concentric capsids with icosahedral symmetry, surrounding an inner core containing the double-stranded RNA genome of about 18–25 kilobases, made up of 10–12 segments, each of which represents a single gene. These viruses do not contain any lipid envelope. The segmented nature of the genome lends itself to a high degree of genetic reassortment and exchange between species of the same genus, particularly when more than one virus co-infects a single host. There are at least nine different genera within the reovirus family, with antigenically distinct coat proteins. Four of these genera, including the rotaviruses, have been found associated with humans: orbivirus and coltivirus, which are pathogenic, and the originally discovered orphans, now called the ortho–reoviruses. This last genus has served as a very important model for molecular biologists interested in learning about the biochemistry and genetics of life forms with double-stranded RNA genomes. Although these viruses were never conclusively linked to any specific disease outbreaks, scientists suspect that they do cause inapparent infections because natural antibodies against these viruses are present in nearly all adults. The virus may be found in sewage and polluted waters. *See also* ROTAVIRUS.

reservoir

A living creature or inanimate object that harbors pathogenic organisms and may serve as an ongoing source of outbreaks of disease. An important distinction between a disease reservoir and a disease vector or carrier is that the latter serve mostly as temporary sites for the organism and function mainly as agents of transmission rather than being the source of the parasite. Lakes and ponds, for instance, often serve as reservoirs for enteropathogenic bacteria and protozoa. Animal reservoirs such as cattle and pigs function as the source for different zoonoses. *See also* HOST; VECTOR; WATER-BORNE INFECTIONS.

resolution limit

The smallest distance between two points that a lens is capable of distinguishing. The naked human eye has an average resolution limit of 0.1 mm, which means that when two points lie closer than 0.1 mm they will be seen as a single blurry spot rather than two separate ones. In microscopy, this limit—also known as the resolving power of a microscope—is the factor that determines the smallest level of observable detail for a given instrument. Theoretically, the smallest distance is equivalent to the wavelength of the rays used for magnification; thus, the resolution limit of optical microscopes falls in the lowest wavelength range of visible light, while electron and X-ray microscopes have smaller resolving powers. See also ELECTRON MICROSCOPE; MAGNIFYING POWER; MICROSCOPE; OPTICAL MICROSCOPE.

respiration

Most of us think of respiration in terms of breathing, the process that enables humans and animals to gain oxygen from the atmosphere so that we may continue to live. Accurate as this idea might be, it is nevertheless an incomplete picture. Indeed, most living things on this planet rely on some kind of respiration. Basically, respiration might be viewed as the biochemical/metabolic process through which most organisms mobilize energy (which they have previously obtained by photosynthesis or other methods) into ATP so as to use it for various synthetic and other functions. In eukaryotic cells, respiration takes place almost exclusively in the mitochondria.

The distinguishing feature of this energy conversion (in contrast to fermentation) is the involvement of an external oxidizing agent. By far the most common oxidizing agent is atmospheric oxygen, although other organic and inorganic molecules might also be used. These latter types of respiratory reactions are called anaerobic respiration. The process may be visualized as a kind of combustion in biological systems: a carbon-based fuel— glucose in most cells—"burns" in the presence of the oxidizing agent (oxygen) to release energy, water, and carbon dioxide. Unlike combustion however, respiration does not take place in a single step or even in a simple series but is broken down into stages so that the energy is released slowly, as needed. The first step is the glycolysis of sugar molecules to pyruvic acid, followed by the Krebs cycle, which might be considered as the true representative of respiration since it is during this phase that the external oxygen is used to release the carbon atoms from the pyruvic acid into carbon dioxide. The conversion of one molecule of glucose results in the net release of 38 ATP molecules, of which only two are generated during glycolysis. *See also* CARBON CYCLE; ELECTRON TRANSPORT CHAIN; METABOLISM; MITOCHONDRION; OXYGEN CYCLE.

respiratory syncytial virus

This virus is the most frequent cause of respiratory infections in infants and is responsible for about half of bronchiolitis cases and one quarter of pneumonia cases in the world. While the exact mechanisms of disease are not yet completely understood, there is some evidence to suggest that the most severe symptoms are likely immunological in origin. Although children are the most susceptible, all age groups can and do contract infections. The most common result is a febrile illness involving inflammation of the upper respiratory passages (rhinitis and pharyngitis) resembling a mild to severe cold. The incubation period is 4–5 days. The most susceptible populations to this virus are the very young and very old, in whom the immune system is weaker. Lower respiratory involvement seems to occur only in very young infants and is often fatal due to the obstruction of the air passages caused by the inflammatory responses.

The virus itself is an enveloped, single-stranded RNA virus in the family para–myxoviridae. The respiratory syncytial virus is classified as a separate genus, *Pneumovirus,* because of its difference from all other viruses in this family—e.g., measles and mumps viruses—in its genomic organization and protein composition. It has a 15–16 kilobase linear genome containing 10 genes that encode 10 distinct structural and functional proteins. Still another distinguishing feature of this virus is the presence of a weakly antigenic surface glycoprotein G. The virus induces the formation of syncytia both *in vivo* and in tissue culture. *See also* PARAMYXOVIRUS; SYNCYTIUM.

retrovirus

Enveloped animal viruses with RNA genomes, associated with a number of diseases including AIDS and leukemia in humans and various cancers in different animals. This name is derived from the fact that despite their RNA genomes, these viruses exist in host cells in the form of DNA that is integrated with the host cell—i.e., their genomes are *reverse-*

*tr*anscribed (from DNA to RNA) to generate new viruses.

The retroviral genomes are different from most other groups of virus in some key respects. First, they are the only known diploid viral genomes, with two identical single-stranded RNA molecules held together by hydrogen bonds in the intact virion. Although the genome is a positive sense RNA molecule—i.e., it is in the form of mRNA—it is never used directly for synthesis. Retroviruses rely completely on the host's transcription machinery for their duplication, and in addition, require specific host cell tRNAs for their replication.

Retroviruses gain entry into a host cell by absorption onto specific receptors on the cell surface. The first step after entry is the conversion of the RNA genome to a double-stranded DNA molecule called the provirus, catalyzed by a special enzyme called reverse transcriptase. Proviruses, which have a few additional sequences over the viral RNA, may be present in either circular or linear forms. Integration of this molecule into the cell genome occurs via a series of cutting and splicing reactions at specific locations of both the provirus and host. With the integration into the host genome, the viral genes come under the host's control mechanisms. While in this state, the genome uses the protein synthetic apparatus of the host cell to produce viral proteins, as well as copies of the viral genome. These components assemble in the cell's cytoplasm to form the nucleocapsids that emerge from the cells by a process of budding from areas of the cell studded with the envelope proteins of the viruses. Finally, the retroviruses undergo a process of maturation outside the host cell. *See also* HIV; HTLV; REVERSE TRANSCRIPTASE.

reverse transcriptase

Enzyme that catalyzes the synthesis of a DNA molecule from an RNA template. This enzyme is used by retroviruses to integrate their genomes into the genes or chromosomes of host cells. The existence of such an enzyme was proven only after Howard Temin and David Baltimore, working independently from one another, isolated it from RNA tumor virus particles in 1970. *See also* BALTIMORE, DAVID; DNA; RETROVIRUS; RNA; TEMIN, HOWARD.

rhabdovirus

Represented by the notorious rabies virus, the rhabdovirus family consists of bullet-shaped, enveloped RNA viruses that infect a number of different animals. Individual viruses may be identified on the basis of their preference for different hosts—e.g., while the rabies virus can infect a number of different warm-blooded animals such as dogs, raccoons, and squirrels; the Lagos bat virus shows a special preference for bats and cats; while a third species, called the Obodhiang virus, prefers to reside in mosquitoes. Different species may also show a tendency to appear in fixed geographical regions—the Kotonkan virus, for instance, has been isolated in Nigeria, while the European bat virus has been found mostly in European countries, particularly France, Germany, Spain, and Finland.

An average rhabdovirus measures about 180 by 75 nm and contains an outer lipid envelope derived from the host cell, which encases a helical nucleocapsid core. The membrane is usually found in association with two viral proteins. The genome is bound tightly to about 2,000 units of the nucleoprotein. Only enveloped viruses are capable of infecting new cells. These viruses replicate in the cytoplasm of the host's cell. During replication, the genome directs the synthesis of certain virus-specific proteins that migrate to specific locations of the cell and become embedded in the cell membrane. New viruses are generated by a process of budding from these protein-studded locations of the cell. During viral synthesis, different components accumulate to form darkly staining patches called negri bodies, the presence of which is a valuable clue in diagnosing rabies virus infections. *See also* RABIES VIRUS; VESICULAR STOMATITIS VIRUS.

rheumatic fevers

A disease condition marked by fever and inflammatory lesions at different locations in the body, occurring as a delayed consequence to throat infections or pharyngitis caused by *Streptococcus pyogenes* group A hemolytic organisms. In acute rheumatic fever (ARF), inflammation may occur in the heart, joints, subcutaneous tissues, and even nerves. It is often seen in young children between the ages of 6 and 15 soon after an attack of strep throat. The inflammation of the heart, a common target, can lead to death or to the development of a chronic condition called rheumatic heart disease. This last condition is marked by a gradual destruction of the tissues of the heart and its blood vessels and eventual heart failure, sometimes several years after the acute episode. While the exact mechanism of rheumatic fever is not completely understood, it is believed to occur due to an immune reaction to *S. pyogenes,* which has antigens similar to antigenic markers in the heart valve tissues. The correct and timely diagnosis of strep throat and proper antibiotic treatment can prevent ARF. A hallmark of the rheumatic fevers is the formation of nodular structures called Aschoff bodies in the myocardium. *See also* STREPTOCOCCUC PYOGENES.

rhinovirus

Responsible for about 50 percent of the cases of the world's most common disease—the common cold—rhinoviruses are naked, single-stranded RNA viruses of the picornavirus family, which show a preference for colonizing the cells of the upper respiratory tract. The virus is highly infective and spreads easily via the aerosols created when sneezing. A highly variable antigenic structure has resulted in upwards of 100 separate serotypes of the rhinovirus. Despite all the advances of modern medicine, a cure for the common cold and the rhinoviruses still appears as out of our reach as ever. *See also* PICORNAVIRUS.

Rhizobium leguminosarum

The much-vaunted beneficial properties of various leguminous crops that restore fertility to soil is largely due to the action of this genus of symbiotic nitrogen-fixing bacteria. The symbiosis is evidenced by the presence of swollen sacs or nodules on the roots of these plants, which on microscopic examination reveal the presence of large numbers of gram-negative rods that have colonized the plant cells. In return for the nitrogen that the plants receive due to bacterial nitrogen fixation, the nodules provide a sheltered habitat for the bacteria, safe from various environmental threats in the soil. The ability to colonize plant cells is encoded in bacterial genes that may be present in either the chromosome or a plasmid. Symbiosis is initiated when the bacteria, which are also capable of living independently in the soil, get attracted to the surface of the root hairs of the legumes. In the soil, the bacteria are able to move freely by a swimming motion due to the presence of a single flagellum near one of its poles. Of all known bacteria, *R. leguminosarum* shows more antigenic diversity among its strains than any other species. *See also* NITROGEN CYCLE; NITROGEN FIXATION; SYMBIOSIS.

ribosome

Cellular component involved in the translation of the mRNA into DNA in the cytoplasm. It consists of complexes of RNA and proteins and interacts with tRNAs, energy molecules, and other components of the protein synthetic machinery to translate the coding sequence of the mRNA into polypeptides. Ribosomes are present in all living things but have markedly different structures and components depending on whether they are present in prokaryotic or eukaryotic cells. It is interesting to note that the eukaryotic organelles such as mitochondria and chloroplasts contain their own ribosomes which are more akin to the prokaryotic ribosomes. During the active or functional phase of protein synthesis, one may find several ribosomes strung along a length of mRNA to form what is called a polyribosome, which is necessary for protein synthesis to proceed properly. *See also* CELL; CYTOPLASM; TRANSLATION.

rickettsiae

The rickettsiae are a group of minute organisms that rank along with the chlamydiae as the smallest genuine life forms known. See Table 2 accompanying the Chlamydiae entry for a summary of differences between these organisms, the Chlamydiae, and viruses. Aside from a single genus (*Rochalimaea*), members of this group are obligate intracellular parasites and are found in a wide variety of vertebrate and invertebrate hosts. They are tiny rods about 2 μm in length and 0.3–0.7 μm in diameter, with a cell wall that displays all the structural and biochemical characteristics of a gram-negative cell, i.e., they contain both the peptidoglycan layer as well as the outer membrane. Metabolically, most rickettsiae differ from other bacteria in that they are incapable of glycolysis and rely entirely upon the Krebs cycle for energy. Although they are capable of producing ATP, they resort to synthesis only when these molecules are unavailable from the host. The synthesis of various molecules is similarly limited to those compounds that are not available from the host.

The rickettsiae have been classified into four main genera: *Rickettsiae*, *Ehrlichia*, *Coxiella*, and *Rochalimaea*. Of these, *Rochalimaea* (type species *R. quintana*) may be distinguished from these others because of its ability to live outside living host cells. The main difference between the remaining rickettsiae lies in the methods they employ to evade destruction by the host enzymes once they have entered the cells. For instance, the genus *Rickettsia* breaks down the membrane of the compartment in which it is enclosed and thus exists free in the cytoplasm. *Ehrlichia* functions in a manner similar to the Chlamydiae and prevents the hosts' lysosomes from releasing their enzymes into the vacuole, while *Coxiella* is naturally resistant to the host enzymes. In addition, *Coxiella* is markedly different from the other rickettsiae in that it has a distinctive life cycle and mode of transmission between hosts.

Various rickettsial species have been implicated in a number of human diseases. With the sole exception of the *Coxiella*-induced Q fever, all rickettsial diseases are transmitted to humans via insect vectors such as ticks, mites, fleas, and lice. The clinical signs and symptoms of various rickettsial diseases, including typhus, scrub typhus, various spotted fevers, and trench fever, are similar and differ mainly with respect to severity of symptoms and their epidemiology—i.e., geographic predominance. See Table 9 accompanying this entry for a summary of rickettsial diseases, specific causative agents, and epidemiological features. The organisms are transmitted to human hosts either through the bite of infected ticks or by the mechanical action of rubbing the site of the bite and introducing the organisms present in the insect or its feces. Once they enter the body, the rickettsiae invade the bloodstream and establish themselves in the endothelial cells (the cells that make up the veins and arteries). Their replication causes the host cells to detach from the walls of the blood vessels and cause obstructions, which leads to the destruction of tissues, hemorrhage, and the occurrence of a rash. These diseases are similar to various viral hemorrhagic fevers in many respects, and care must be taken during diagnosis to identify the culprit properly so that the correct treatment may be administered. For instance, rickettsial diseases are easily treated with such broad-spectrum antibiotics as tetracycline and chloramphenicol, which are completely ineffective against viruses. Early diagnosis of these diseases is important so as to prevent complications, such as endocarditis and shock, which develop rather dramatically if the disease is neglected. Immunity to these organisms is poor, and thus far, no adequate human vaccines have been developed. *See also* CHLAMYDIAE; *COXIELLA BURNETTI*; Q FEVER.

rifamycin

Relatively broad-spectrum antibiotic that acts by inhibiting RNA synthesis in bacteria. *See also* ANTIBIOTIC.

Table 9. Rickettsial Diseases in Humans

Disease	Causative agent[1]	Primary host	Transmission	Clinical features	Occurrence
Boutonneuse fever[2]	Rickettsia conorii	Rodents and dogs	Tick bite	Fever, headaches, and rash with lymphatic involvement; A characteristic necrotic lesion or eschar develops at the site of the tick bite.	Mediterranean, Eastern Europe. Middle East. Africa and India
Ehrlichioses	Ehrlichia canis	Dogs	Unknown[3]	Fever, headaches, swollen lymph nodes etc., with only occasional development of a rash; The absence of a primary lesion is one of the reasons to doubt its mode of transmission to humans.	United States.
Epidemic typhus.[4] (Also known as louse-borne or sylvatic typhus.)	R. prowazekii	Humans and flying squirrels	Louse feces rubbed into the skin due to scratching around the region of the insect bite.	High fever, headaches and characteristic rash on trunk and extremities but seldom on face; Usually occurs in epidemic episodes.	Worldwide, with higher incidence in Africa and Asia.
Murine typhus.[4] (Also known as endemic typhus.)	R. typhi	Rodents, especially mice.	Fleas' feces rubbed into the site of the insect bite during scratching.	Very similar to epidemic typhus but milder.	Worldwide; In the US appears confined to regions of Texas.
Q fever	Coxiella burnetii	Farm animals—cattle, sheep and goats; Rodents.	Inhalation of organisms from contaminated air; Occasionally by tick bites.	Respiratory disease with symptoms very similar to influenza.	Worldwide, with a correlation to slaughterhouses and dairy farms.
Queensland tick typhus[2]	R. australis	Marsupials and rodents.	Tick bites	Symptoms of fever and rash, similar to but milder than most other spotted fevers.	Australia
Rickettsial pox[2]	R. akari	Mice	Mite bites	Symptoms of fever and rash, similar to but milder than most other spotted fevers.	North America, Russia, South Africa and Korea
Rocky Mountain spotted fever (RMSF)[2]	R. rickettsii	Rodents, rabbits, raccoons, foxes, woodchuck and deer.	Tick bites	Fever, headache, lymphatic swelling and rash; There is no lesion at the site of the initial lesion.	Western Hemisphere with predominance in the United States.
Scrub typhus	R. tsutsugamushi	Rodents	Mite bites	Eschar at site of bite and a maculopapular rash with fever.	Eastern Asia, Australia, and South Pacific islands.
Sennetsu fever (glandular fever)	Ehrlichia sennetsu	Not known.	Unknown, presumably via insects.	Remittent fever, chills, joint pains and lymphatic enlargement.	Western Japan and Malaysia.
Siberian tick typhus[2]	R. sibirica	Rodents	Tick bites	Spotted fever symptoms of milder variety than RMSF.	Central Asia, Siberia, Mongolia and Central Europe.
Trench fever	Rochalimaea quintana[5]	Humans	Body louse, via feces rubbed into bite wounds.	Relatively mild symptoms resembling rickettsial spotted fevers.	Europe, Africa and North America.

Notes:

[1]Individual species of the genus *Rickettsia* , will be identified as *R. species name,* after the first mention.

[2]Part of the "spotted fever" group of rickettsial diseases.

[3]This organism is known to be transmitted from one dog to another via ticks, but the mode of transmission to humans has not yet been confirmed.

[4]Part of the "typhus fever" group of rickettsial diseases.

[5]This organisms has been reclassified into the non-rickettsial genus *Bartonella,* although the disease discussion is limited to this section because of its common features with other rickettsial diseases.

risk assessment

In the epidemiological context, risk assessment is the systematic approach to identifying and evaluating—both qualitatively and quantitatively—the risks posed to a population by a potentially dangerous entity. Relative to microbes, for instance, risk assessment might be applied to the adverse effects of a pathogenic bacterium or virus. According to a seminal report by the National Research Council, risk assessment comprises four steps. The first step is the identification of the potential hazard and the collection of all known information about it. This step would also include a qualitative evaluation of the agent. For instance, if the agent in question is a virus, then one might try to identify its primary hosts and mode of transmission, as well as the specific site of infection in a human. The second and third steps, which go hand-in-hand, involve a quantitative assessment of the agent, namely, dose response and the population's exposure to the agent. The final step is the integration of the information obtained in the first three steps to formulate a plan of action to deal with the agent—namely, to estimate the risk the agent poses and offer suggestions for future courses of action. *See also* RISK RATIO.

risk ratio

This ratio, also known as an odds ratio, is used in public health for assessing the risk of exposure or occurrence of a disease. For instance, the disease-odds ratio for a population is the ratio of the odds in favor of the occurrence of a disease among the exposed fraction of the population to the odds in favor of disease occurrence among the unexposed. Similarly, the exposure-odds ratio is the odds in favor of exposure to a disease among cases to the odds of exposure among non-cases. *See also* RISK ASSESSMENT.

RNA

One of the two major nucleic acid species present in living organisms, this molecule is a polymer of nucleotides composed of a back-bone of ribose sugar and phosphate with side chains of four nitrogenous bases—adenine, guanine, cytosine, and uracil (A, G, C, and U). Like DNA, RNA molecules also encode information in the linear sequences of their bases, but with the exception of the RNA viruses, these molecules do not carry genes. Rather, they function as a shuttle between DNA and the rest of the cell. With the exception of one category of viruses that contains double-stranded RNA as genomes, this nucleic acid is present in a single-stranded form. However, short complementary portions of the molecule may fold over on themselves and form double-stranded tracts with hydrogen bonds between A and U and between G and C, respectively.

Most living cells contain three main species of RNA, which perform distinct functions in the translation of information from DNA to protein. Ribosomal RNA (rRNA) is one of the major components of the ribosomes. Messenger RNA (mRNA) molecules are "transcripts" of the information carried in a gene, and they constitute the intermediary form between DNA and protein during gene expression and protein synthesis. Transfer RNA (tRNA) molecules carry specific amino acids to the site of synthesis for incorporation into the growing protein. In addition to these three species, a cell may contain special regulatory molecules called antisense RNA, which can bind to complementary sequences of nucleic acid (usually mRNA or a gene) and inhibit its activity. *See also* DNA; TRANSCRIPTION; TRANSLATION.

Ross, Ronald (1857–1932)

A British Army surgeon who provided the experimental evidence for the theory that malarial parasites were transmitted by mosquitoes, specifically, the *Anopheles* mosquito. Ross received the Nobel Prize for this work in 1902 and was knighted in 1911. Ross also made valuable contributions to the field of applied mathematics (as it related to epidemiology) and was an accomplished poet. *See also* MALARIA.

rotavirus

Now known to be one of the world's most common causes of diarrhea and gastroenteritis in infants, the rotaviruses were first discovered in 1973 in Australia in duodenal biopsies and the feces of children suffering from acute gastroenteritis. They are classified in the reovirus family along with other viruses known to contain a double-stranded RNA genome. When viewed under the microscope, they appear as smooth round particles with two concentric shells. The genome consists of 11 segments with a total size of 18 base pairs. Members of this group are further classified into six different serotypes, designated as A–F and based on antigenic differences. The majority of the human species belong to the A serotype.

Gastroenteritis occurs as a result of the multiplication of the virus in the cell lining the intestinal tract. Viruses gain entrance into the body through contaminated water or food. They enter the cells of the tiny folds (the villi) of the inner wall of intestines, and induce the cells to shrink and die, which reduces the surface area of the intestines that is usually available for the absorption of fluids. As less fluids are absorbed back into the body, they accumulate in the lumen and are discharged with the feces, resulting in watery stools. It should be noted that this mechanism of inducing diarrhea is markedly different from the mechanisms by which bacterial enterotoxins induce fluid loss. Clinically, the diarrhea, which lasts for 4–5 days, becomes apparent 1–3 days after infection and is typically preceded by nausea and vomiting. The infection is often self-limiting, and well-nourished children recover spontaneously. However, the morbidity and mortality rates for rotavirus infections is considerably higher among lower socioeconomic groups. As in most cases of diarrhea, dehydration is the most frequent cause of death. Diagnosis is best achieved by the identification of viruses by electron microscopy of fecal samples. Control measures are aimed at improving the general nutrition of affected populations and administering rehydration therapy to the severely affected. As yet, no vaccines have been developed for general use, although these are under investigation. An important consideration in treatment is the prevention of secondary bacterial infections. *See also* DIARRHEA.

Rothia

A member of the group of actinomycetes, the genus *Rothia* consists of gram-positive bacteria ranging from small rods to filamentous forms. The generic name is derived from the name of the investigator, Roth, who first described their characteristics in detail. A single species, *R. dentacariosa,* has been identified; it is mainly found in the oral cavity and carious lesions. These organisms are proteolytic; they demonstrate a high tolerance (a preference even) for oxygen, while maintaining the capability for anaerobic growth, and they produce the enzyme catalase. *Rothia* strains are serologically distinct from other genera in the family. *See also* ACTINOMYCES.

Rous, Peyton (1879–1970)

The first scientist to recover a filterable agent from chicken tumors (sarcomas) that could induce tumors when injected into cancer-free chickens. When he first presented his results in 1911, the scientific community was extremely reluctant to accept the idea that viruses could cause cancer and did not act upon this evidence in any meaningful manner for some decades. Rous himself switched fields and went on to make important contributions to the study of blood, although he was finally recognized with a Nobel prize in Medicine in 1966 for his seminal discovery. *See also* AVIAN SARCOMA VIRUS; CANCER; ONCOGENIC VIRUSES.

rubella virus

The causative agent of German measles, the rubella virus is classified all by itself in a unique genus, *Rubivirus*, in the family of Togaviruses. Like other members of this group, it consists of a single-stranded, positive sense RNA genome enclosed in an icosahedral capsid surrounded by a lipoprotein envelope. However, it differs from all other viruses in this family in that it alone has no

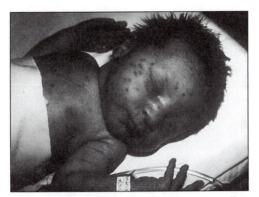

Infant with congenital rubella and "blueberry muffin" skin lesions. *CDC.*

insect vector and is transmitted from one human host to another via aerosols that are released while sneezing or coughing. The structure of the envelope makes this virus susceptible to solvents such as ether and chloroform. The antigenic structure of the virus is stable, and despite the presence of several proteins—including a surface hemagglutinin—the rubella virus has just one serotype.

In addition to German measles, which is a relatively mild disease in children and young adults (though somewhat more severe in adults), the rubella virus is teratogenic, i.e., it is capable of passing from an infected pregnant woman to the fetus and can cause severe congenital abnormalities and even death. The infection, which occurs via the placenta, has its worst effects during the early stages of pregnancy, when it multiplies within the embryonic cells and inhibits proper cell divi-

sion. The teratogenic potential reduces drastically after the first trimester, and infection of the mother after the fourth month may not even result in the passage of the virus into the fetus. Common teratogenic effects include mental retardation, eye defects such as cataracts and glaucoma, congenital heart disease, and deafness. Some of these effects may not even become apparent until months or years after birth. The MMR vaccine has been successful in keeping the disease at bay, but it is vital that it not be given to pregnant women. *See also* GERMAN MEASLES; MMR VACCINE; TOGAVIRUS.

Ruminococcus

These bacteria are frequently found residing in symbiotic relationships in the rumen of various animals, such as cows and sheep, where they help break down cellulose by the production of extracellular enzymes called cellulases. A common example is *Ruminococcus flavefaciens*. Organisms are gram-positive and anaerobic. The sugars produced by the degradation of cellulose may be used either by the host animal or broken down into smaller molecules by bacterial fermentation. These bacteria are typically heterofermentative and produce a number of different small organic molecules such as acetic acid and formic acid in addition to the usual products of fermentation. *See also* CELLULASE.

S

Sabin, Albert B. (1906–1993)

Scientist who developed a live, attenuated form of a vaccine against poliomyelitis during the mid-1950s. Administered orally, this vaccine was introduced only after the successful release of Jonas Salk's injectible vaccine, but it proved much safer for use. It passed large-scale field tests all over the world from 1958 to 1960, and it is the vaccine of choice against the disease in most countries today. *See also* ATTENUATED VACCINE; POLIOMYELITIS; POLIOVIRUS; SALK, JONAS.

Saccharomyces cerevisiae

Familiar to most of us as "baker's" or "brewer's" yeast used for leavening breads and for making beer and wine, *Saccharomyces cerevisiae* is the prototypical yeast. In addition to the commercial uses mentioned, this yeast is also a favorite subject of scrutiny in the laboratory as a relatively simple model system for understanding eukaryotic life at a cellular and molecular level. It is a non-photosynthetic, saprophytic organism that is unicellular and multiples by a process of budding. Cells respond to the Gram stain as if they were gram-positive, although the reason for their ability to retain the dye is not a reflection of peptidoglycan content because it is in the bacteria. *S. cerevisiae* is dimorphic in nature and forms filaments (fungal hyphae) under conditions of nitrogen depletion. As a

matter of interest, it is worth mentioning that despite working with this organism in the laboratory for the better part of the century, scientists did not discover this dual nature of its life cycle until 1992. Thus far, *S. cerevisiae* has not been linked to any human disease. *See also* BEER; BREAD; WINE.

Salk, Jonas (1914–1995)

Few scientists have experienced the simultaneous public praise and professional controversy of Jonas Salk, the developer of the first successful vaccine against poliomyelitis. At a time when most of the scientific community was attempting to produce vaccines by attenuating live viruses, Salk experimented with an alternative: treating the viruses with formaldehyde, which inactivated or "killed" them while leaving their antigenicity intact. The success of this vaccine in limited lab trials prompted him to develop it for public usage rather quickly, which led to censure from the scientific community. However, coming when it did—in the wake of one of the country's worst polio epidemics—the effectiveness of the vaccine won him public fame and appreciation. It should be noted that Salk received no significant scientific honors—such as the Nobel Prize—or monetary gain from his discovery. *See also* POLIOMYELITIS; POLIOVIRUS; SABIN, ALBERT.

Salmonella

This genus in the Enterobacteriaceae family of gram-negative rods consists of a number of species of importance as human pathogens. Infections with these bacteria typically occur as a result of ingesting contaminated water or food, although inanimate fomites are also frequently implicated. The salmonellae are hardy and capable of surviving in both frozen and moist environments for up to several months. This property undoubtedly plays a major role in causing epidemics of *Salmonella* infections. Most strains of *Salmonella* are motile due to the presence of peritrichous flagella, and motility appears closely linked to the pathogenicity of these organisms. *See also* IMViC TESTS.

Salmonella enteritidis

With more than 1,500 known serotypes in existence, this organism is perhaps the most widespread of the salmonellae in the world, besides being the most prevalent species of *Salmonella* in the United States. These organisms may reside in a large number of domestic and wild animals, including poultry, swine, cattle, dogs, and cats. *S. enteritidis* and *S. typhimurium* (a common serotype often classified as a separate species) are common causes of food-borne gastrointestinal infections, generally referred to salmonellosis. The signs and symptoms of disease include an acute enterocolitis—inflammation in the colon—headaches, fevers, diarrhea, nausea, and occasional vomiting, within a few hours to a day of ingesting the contaminated food or water. The disease is a result of the colonization and subsequent erosion of intestinal tissue by the bacteria, rather than the production of enterotoxin, although some strains may produce this as well. The infection may be self-limiting or may spread into the bloodstream, giving rise to several complications, including septicemia and destruction of other tissues and organs. Complications and death are relatively rare outcomes that tend to occur in people with compromised immune systems, such as the very young, the very old,

AIDS patients, and organ transplant recipients. Salmonellosis may be diagnosed by the isolation and identification of bacteria from stool samples. Tests such as the IMViC panel, salt tolerance, and growth on differential media are required to differentiate the culprit from other enteric bacteria such as *E. coli* and *Shigella*. Once identified, antibiotic therapy should be administered to clear the infection and proper precautionary methods should be adopted to limit the spread of infection. No immunization is available. *See also* FOOD-BORNE DISEASES.

Salmonella typhi

This bacterial species is the cause of typhoid fever, an important food- and water-borne infection that affects several million people every year. These bacteria cause the characteristic symptoms of typhoid fever by their ability to invade the bloodstream of infected individuals soon after they colonize the intestines. The spread of the bacteria to different parts of the body results in the destruction of cells in remote locations, which manifests itself in an overall loss of resistance to other infections, as well as extreme weakness. This process is different from most other salmonella infections, where bacteremia is a complication rather than a matter of course. While many such infections are completely resolved (either due to antibiotics or the body's own abilities to clear infections) *S. typhi* may also settle down to colonize the gall bladder, where it causes no further damage to the infected individual but is periodically shed into the feces and thus serves as a continuous source of infection for others. Such was the case with Mary Mallon, the notorious "Typhoid Mary." *S. typhi* may be differentiated from other *Salmonella* species on the basis of phage typing tests as well as specific biochemistry and pathogenicity. *S. paratyphi* is a closely related but antigenically distinct species that induces a much milder disease than typhoid fever. *See also* CARRIER STATE; TYPHOID FEVER; TYPHOID MARY.

sanitation

The development and use of various measures to obtain and maintain environmental conditions favorable to good health.

saprophyte

An organism that derives nutrition—i.e., energy and carbon—from complex organic molecules. Because it uses organic substrates, a saprophyte is by definition heterotrophic. The difference between this and the parasitic heterotrophs is that the saprophytes do not require living cells or organisms. Such organisms typically live off dead and decaying plant and animal matter. Many fungi, as well as a large number of bacteria, are saprophytic in nature. *See also* HETEROTROPH.

Sarcina ventriculi

Little-studied genus of gram-positive, obligately anaerobic, spore-forming bacteria that typically occur in nature in packets of eight cells. In terms of ribosomal sequence and GC ratio (falling within a narrow range of 28–31 moles percent) *Sarcina* appears to be most closely related to *Clostridium perfringens*, but is not associated with the kind of severe gastrointestinal disease caused by the latter. In fact, *Sarcina* has been isolated from the gastrointestinal tract of humans without any apparent disease.

scalded skin syndrome

This is typically a disease of newborn babies caused by *Staphylococcus aureus*. It is also known as Ritter's disease, exfoliative dermatitis, or bullous impetigo. Strains of *S. aureus* that cause this condition produce an enzyme called exfoliatin which causes the skin to peel. Organisms from the environment gain entry into infants' bodies at the umbilical cord or near the site of circumcision. The infection is marked by the formation of small blisters filled with fluid and bacteria, which rupture to form a thin crust and eventually give the appearance of a ringworm infection due to the peeling of surface layers of the skin. Sometimes the rupturing of the blisters spreads the infection over a more extensive area. The medical term in this case is pemphigus neonatorum. *See also* STAPHYLOCOCCUS.

scanning electron microscope (SEM)

Adaptation of the electron microscope in which the image of the object is obtained by the reflection of electrons from the surface of the specimen only. Reflection of this type is facilitated by coating the specimen in some electron-opaque material such as metal atoms. The chief advantages of the scanning electron microscope (SEM) is the ease of sample preparation, the enormous depth of focus, and the ability to glean surface details about a specimen. The method is also used to determine internal structures with respect to a single plane: by freezing the structure, cracking it open, and coating the cracked surface. *See also* ELECTRON MICROSCOPE; METAL DECORATING.

scanning transmission electron microscope (STEM)

Adaptation of a conventional transmission electron microscope (TEM), which relies on a different method of image processing after the electrons have passed and refracted through a specimen. Rather than process the electrons on a photographic plate, a scanning transmission electron microscope (STEM) analyses the energy patterns of the transmitted electrons to compose an image. The resolving power of such microscopes is comparable to TEM and images are probably sharper. *See also* ELECTRON MICROSCOPE.

scarlet fever

Disease caused specifically by those strains of *Streptococcus pyogenes* that harbor a bacteriophage containing a gene for a pyrogenic (fever causing) or erythrogenic exotoxin. This disease usually occurs along with streptococcal pharyngitis on the second day after infection. It is characterized by the development of a rash—erythema—which first appears on the chest and spreads to other parts of the body, in particular, the neck, armpit area, elbows, groin, and inner surfaces of the thighs. The face is not usually affected, although there may be redness there due to fever and flushing. In addition to the rash, scarlet fever is distinguished by the development of red papillae on a fuzzy, yellowish-white, coated tongue, known as "strawberry tongue." Over time, this develops into a moistly red or beefy-looking "raspberry tongue" due to the shedding of the upper fuzzy layers. Fever, nausea, and vomiting may accompany severe infections. If left untreated, scarlet fever is likely to give rise to the same sequelae as other streptococcal infections. The disease is diagnosed by its symptoms and treated with antibiotics normally used against the organism. *See also* STREPTOCOCCUS PYOGENES.

Schick test

Epidemiological test used to monitor the status of a person's immune system and assess susceptibility to diphtheria. Named for the Viennese physician who first developed it in 1913, it involves injecting small amounts of the diphtheria toxin into the skin of the patient (not subcutaneous but intracutaneous) and monitoring the reaction. People with no circulating antitoxin in their blood will develop a small red lesion at the site of injection within 24 hours of receiving the injection, which will gradually grow to about one centimeter in diameter, become slightly pigmented, and gradually scale off. This reaction is the result of the response of the local skin cells to the toxin. To avoid confusion with a local allergic reaction, which is caused by the injection of virtually every foreign substance, the test should not be read earlier than five days, by which time the nonspecific reaction wears off. People who have adequate levels of antitoxin in their bloodstream will not develop the lesion. Such individuals would be at low risk of contracting diphtheria under ordinary conditions of exposure. *See also* ANTITOXIN; DIPHTHERIA.

schizogony

Form of asexual cell division, seen in certain stages of protozoan organisms such as *Plasmodium*, in which multiple nuclear divisions take place within the cell before the cytoplasm splits to form multiple progeny, which are then released from the host cells. *See also* CELL DIVISION; *PLASMODIUM*.

scrapie

Named for the peculiar behavior demonstrated by afflicted sheep and goats, in which they develop an intense itch and attempt to "scrape" off their wool and hair, scrapie is a fatal neurodegenerative disease caused by the proteinaceous disease agent called a prion. This disease is perhaps the most exhaustively studied of all prion diseases and is the model scientists have used to understand the disease process and isolate the culprit.

Histologically, scrapie is similar to other spongiform encephalopathies in animals as well as humans, being marked by the development of vacuoles in the brain from which infectious prion material may be isolated. The nerve-cell degeneration has a marked impact on the behavior of the animals; they become irritable and gradually loose coordination, reaching a point where they are unable even to stand without collapsing. Both sporadic and familial cases have been identified. Animal feed for cattle and horses prepared from improperly processed sheep's meat appears to have been the source of infection that set off the alarm on mad-cow disease. The chronic wasting disease of elk and mule deer, and transmissible mink encephalopathy, are examples of scrapie-like conditions in other animal species. *See also* GAJDUSEK, CARLETON; MAD-COW DISEASE; PRION; PRUSINER, STANLEY; SPONGIFORM ENCEPHALOPATHY.

septicemia

Systemic illness characterized by symptoms of high fever and shock caused by the presence of infecting organisms, usually bacteria, in the blood. *See also* BACTEREMIA.

Serratia

This genus is one of the free-living members of the family Enterobacteriaceae, which consists mostly of gram-negative rods that inhabit the intestinal tracts of various animals. The most common species in this group is *S. marcescens*, an opportunistic pathogen of the urinary and upper respiratory tracts. It has also been implicated in cases of septicemia following surgery. Infections may be diagnosed by the isolation of the organism by culturing material—urine, mucus, or blood—from the site of infection. A related species, *S. liquifaciens,* is often associated with the spoilage of refrigerated vegetables and meats. Infections may be treated with antibiotics, although the development of multiple drug resistance in this and other coliforms is a fast-growing problem.

In the laboratory, *Serratia* colonies are easily identifiable due to the production of a bright pink or crimson pigment. In addition, it may be distinguished from other coliforms on the basis of the IMViC tests. Organisms are typically motile due to peritrichous flagella and may be defined in terms of their somatic, flagellar, and capsular antigens. They are microaerophillic, facultative anaerobes that ferment glucose and lactose and are proteolytic. *See also* IMViC TESTS.

sewage disposal

Centuries ago, when Earth had fewer human beings and people led more or less nomadic lives, it was possible to leave the disposal of various wastes to nature. With people constantly on the move, human wastes such as feces, urine, and spoiled food did not accumulate in significant amounts. A variety of natural processes, predominantly microbial but also mediated by other scavengers and even some non-biological means, would degrade waste materials over time. But, as populations began to establish themselves in pockets, increase in size, and spread out, the rates of both the consumption of resources and generation of wastes rose to such an extent that nature's methods were no longer adequate. The efficient treatment and dis-posal of wastes became imperative, not only because of their negative impact on the quality of life but also because of the hazard they posed for health and the environment. People devised drains and sewage systems, which diluted the wastes in water and disposed of them in remote locations beyond their living quarters. In time, these measures, too, became inadequate. Waste management today cannot be merely a process of dilution and disposal of sewage; rather, it involves treating the material to render it less harmful.

Sewage treatment processes use microorganisms, borrowing from nature's own devices. The first step in sewage treatment is to separate the solids and suspended matter from liquid sewage and treat each independently. Liquid or effluent wastes are treated with aerobic heterotrophs, which remove most of the dissolved organic material and consequently lower the biological oxygen demand of water to sufficiently low levels. After a disinfecting treatment to remove pathogens, the aerobically treated water may be discarded safely into the environment and is even used for irrigating farmlands.

The treatment of sewage solids, also known as sludge, is carried out in tanks at about 35°C by anaerobic bacteria (e.g., *Clostridium, Bacteroides,* and certain methanogens), which digest various complex organic molecules to simpler compounds. A large proportion of the material is eventually converted by fermentation to acids and gases. The solid end product, which is relatively odorless, may be used as fertilizer. Biogas or methane is another useful byproduct of this process. *See also* ACTIVATED SLUDGE; BIOLOGICALLY AERATED FILTER.

sexually transmitted disease (STD)

Synonymous with venereal disease, this term is used more widely today to denote the class of diseases whose primary mode of transmission is through sexual contact. The subject of sexually transmitted diseases (STDs) is one of the most important issues, alongside personal hygiene and food safety, about which public health officials and community health

Sign promoting condom use in Helsinki, Finland. *CDC/ Dr. Edwin P. Ewing, Jr.*

workers seek to educate the public. *See also* AIDS; C*HLAMYDIA TRACHOMATIS*; GONORRHEA; HEPATITIS B; SYPHILIS; VENEREAL DISEASE.

Shigella dysenteriae

This intestinal bacterium, frequently found in feces-contaminated water, is best known as the cause of bacillary dysentery in humans. It is a highly potent organism, as evidenced by observations that as few as 10 cells are capable of invading the intestine and inducing symptoms. Major symptoms include an abrupt onset of severe abdominal pains, cramps, diarrhea, fever, vomiting, and the presence of blood, pus, and mucus in the stools, usually several hours after the ingestion of contaminated water or food. Most of these symptoms are due to the invasion of the intestines by these bacteria rather than from the effects of an enterotoxin, although some strains do produce an enterotoxin called shiga toxin. Diagnosis is made by the isolation of the organisms from stool samples. The bacteria are gram-negative rods that can ferment sugars without producing gas. They may be differentiated from other members of the Enterobacteriaceae by the IMViC panel of tests. *See also* DYSENTERY; IMViC TESTS.

shingles

Also known as herpes zoster, shingles is an acute disease of viral origin characterized by severe pain and the formation of chickenpox-like lesions along areas enervated by sensory nerves, particularly in the region of the face, neck, and thorax. While the exact mechanism of pathogenesis is not clearly understood, the disease appears to be caused by the reactivation of a latent Varicella-Zoster virus several years after a primary episode of chickenpox. It is seen to occur most frequently among elderly or immuno-compromised individuals, although it may occur at any age. There is some evidence to suggest that viral latency is maintained by mechanisms of cell-mediated immunity, and that the eruption of shingles is due to the weakening of this system. Reactivation is often associated with surgical or other trauma to the nerves harboring latent Varicella-Zoster virus; presumably, this releases the virus, enabling it to travel along the nerve back to the skin and induce the symptoms of shingles. The disease is mostly self-limiting, although it may be complicated by conditions such as postherpetic neuralgia.

The projected incidence of shingles is related to the incidence of chickenpox, against which there were no vaccines until 1995. According to CDC estimates, about 10–20 percent of the population in the United States is likely to develop the disease. Diagnosis is achieved by the evaluation of clinical symptoms as well as identification of the virus in material from the lesions. The differentiation between this virus and the herpes simplex virus is especially important because they are similar but require different treatments. Acyclovir and other drugs of this class may be used to reduce healing times and prevent the formation of new lesions after shingles has been diagnosed. *See also* HERPESVIRUS; POSTHERPETIC NEURAGLIA; VARICELLA-ZOSTER VIRUS.

smallpox

Although smallpox has been eradicated and is no longer a threat to humans, this viral dis-

Smallpox lesions on abdomen. *CDC/James Hicks.*

The last known person in the world to have smallpox, 1977. *World Health Organization.*

ease is worth reviewing because it has played an important role in history, both as an agent of mass devastation as well as a model for the control of other infectious diseases. That the disease was known in antiquity is borne out by accounts dating back to the time of the Egyptian pharaohs (1100 B.C.). Because the infection has such distinctive characteristics and has been so well documented in the past, medical historians have been able

to identify it as the cause of various past epidemics with a high degree of certainty.

Primary smallpox infection occurs via the upper respiratory tract, usually from aerosols created by infected patients in the acute phase of the disease. Following this primary infection, there is a 10- to 12-day incubation period during which the virus multiplies and spreads to various internal organs and the skin. Disease symptoms begin with a fever for 2–4 days, followed by the appearance of the smallpox lesions or pocks, which are abundant on the face and limbs and less so on the trunk. The lesions progress through multiple stages within the first few days— starting as a macular rash which develops into fluid-filled pustules, eventually becoming encrusted 8–10 days after their first appearance, and later falling off, leaving behind characteristic scars.

The prognosis for smallpox often depended on the virulence of the infecting strains. Variola major caused the most severe infections and had a mortality rate of 20–50 percent. Variola minor was a far less virulent strain, with less than 1 percent mortality even in unvaccinated populations.

A typical case of smallpox is easily identifiable by the rash; however, by the time this appears, the internal organs are already damaged and the disease usually runs its course. There are no known effective antiviral drugs, and treatment was usually directed at alleviating the symptoms and—perhaps more important—limiting the spread of infection by minimizing the exposure of patients. When outbreaks were suspected, patients exhibiting initial symptoms of a cold and fever were confined because they were at their most infective just before the rash broke out.

Perhaps the most effective preventive measure against smallpox has been immunization. The story of the country physician Edward Jenner and his ingenious use of infectious material from a related disease called cowpox to render people resistant to smallpox is well known. But even before Jenner invented his vaccination technique, the practice of raising immunity to smallpox by inducing a milder form of the disease, by inoculation via aerosols, was widespread in China, India, and the Middle East. This method was first brought to the West during the seventeenth century by Lady Mary Wortley Montagu, a British noblewoman who visited Constantinople and learned of the practice there. In the modern context, the vaccine is important because of its key role in eradicating the disease. The last naturally occurring outbreak of smallpox was reported in Somalia in October 1977, and the World Health Organization announced the complete eradication of smallpox in 1979. *See also* POXVIRUS; VARIOLA VIRUS.

Snow, John (1813–1858)

This nineteenth-century British physician/surgeon is perhaps best known for pinpointing the source of an outbreak of cholera, during the 1854 epidemic in London, to a single water pump—the "Broad Street pump." His finding was yet another piece of evidence to support the idea that cholera was a waterborne rather than miasmal disease. Modern day epidemiologists regard him as the "father of shoe-leather epidemiology." *See also* CHOLERA.

Spallanzani, Lazzaro (1729–1799)

Italian scientist who is credited with providing convincing evidence against the idea of spontaneous generation. His experimental method consisted of boiling hay infusion broth and leaving it in flasks with necks that had been melted shut, thus cutting off any contact with the external environment. Broth treated in this manner remained sterile indefinitely, in contrast to broth that was exposed to the air, which carried the "seeds" of microbial infection. *See also* ORIGIN OF LIFE.

species

Taxonomic category that is subordinate to genus and always used in conjunction with the generic name to identify an individual organism. In higher forms of life, a species represents its separate identity from other species by not interbreeding with them. This

distinction does not hold together in case of prokaryotes and other haploid organisms because true "breeding" does not occur in these forms of life. *See also* BINOMIAL NOMENCLATURE; TAXONOMY.

spheroplast

Cell wall-less but otherwise intact (i.e., not-lysed) form of a bacterium, yeast, or fungal cell. These microbes are called spheroplasts because they assume a spherical shape in isotonic solutions regardless of the shape of the organism in its walled state. The lack of the

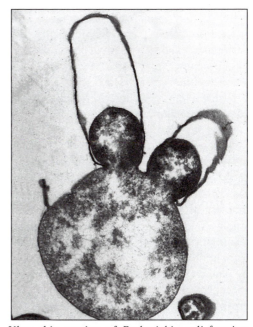

Ultra thin section of *Escherichia coli* forming spheroplasts in a medium containing penicillin. © *Science Source/Photo Researchers.*

cell wall renders these structures extremely sensitive to osmotic changes in their surrounding media, and they can only maintain their integrity under isotonic or hypertonic conditions (i.e., solutions with higher salt concentrations than the interior of the cell). In hypotonic solutions, the spheroplasts swell up and burst, due to the indiscriminate passage (osmosis) of fluid into the cell. One way to produce spheroplasts is to grow gram-positive organisms in isotonic solutions in the presence of penicillin or other β-lactam antibiotics, which prevent cell-wall synthesis. *See also* CELL WALL; β-LACTAM ANTIBIOTIC.

spinal meningitis

An often fatal disease caused by *Niesseria meningitidis*, characterized by an inflammation of the meninges (i.e., meningitis) or the outer layers of the brain tissue in response to the presence of bacteria in the brain. Primary infection with the niesseriae occurs in the nasopharyngeal area, where bacteria may reside for months without causing any outward symptoms. Under weakened circumstances, the infection spreads to the lymph canal, from which it may enter the bloodstream and spread to other parts of the body. *N. meningitidis* shows a particular predilection for the meninges and is small enough to evade the blood-brain barrier. The disease is marked by a sudden onset with intense headaches and vomiting accompanied by a sore neck, progressing rapidly to a coma.

Because the primary site of infection is the respiratory tract, *N. meningitidis* is spread easily via respiratory secretions and is thus associated with epidemic outbreaks rather than isolated cases. Incidence is highest among children under 5 years of age. Spinal meningitis may be treated with broad-spectrum antibiotics. Early diagnosis is imperative and can be achieved by culturing clinical specimens such as nasal exudates and cerebrospinal fluid in blood agar. *See also* NEIS-SERIA.

Spirillum minus

Spiral-shaped bacteria of uncertain taxonomy that is one of the etiological agents of rat bite fever in humans. The natural habitat of these bacteria is the body (bloodstream and internal surfaces) of animals such as rats, mice, cats, guinea pigs and some other domestic carnivores. Little is known about the physiology or biochemistry of these bacteria because they have not been isolated as pure cultures or grown on solid media. *Spirillum minus* is not amenable to the Gram stain and is best visualized by using dyes like the Giemsa stain. The bacteria appear to be sus-

ceptible to a broad range of antibiotics. *See also* RAT BITE FEVER.

spongiform encephalopathy

Group of neurodegenerative diseases caused by prions, characterized by the formation of large vacuoles or holes in the cortex and cerebellum of the brain, giving the organ a sponge-like appearance. The destruction of the brain tissue leads to severe neurological symptoms, including loss of muscular control, dementia, and wasting paralysis, eventually ending in death. Besides the brain vacuoles, typical histological findings in the nervous tissues include the deposition of amyloid, noninflammatory lesions and the destruction of astroglial cells of the brain. Most of these diseases appear slowly, although there is also a possibility of a sudden onset. *See also* CREUTSFELD-JACOB DISEASE; FATAL FAMILIAL INSOMNIA; GAJDUSEK, CARLTON; MAD COW DISEASE; PRION; PRUSINER, STANLEY; KURU.

spore

Metabolically inactive or sluggish form of a bacterial cell, usually present as a highly dehydrated cytoplasm encased in a thick resistant coating so as to be able to withstand adverse environmental conditions such as temperature extremes (usually heat), desiccation, and harsh chemicals. Only specific groups of bacteria, such as species of *Bacillus* and *Clostridium*, are able to form spores, and this ability is an important criterion in the classification of these organisms. Spores are usually formed in response to environmental stress and can give rise to a new vegetative cell (i.e., germinate) when ambient conditions are restored. Contamination with spores is often a cause for the delayed spoilage of different kinds of foods, such as those treated with pasteurization, which kills the vegetative cells.

Sporothrix schenckii

Dimorphic fungus associated with a mycosis called sporotrichosis in humans. Found in various kinds of soils, it is widely distributed in the world and appears to cause disease only sporadically among people who come in frequent contact with the infected soils, such as gardeners, farmers, and miners. Sporotrichosis usually begins with the development of a small nodule in one of the extremities, presumably at the site of a scratch or previous injury through which the fungal spores gain entrance into the body. The nodule grows and involves the regional lymphatic vessels, which become firm and cord-like and proceed to develop a series of nodules in the area. Eventually the nodules grow soft and ulcerate. This disease does not usually progress to a disseminated, systemic form, and fatalities are uncommon. If spores are inhaled in large quantities, sporotrichosis may present as a pulmonary disease with respiratory distress and various signs of lung involvement. Diagnosis is made by examining a tissue biopsy or exudate for the presence of fungi. The disease may be treated with various azole fungicides and oral iodide drugs.

staining

For the purposes of this volume, we may define staining as a technique for achieving contrast between an object and its surrounding medium to enhance the visibility of the object in microscopy. Because most microorganisms are nearly transparent, coating them with a dye greatly helps us see them under an ordinary optical microscope. Simple staining refers to methods in which a microbe is simply coated with a dye (e.g., methylene blue and safranin) so that the organisms are seen as colored objects against a light background. There are also a number of specialized staining procedures, such as negative stains, differential stains, and chemicals with a special affinity for one or more cellular component, which help highlight specific features of an organism or differentiate between different organisms on the basis of structure or chemical composition. *See also* ACID FAST STAIN; CONTRAST; GRAM STAIN; MICROSCOPE; OPTICAL MICROSCOPE; NEGATIVE STAIN.

Staphylococcus

Genus of commom bacteria found free-living in soil and water and as part of the indigenous microflora of the upper respiratory tract, different species of which may or may not be pathogenic in humans. Morphologically, these bacteria are small cocci that exist in irregular clumps or grape-like clusters for which they are named (*staphylo* is Greek for cluster). From the human perspective, there are three main species of *Staphylococcus* that share morphological and biochemical features and differ mainly with respect to their pathogenicity. *S. aureus* is the most prevalent and also the most pathogenic species. It is pyogenic—pus-forming—in nature and has been implicated in a number of diseases ranging from simple boils and carbuncles to serious conditions such as osteomyelitis, toxic shock syndrome, scalded skin syndrome in infants, and food poisoning. *S. epidermidis* is normally present on our skin as a commensal. The third species, S. *saprophyticus*, has been found growing on dead tissues. When grown on nutrient agar plates, *S. aureus* colonies may range in color from white to golden yellow, corresponding to the color of the pigment produced. Although these pigments were once used for the identification of different staphylococci (the species name *aureus* is derived from the word for gold), they are no longer considered a reliable indicator of species type.

While staphylococci do not form spores, their compact size and shape makes them resistant to drying and enables them to survive for several months in air and dust. They are facultative anaerobes that ferment a variety of sugars—with the highest preference for glucose—to produce lactic acid and no gas. Mannitol fermentation is a useful marker for differentiating among species and strains. The pathogenic organisms produce various extracellular products known to have adverse effects on humans. These effects include the enzyme coagulase, which functions as a factor in blood coagulation; hemolysins (or staphylolysins), designated as alpha, beta, gamma, and delta, which lyse red blood cells; and leukocidins, which attack the phagocytic white blood cells. In addition, some strains also produce a type of enterotoxin (i.e., a substance that is toxic in the intestine). The staphylococci show an affinity for bacteriophages due to the presence of specific receptors on their surface. In fact, the ability to be lysed by a specific type of phage is a basis for differentiating among different strains of *S. aureus* and is a useful tool for epidemiological purposes. *See also* COMMENSAL; FOOD-BORNE DISEASES; SCALDED SKIN SYNDROME; TOXIC SHOCK SYNDROME.

sterile technique

This term is often used interchangeably with aseptic technique, which is a more accurate description of the methods because complete sterility in a working environment is not really achievable. *See also* ASEPTIC TECHNIQUE; STERILIZATION.

sterilization

In a microbial context, sterilization refers to any process that completely eliminates all living organisms from the object that is treated. Unlike disinfection, sterilization is a discrete property—i.e., something is either sterile or it isn't. Thus, the presence of even a single living organism on the surface of an instrument or in a batch of medium would render them asterile.

Sterilization per se is an impractical objective except in preparing materials and equipment for growing pure cultures of various microbes in the laboratory, or in the preparation of equipment for certain medical and surgical products. In most cases, disinfection is the more attainable and practical goal. Effective sterilization techniques take advantage of one or more of the known properties of normal environmental organisms—such as heat sensitivity or their small size—to destroy them or, alternatively, to physically remove them. The actual method used depends on the nature of the material to be sterilized. For example, either dry heat or steam (in an autoclave) may be used to sterilize surgical equipment and glassware, while only the latter method is suitable for fluids. Filtra-

tion methods, either alone or in combination with centrifugation, are also useful for removing living organisms from liquids. Such methods are particularly useful for sterilizing liquids such as serum whose properties would be significantly altered by heat sterilization. UV irradiation and the use of chemicals may be useful for sterilizing surfaces, but sterility lasts only until the first exposure of the surface to the ordinary environment. These latter methods are, therefore, disinfectant measures rather than sterilization. *See also* AUTOCLAVE; DISINFECTION; STERILE TECHNIQUE.

strain (bacterial)

A single isolate of a bacterial species, which differs from other known organisms in just one or a few characteristics, usually relatively minor features from a taxonomical standpoint. *See also* SPECIES; TAXOMONY.

Streptobacillus moniliformis

A causative agent of rat bite fever in humans, this organism is a gram-negative, highly pleomorphic bacterium, occurring as short rods that may form filaments or chains. Bacterial cells often give rise to yeast-like swellings and readily form L-forms, which often give the chains a beads-on-a-string appearance. *S. moniliformis* is a fastidious organism that requires media enriched with blood or scrum for its growth. The bacteria are either aerobes or facultative anaerobes and non-motile. They may be found as commensals in healthy rats and have also been isolated from infected mice and other rodents. *See also* RAT BITE FEVER.

Streptococcal toxic shock syndrome.
See TOXIC SHOCK SYNDROME

Streptococcus

Streptococcus is a large, heterogeneous bacterial genus, comprising some of the most important human pathogens, a variety of commensals, and numerous other free-living species found widely distributed in the soil and air. A number of organisms that were originally classified within this genus have since been reclassified in other genera (e.g., *Lactococcus, Leuconostoc,* and *Pediococcus)*, but even so, the number of bacteria classified in the genus *Streptococcus* is large. Morphologically, the streptococci are spherical organisms that usually occur in chains or sometimes in pairs (diplococci) and lack flagella or any other means of motility. They stain gram-positive and different species may or may contain a capsule. Physiologically, they tend to be facultative aerobes that ferment sugars to produce acid.

Because of their vast diversity, the classification of the streptococci has been a subject of considerable importance to microbiologists. One of the first methods used to classify streptococci was to differentiate them on the basis of their growth characteristics on blood agar, which in turn was a reflection of the type of hemolysins produced by different species. To this day, hemolysis is used as a preliminary means of identification of streptococci and related organisms that are isolated in the laboratory. The α-hemolytic species produce enzymes that partially lyse sheep red blood corpuscles, resulting in the development of cloudy, greenish, or brownish zones of hemolysis around the bacterial colonies. *S. pneumoniae, S. salivarius,* as well as species of the newer genus *Enterococcus* belong to this group. The β-hemolytic streptococci, which are among the most pathogenic species, produce hemolysins that completely lyse the blood cells, resulting in the formation of clear zones around the colonies. *S. pyogenes* is the representative species. The γ-hemolytic organisms do not actually produce any hemolysins and are typically nonpathogenic. Strains of both α and β streptococci may appear to belong this last category due to mutations resulting in the loss or inactivation of their hemolysins. A second method for classifying the streptococci was developed by Rebecca Lancefield on the basis of the antigenic structure of the Protein M component of the cell wall. In this scheme, the streptococci are subdivided into some 15 categories (Groups A–O) of which A–G comprise the β-hemolytic organisms. Species of human importance and interest are discussed in the following sections. *See also* HEMOLYSIN; LANCEFIELD, REBECCA; PROTEIN M.

Streptococcus pneumoniae

Originally known as *Diplococcus pneumoniae,* his organism is the most common cause of bacterial pneumonia in the world and is also responsible for nearly half of all ear infections. One of the major virulence factors is the thick polysaccharide capsule, as evidenced by the fact that capsule-less mutants of *S. pneumoniae* (i.e., the R-strains) are avirulent. These capsules help the bacteria colonize tissues and avoid destruction by phagocytosis.

Bacteria gain entry into the body via the respiratory tract and colonize the lungs to cause pneumonia. Other systemic sequelae, such as bacteremia and meningitis, may result if the infection does not resolve itself or is left untreated, and also in cases where the infected individual is highly immuno-compromised. Antibiotic therapy, which had proven useful in the past, is now losing efficacy because these bacteria have developed resistance to many of the widely used types. Prevention of these infections is, therefore, very important. Vaccinations for 23 existing serotypes do exist, but because of the diversity and possible confusions regarding exposure, these are underused. Also, the vaccines are not recommended for use in children under two years. With the recent sequencing of the 2.2 kilobase genome of *S. pneumoniae,* scientists hope to gain more insights into its pathogenesis, virulence factors, and antibiotic sensitivities to devise better strategies to prevent infections. *S. pneumoniae* occupies a special place in the history of microbiology as the first organism in which bacterial transformation was observed and as the model system that led to the discovery of the genetic role of DNA by Oswald T. Avery, Colin MacLeod, and Maclyn McCarty in 1944. *See also* AVERY, OSWALD; CAPSULE; MACLEOD, COLIN; MCCARTY, MACLYN; TRANSFORMATION (BACTERIAL).

Streptococcus pyogenes

This streptococcal species, comprising more than 70 different serotypes or strains, is responsible for the majority of the well-known streptococcal infections, including both skin and respiratory disorders. The most frequent upper respiratory tract disease is a type of pharyngitis called strep throat. Although this is often self-limiting with nonspecific symptoms, proper diagnosis is important for avoiding streptococci-specific complications such as acute rheumatic fever and acute glomerulonephritis, which can have fatal consequences. Pharyngitis is caused by strains of *S. pyogenes* that carry a bacteriophage with a gene for a special pyrogenic or erythrogenic toxin responsible for scarlet fever. These toxin-producing strains have also been implicated in outbreaks of toxic shock syndrome similar to the type caused by staphylococci. Primary pharyngeal infection may also give rise to suppurative (pus- or mucus-forming) sequelae such as abscesses on the tonsils and throat, sinusitis, and middle-ear infections. Skin and soft-tissue infections of *S. pyogenes* include impetigo, erysipelas, and cellulitis. In general, there is some correlation between the strain or serotype of the organism and the type of disease caused. *See also* IMPETIGO; SCARLET FEVER; TOXIC SHOCK SYNDROME.

Streptomyces

One of the most extensive genera among the prokaryotes, *Streptomyces* consists of over 500 individual species of bacteria, many of which have garnered the interest of the pharmaceutical industry because of their ability to produce antibiotics. Perhaps the best known example is *S. griseus*, which produces streptomycin. Other commercially important species include *S. coelicolor* and *S. clavuligerus*. *Streptomyces* are not known to be pathogenic in humans but have been found to cause certain plant diseases. They are gram-positive, mostly filamentous bacteria that resemble the fungi in most morphological characteristics (save their nuclear organization), including the presence of extensive, well-developed mycelia, with aerial hyphae,

for the formation of spores. Horizontal hyphae may also fragment to form chains of spores. Both types of spores are asexual structures that are metabolically less active than the vegetative cells and can germinate under favorable conditions to give rise to new mycelia. Metabolically, *Streptomyces* is a heterotroph found in the soil, capable of using compounds such as glucose, starch, and lactose. Antibiotics are not primary metabolites produced during the period of active growth of the bacteria, but rather secondary metabolites that appear as the growth rate of the bacteria slows down. The exact genetic mechanisms that regulate antibiotic production are still being worked out. *See also* ANTIBIOTIC; STREPTOMYCIN.

streptomycin

Natural aminoglycoside antibiotic that is produced by *Streptomyces griseus*. This was the next antibiotic to be discovered after penicillin. Although it has some toxic effects on human hosts—because of its effects on protein synthesis—it has proved useful in the past for treating tuberculosis. *See also* AMINO–GLYCOSIDE ANTIBITIC; ANTIBIOTIC; *STREPTOMYCES*.

subacute sclerosing panencephalitis

Chronic progressive and ultimately fatal disease of the central nervous system in children or adults seen as a delayed sequelae of a measles infection. The exact mechanisms by which this disease develops are not precisely understood, but it appears that the virus goes into a latent state and causes the disease upon reactivation, which may occur years after the initial measles infection. Patients show high titres of anti-measles antibodies in both the blood and the cerebrospinal fluid. Although intact viruses are not found, it is possible to isolate them in tissue culture with the help of some additional viruses. Scientists speculate that disease occurrence may correlate with the amount of the virus that reaches the brain tissue during the initial infection. Once in the brain tissue, the viruses are unable to multiply completely because of the lack of an envelope protein, which leads to the accumulation of viral proteins and incomplete viral particles in the cells. Over time, the accumulation of this material leads to the destruction of the cells and the sclerosis in the brain. Diseases of this type are also known as "slow virus" diseases to reflect the long period of latency between initial infection and the appearance of symptoms. *See also* MEASLES.

sulfonamides

Synthetic antibiotics that are bacteriostatic for many gram-positive bacteria and which act by interfering with the synthesis of folic acid, an important participant in many essential metabolic processes. The sulfonamides are similar in structure to one of the normal components of folic acid and thus can be mistakenly incorporated into the molecule during synthesis. This results in the formation of an inactive compound that is unable to replace folic acid in metabolism, thereby inhibiting bacterial growth. *See also* ANTIBIOTICS.

surveillance

Ongoing scrutiny of a specified population in terms of health data—i.e., the occurrence and frequency of different diseases, or the change in various factors that may affect the diseases—so as to maintain the health of the overall population and efficiently mobilize and implement adequate measures to deal with any health crises that may arise. To this end, surveillance is an ongoing process of systematic collection, analysis, and interpretation of epidemiological data, integrated with the dissemination of the results of these data to the interested parties, such as local health care providers and public health officials. Epidemiological surveillance tasks are usually carried out at a national or regional level by various government and private agencies. The official organization for this purpose in the United States is the Centers for Disease Control (CDC), while the World Health Organization (WHO) performs many of the same tasks at the international level.

symbiosis

A stable, physical association between two living species—called the symbionts—usually from different taxonomic groups. The term is often used to denote a mutually beneficial relationship between organisms, but parasitic and commensal (one-sided) relationships are also forms of symbiosis. *See also* COMMENSALISM; MUTUALISM.

syncytium

A continuum of cells, or a multinucleate mass formed by the fusion of several cells together, usually due to the action of a virus. The infection of a cell with a virus typically introduces certain viral proteins into the membrane of the host cell, which recognize and bind to molecules on the surface of other cells, resulting in fusion. The end result of this fusion is cell death. The formation of syncytia is a characteristic cytopathic effect seen in tissue cultures due to infection with the measles virus or the respiratory syncytial virus. The induction of syncytia by HIV appears to be a major mechanism by which these viruses kill T cells. *See also* CYTOPATHIC EFFECT; RESPIRATORY SYNCYTIAL VIRUS.

syphilis

A venereal or sexually transmitted disease caused by the spirochete *Treponema pallidum*. The disease has a long history, and over time has acquired many nicknames, including the "French pox" and the "Englishmen's disease," during the war between the two countries in the late 1400s. The modern name for the disease comes from a poem written in 1530 by a scientist called Fracastorius, in which a shepherd was punished by the gods with the disease for being disrespectful. This same scientist was also the first to speculate—in a scientific article—that the disease was passed from one person to another in some unseen germ via sexual contact. The disease later gained even more notoriety as its spread came to be associated with sailors who traveled around the world from port to port.

Untreated syphilis progresses through three stages. The primary phase of the disease begins with the entry of *T. pallidum* into the body via the skin or mucous membranes following sexual contact. The most common site for infection is the genital area, although it is occasionally transmitted orally. Organisms begin to multiply almost immediately and produce a local lesion called a hard chancre at the site of entry. This is usually a rather small, inconspicuous lesion that may go unnoticed and may even heal completely without treatment. However, the organisms continue to multiply and enter the local lymph nodes and the bloodstream, from which they are able to invade other tissues.

Secondary syphilis, which coincides with the spread of the organism, sets in sometime between four and eight weeks after the initial infection. Often the body reacts with an allergic reaction resulting in a widespread rash. In addition, the patient may develop lesions or sores in the mucous membranes of the mouth and genitalia and experience mild systemic symptoms such as fever, sore throat, and headaches. Although not debilitating, the secondary stage of syphilis is extremely infectious because the sores and rashes are full of organisms that may spread via direct contact or though articles containing secretions from these sores.

The external symptoms of secondary syphilis may recede or disappear spontaneously like the primary lesion, but inside the body the organism continues to multiply and spread. This leads to the development of a chronic stage called tertiary syphilis, during which virtually any organ may be the target for attack. This phase is characterized by the formation of noninfective, ulcerative lesions called gummae on multiple sites both externally and internally. Gummae on the skin may simultaneously spread and heal, which causes disfigurement due to the formation of scar tissue. The internal spread and multiplication of the organisms results in the massive destruction of a variety of tissues, in part by the organism and also due to the inflammation of the regions. The most serious outcomes are associated with damage to the heart, arteries, and nervous system. In the final stages, syphilis may be associated with aneurisms of the aorta, arteriosclerosis, and, when the nervous system is damaged, general paralysis and even various forms of insanity.

In addition to sexual contact, *T. pallidum* may be spread from an infected pregnant mother to her child via the placenta, especially during the secondary phase of the disease. This often leads to fetal death, or the baby may be born with a form of congenital syphilis. Babies may have rashes or sores that are infectious, although often the symptoms are mild and may go undiagnosed. *See also* SEXUALLY TRANSMITTED DISEASE; *TREPONEMA*; VENEREAL DISEASE.

T

T cell

Type of lymphocyte involved in the body's cell-mediated immune response to various harmful agents such as viruses and cancer antigens, as well as unfamiliar antigens such as those on transplanted tissues or organs. The T cells undergo maturation in the thymus gland, which is the reason for their name. They constitute the major proportion of circulating lymphocytes—anywhere from 60–70 percent. Although morphologically identical, T cells are divided into three subgroups according to differences in their function and in their antigenicity, which in turn is a reflection of protein receptors on the cell surface. The main types of T cells include helper T cells and cytotoxic T cells. *See also* B CELL; IMMUNE SYSTEM; LYMPHOCYTE.

taxonomy

The branch of science that groups living organisms into hierarchical groups or "taxa" according to their similarities and differences. Ideally, taxonomic groups should reflect the phylogenetic or evolutionary relationships between the different organisms being classified. Within a group of organisms (e.g., the bacteria), it may be possible to gauge phylogeny on the basis of the sequence of the ribosomal genes or other genetic sequences that are present across species. However, this is not always feasible, and generally classification is based on genotypic and phenotypic features, such as morphology, colony characteristics, structural features, and biochemical/metabolic properties. *See also* BINOMIAL NOMENCLATURE; LINNEAUS, CARL.

Temin, Howard M. (1934–1994)

This American scientist made fundamental contributions to our modern understanding of the oncogenic or tumor-inducing viruses. Working in the 1950s along with his colleague Harry Rubin, Temin developed *in vitro* procedures for growing, monitoring, and quantifying animal viruses. Temin was also one of the pioneering virologists to study oncogenic RNA viruses and to postulate a mechanism for their action in human cells. He was able to furnish proof for his hypothesis by isolating the enzyme reverse-transcriptase, for which he shared the 1975 Nobel Prize with David Baltimore and Renato Dulbecco. *See also* BALTIMORE, DAVID; DULBECCO, RENATO; ONCOGENIC VIRUS; REVERSE TRANSCRIPTASE.

tetanus

Acute spasmodic, disease—usually fatal—caused by infection with *Clostridium tetani*. Infection usually occurs by the contamination of wounds with spores from the air, dust, or soil. Clinical symptoms typically appear after an incubation period of 6–15 days (corresponding to germination of the spores and multiplication of the vegetative cells) although onset might occur quickly—in one day—or after long periods of up to two months. Because this species does not produce many protein- and sugar-degrading enzymes like the gas gangrene clostridia, the initial infection may often go unnoticed, giving rise to what is know as a cryptogenic tetanus.

The major symptoms of tetanus are due to the effect of the spasmogenic exotoxin. At first, the patient experiences muscular tension and cramps in the area around the infected wound, but as the infection and toxin spread in the body, symptoms become more generalized, including backaches, headaches, irregular heart rhythms, generalized irritability, and facial twitches. A prolonged spasm of the jaw muscles may restrict the patient's ability to move the mouth, which is why this disease came to be known as lockjaw. Tetanus is usually fatal due to the involvement of respiratory muscles.

The treatment of tetanus is aimed at neutralizing the toxin with the help of an antitoxin, preventing the proliferation of the clostridia with appropriate antibiotics. The incidence of this disease in developed countries is low because of the ability to immunize patients using a toxoid. It should be noted that clinical tetanus does not confer immunity against future episodes of the disease. Thus, people who are likely to be exposed should be treated periodically with booster shots. *See also* CLOSTRIDIUM TETANI; LOCKJAW; TOXOID.

tetracycline

Broad-spectrum antibiotic that acts by binding to transfer RNA (t-RNA) molecules and inhibiting protein synthesis. In addition to treating a wide variety of human and animal diseases, this group of antibiotics has also proven useful in treating certain plant diseases such as coconut lethal yellowing. *See also* ANTIBIOTIC.

Theiler, Max (1899–1972)

A member of the research group that pinpointed the cause of yellow fever to a virus, overturning the original belief that it was caused by bacteria. Theiler was also respon-

sible for demonstrating that this disease could be experimentally transmitted to white mice. This knowledge provided scientists with an easy and more economical system for cultivating and studying the yellow fever virus in the laboratory, and it enabled them to develop a successful vaccine against a disease that had plagued humankind at least since the seventeenth century. Theiler was awarded the Nobel Prize in 1951. *See also* YELLOW FEVER.

thermophile

Organism that not only can tolerate, but in fact has a preference for, living at extremely high temperatures.

Thermotoga maritima

This thermophilic, rod-shaped bacterium is of special evolutionary significance because the DNA sequence of its ribosomal genes has revealed it to be one of the most slowly evolving lineages among the true bacteria (Eubacteria). The organism was originally isolated from marine geothermal sediments in Italy and has an optimal growth temperature of 80°C. It can use a great many simple carbohydrates, as well as polymers such as cellulose and xylan, as the primary carbon and energy source (by conversion to hydrogen). The complete genome of this organism has been sequenced, and although the core is eubacterial in origin, about one-fourth of it contains archaebacterial sequences. *See also* ARCHAEBACTERIA.

Thiobacillus ferrooxidans

The bacteria of this genus are distinguished by their ability to derive energy by oxidizing reduced forms of either sulfur or iron (ferrous). An interesting feature of the metabolism of *T. ferrooxidans* is that while it is generally an obligate aerobe with an absolute requirement for oxygen, it will grow under anaerobic conditions in the presence of ferric ions (the oxidized form of iron), which it will use during respiration as the terminal electron acceptors in lieu of oxygen. Thus, the same organism can either oxidize or reduce iron based on its microenvironment. *T. ferrooxidans* abounds in mine tailings and coal deposits where it is believed to play an important role in oxidizing iron and inorganic sulfur compounds. While these bacteria have been implicated as major players in the pollution of streams in mining areas, their metal-utilizing abilities and acid tolerance may also be channeled to beneficial purposes. For example *T. ferrooxidans* may be used to treat acid-mine drainage by accelerating the release of heavy metals into solution where they may be disposed of more easily. It is also possible to use these bacteria in bioleaching processes to extract metals from low grade ore. They are gram-negative, motile rods that, because of their metal and sulfur utilization, have a preference for acidic environments.

thrush

A superficial fungal infection in children—caused by a species of *Candida*—characterized by the appearance of white patches in the mouth and throat. Known since ancient times, thrush was described in the writings of Hippocrates and Galen. Because the fungus is widespread in the environment, exposure to these organisms is fairly common, and only children with weak immune systems unable to fight off infections contract the disease. In addition, women with vaginal yeast infections may pass the fungus on to the mouths of their babies during birth. *See also* CANDIDA.

thylakoid

The site of photosynthesis, and sometimes of respiration, in cyanobacteria. Thylakoids are chlorophyll-containing, membrane-bound sacs that lie close to, but distinct from, the cell wall in a parallel arrangement. *See also* CHLOROPHYLL; CYANOBACTERIA; PHOTOSYNTHESIS.

tinea pedia. See ATHLETE'S FOOT

tissue culture

Laboratory technique of the *in vitro* culturing of cells, organs, or parts of organs independent of their source. The term is often used interchangeably with cell culture, which refers specifically to the growth and maintenance of cells dispersed from their tissues of origin. Removed from their normal environment, these cells lose their resemblance to the parent tissue within a few cycles of replication. Certain cells types lose their ability to proliferate beyond a few generations, apparently requiring some sort of "master switch" to regulate their activities. These are called the primary cell cultures. Other cell types become transformed in such a manner as to become "immortal" and may be subcultured indefinitely. The cultures derived from these types of cells are called continuous or established cell lines. An example is the HeLa cell line, now distributed worldwide by ATCC, which was originally derived from the epithelial cells in a cervical carcinoma, believed to come from a woman named Helen Lane or Lange. Both primary and continuous cell cultures are used for the laboratory cultivation of various viruses and of other obligate intracellular parasites. Organ culture refers to the maintenance of part or whole organs *in vitro*, in a manner so as to preserve at least part of the integrity of the three-dimensional shape and intercellular relationships of different cells in that organ *in vivo*. *See also* CYTOPATHIC EFFECT; ENDERS, JOHN; TRANS—FORMATION (CELLULAR).

togavirus

This is a family of mostly arthropod-borne RNA viruses that have gained importance in recent decades as emerging pathogens, causing various viral encephalitides and arthritic fevers. A notable exception to the arboviruses in this family is the rubella virus, the causative agent of German measles, which does not infect any arthropods and which is classified into a separate genus from all other togaviruses. The remainder of the members of this family are grouped into a single genus called alphavirus. Both the alpha and rubella viruses are spherical particles consisting of an icosahedral capsid with a tightly bound lipid envelope. The latter also contains proteins and confers antigenic properties upon the viruses. The genome is a single-stranded RNA molecule of the mRNA type and is capable of infecting cells even when divested of its protein and lipid coverings. The togaviruses replicate in the cytoplasm of the host cells, and acquire their envelope as they emerge from the host cell by budding. The rubella viruses are unique among this family as the only viruses that can acquire their envelopes from membranous structures inside the host cell, e.g., the endoplasmic reticulum.

Alphaviruses are typically maintained in nature in a cycle between mosquitoes and a reservoir host, usually small birds or rodents. The spread to humans and other large mammals is relatively rare, although it is becoming more common with the rise in immuno-suppressed populations due to AIDS. Three main species of the alphaviruses have been identified as causes of encephalitides (see Table 4 accompanying the Encephalitis entry for details), which are also called the equine encephalitides because they are often associated with horses and appear to be spread to humans from this source via mosquitoes. Several other species of the alphaviruses have been identified as the causes of a triad of symptoms, including fever, severe joint pains, and rashes, occurring in various tropical regions of Africa, southern Asia, the Philippines, South Pacific islands, Australia, South America, and, occasionally, parts of Europe. These viruses have been named for the African terms in which the diseases are described, e.g., Chikungunya and O'nyong-nyong viruses, or after the areas of the original outbreaks, e.g., the Mayaro, Ross River, and Sindbis viruses. The disease may either resolve itself or persist for several months. Control measures are aimed at insect control to prevent spread to large human populations. Vaccines have been developed for the eastern and Venezuelan equine viruses and are administered to those populations perceived at risk, including horses. An experimental vaccine against the Chikungunya virus, which is widespread in Africa and Asia, is being tested in humans. *See also* ARBOVIRUS; ENCEPHALITIS; RUBELLA VIRUS.

toxemia

Toxemia refers to the presence of a toxic substance—usually produced by some infecting agent—in the bloodstream. Because the substance is easily spread to all parts of the body along with the blood, it causes generalized damage in various tissues and organs.

toxic shock syndrome (TSS)

Disease, predominantly of women and postpubertal girls, caused by infection with a strain of *Staphylococcus aureus* that produces a unique exotoxin capable of attacking the epidermis. Symptoms of toxic shock syndrome (TSS) are brought on by septicemia—i.e., the proliferation of bacteria in the bloodstream. The disease begins abruptly with symptoms of fever, diarrhea, vomiting, and abdominal cramps that escalate into hypotension (low blood pressure) and shock within 72 hours. Other symptoms, which may often go unnoticed, include the development of a rash and a thick vaginal discharge. News of TSS hit the headlines during the late 1970s when it broke out in almost epidemic proportions among women. Disease incidence was found to be associated with the growing rate of tampon use. This led to the suspicion that the infecting organism entered the bloodstream from tiny lesions in the vagina where it was introduced during the process of tampon insertion. Even today, tampon users are at the highest risk of contracting the disease. However, TSS may develop in males and preadolescent girls if the exotoxin-producing strains of *S. aureus* enter the bloodstream via other lesions in the body.

During the 1980s, a new group of agents came to be associated with TSS-like symptoms, although not necessarily of vaginal origin. Most often this new disease was linked to a cutaneous lesion and was characterized by acute pain in the extremities as well as other parts of the body, fevers, and shock, and a case-fatality rate often exceeding 50 percent. The causative agents were found to be the group A streptococci, i.e., strains of *Streptococcus pyogenes* in a new dramatically virulent guise. The early association of this syndrome with necrotic wounds gave these bacteria the popular label of the "flesh-eating bacteria," considerably heightening the public's terror of the agent and disease. Timely intervention with the appropriate antibiotics is important, but equally important in the treatment of streptococcal toxic shock syndrome (STSS) is the exploration of the body and appropriate measures to remove bacteria. *See also* STAPHYLOCOCCUS; STREPTOCOCCUS PYOGENES.

toxoid

A molecule that is similar, but not identical, to a toxin. The tetanus toxoid, for instance, shares the antigenic structure of the parent toxin but lacks its toxic properties, which include the ability to induce a prolonged nervous stimulation of muscles. Thus, the toxoid is a useful substance for preparing a vaccine against tetanus—it raises the appropriate neutralizing antibodies without causing any nervous damage. *See also* ANTIBODY; ANTIGEN; TETANUS; VACCINE.

Toxoplasma gondii

This protozoan is an obligate intracellular parasite that exhibits a wide host range and low host specificity within the mammals. The primary hosts appear to be cats and other feline species, while humans, rodents, dogs, and a variety of other mammals, as well as birds, may function as the intermediate or secondary hosts. The life cycle of *T. gondii* consists of two distinct phases—an intestinal phase, during which the organism multiplies within the epithelial cells of the definitive host only, and an extraintestinal phase, which may take place in cats or any of the other intermediate hosts mentioned above. Ingestion of material contaminated with parasitic oocysts is the major method of transfer from host to host. Pregnant mothers may pass the protozoa to their offspring via the placenta.

The intestinal phase of the parasite consists of both sexual and asexual reproductive cycles. Primary infection occurs via the ingestion of the oocysts, which upon entry into the intestine release infective, vegetative cells that immediately invade the epithelial cells. These cells, called the sporozoites, rapidly undergo a few cycles of asexual reproduction inside the epithelial cells and, in a matter of days, differentiate to form sexual gametes that then fuse to form diploid cells called the oocytes. This phase of the cycle is self-limiting and lasts only for a few weeks, after which the sporozoites and gametes of *T. gondii* are no longer found in the intestinal epithelium. The oocytes are released into the intestinal lumen, where they mature into oocysts within 2–3 days by undergoing meiosis such that they then contain haploid parasites encased in a thick cyst wall. Oocysts that are shed with the feces are metabolically inactive and may survive in the soil for up to one year, provided the environmental conditions are congenial.

The extra-intestinal phase of the life cycle of *Toxoplasma* is completely asexual and haploid. It begins with the ingestion of the oocyst by various intermediary hosts or, in some instances, with the release of the haploid parasites into the cat feces. In case of the latter, the parasite proceeds to treat the cat as the intermediate host as well. Two major forms of the parasitic cell are seen to develop during this phase: present early on, during the acute phase of infection, are the actively proliferating trophozoites, which indiscriminately invade the host cells, multiply asexually within them, and then lyse the cells to release more infective parasites. As their numbers increase, the trophozoites spread to various parts of the body via circulating cells. The second form of the parasite, which is a resting cyst, is seen to develop in tissues such as the muscle and brain, presumably aided by the host's immune system. This encysted form is more resistant than normal trophozoites to digestive enzymes as well as being metabolically slower. The disintegration of the cyst wall, which happens slowly over time, can lead to the proliferation of the parasites and corresponding tissue damage. Because the encysted parasites can survive for long periods, meat from infected animals is a potential source of infection.

Given the high rate of human exposure to *Toxoplasma*, the rate of disease—called toxoplasmosis—is relatively rare. Many infected individuals may be completely asymptomatic or exhibit only mild, nonspecific flu-like symptoms. Rare but serious complications include encephalopathy due to the proliferation of organisms from cysts in the brain. Congenital infections are also very severe, particularly if the mother is infected in the early stages of pregnancy. In general, the disease is seen to affect immuno-compromised or suppressed individuals more easily and severely, and it is consequently seen at a much higher rate and with more complications among AIDS patients.

Diagnosis of the disease is achieved by the identification of *T. gondii* in biopsy specimens of encysted tissues, white blood cells, and spinal fluid. Serological tests to detect the antigens of both trophozoites and cysts are also recommended, particularly if the patient history indicates exposure to cats (and cat feces) or contaminated meat. Drug therapy is available to combat this disease. Thus far, no vaccines have been developed either for humans or to control the proliferation of *Toxoplasma* in cats.

trachoma

Infectious disease of the eyes caused by *Chlamydia trachomatis*. The disease often results in blindness as well as scarring of the corneal and other ocular tissues. Endemic in south-central regions of the United States, parts of northern India, and in remote areas of Africa, trachoma is nevertheless known to occur worldwide, and affects nearly 500 million people. In endemic areas, the clinical symptoms are seen most often in young children.

Infection with chlamydia occurs through direct contact. Common sources include fingers—from touching infected eyes—fomites such as face cloths and other linens, or bath-room articles and swimming pools. Newborn babies may often be infected from the birth canal of infected mothers—thus trachoma is not necessarily brought on by preexisting cases of ocular infections but may also originate from individuals with other chlamydial infections, notably cervical infections. The disease begins as a chronic inflammation of the conjunctival membrane (conjunctivitis) in response to the chlamydial infection. An accumulation of lymphocytes and other immune cells results in the formation of large red or yellow follicles, which may also coalesce to form larger lesions, which if not treated properly, will become necrotic and cause scarring of the conjunctiva. Regardless of the status of the chlamydial infection, conjunctival scarring has serious long-term effects. First, the scars contract and turn in the eyelids, leading to considerable friction between the eyelashes and cornea, which in turn, damages the corneal tissues. Trachoma-induced blindness is thus as much a product of mechanical damage of the tissues as it is a direct consequence of inflammation. The prolonged injury also leaves the eyes more susceptible to secondary bacterial infections with organisms such as *Hemophilus*.

If diagnosed in time, trachoma may be treated with broad-spectrum antibiotics such as tetracycline or erythromycin, to clear the chlamydial infections. However, because the most serious consequences of trachoma are post infective, proper prevention is vital. Public education regarding sanitary habits; disinfection of articles that have the potential to function as fomites, especially within households; and vigilant monitoring of the microbial status of various water sources, (particularly publicly accessible sites such as swimming pools) are all prudent measures. Sociocultural practices that reduce the contact between infected and uninfected family members or members of the community go a long way towards reducing severity and incidence of the disease. *See also* CHLAMYDIA TRACHOMATIS.

transcription

The synthesis of RNA molecules from their DNA templates. Transcription is a relatively simple process, that involves the stepwise synthesis of single-stranded RNA chains from the corresponding template segment, catalyzed by appropriate RNA polymerases. Uracil (U) replaces thymine (T) in the AT base pairs. An RNA molecule is always the first product to emerge from a gene, regardless of whether the final product is to be RNA or protein. Transcription is thus the terminal process in t-RNA and ribosomal RNA synthesis, but only an intermediary step in protein synthesis. The RNA species that will be processed further to make proteins is called messenger RNA. This stepwise synthesis is necessary to minimize direct participation, and therefore wear and tear, of DNA during the lifetime of the organism. *See also* DNA; RNA; TRANSLATION.

transduction

Process of genetic transfer in bacteria where the vehicle for carrying the gene is a bacteriophage. The virus contains a piece of bacterial DNA in its genome and when it infects a new cell the properties encoded in this DNA are expressed in this new cell. Two modes of transduction occur in nature. Generalized transduction occurs due to lytic phages, while specialized transduction occurs as a result of a different class of viruses called lysogenic phages. Both types of transduction have been successfully used as tools in genetic engineering to introduce specific genes into target cells. *See also* BACTERIOPHAGE; GENETIC ENGINEERING.

transformation (bacterial)

Process of genetic transfer in bacteria, where one organism takes in a piece of naked DNA of another from the environment and acquires the traits encoded in this piece. The ability to undergo transformation is not a property of all bacterial cells but is restricted to what are called competent cells at certain stages of their life cycle. Transformation was first observed by British physician Fred Griffith, who noticed that certain avirulent mutants of *Streptococcus pneumoniae* became virulent (i.e., were "transformed") when mixed with dead organisms of a disease-causing strain. Some years later, at the Rockefeller University in New York City, Oswald Avery, Colin MacLeod, and Maclyn McCarty investigated this phenomenon further and in the course of the search for the "transforming principle" discovered that DNA was the carrier of genetic material. *See also* AVERY, OSWALD; DNA; GENE; MACLEOD, COLIN; MCCARTY, MACLYN; *STREPTOCOCCUS PNEUMONIAE*.

transformation (cellular)

The manifestation of changes in the behavior of a cell, such as improperly regulated cell division, unchecked growth, and alterations in the surface properties, e.g., shape, contact inhibition, and the ability to communicate with neighboring cells or the extracellular medium. The underlying cause for these changes is the mutation of one or more regulatory genes. Different agents that may induce these mutations include viruses, chemicals, or radiation. Cellular changes induced by a transforming virus are fairly specific, and may be observed in the laboratory by infecting tissue culture cells with the virus. Similar changes *in vivo* may lead to cancer. *See also* CANCER.

translation

The synthesis of protein chains from messenger RNA in the presence of ribosomes and the appropriate transfer RNA molecules. This process involves a stepwise addition of amino acids, carried by cognate t-RNAs, on a growing peptide chain, using the ribosomes as the site for peptide bond formation. The correct order of amino acids is encoded in the DNA or m-RNA sequence in a special triplet code, in which each amino acid is specified by a sequence of three nucleotides called a codon. The end result of translation is the formation of a linear peptide chain, which is quickly folded into a specific three-dimensional conformation and transformed into a functional protein through a series of post-translational events. *See also* PROTEIN; RIBOSOME; RNA.

transmission electron microscope (TEM)

Another term used for a conventional electron microscope in which images of the specimen are formed by the diffraction of electrons that are transmitted through the specimen. This approach has a higher resolving power over scanning electron microscopy, and is more useful for discerning details of interiors of structures because its images are of three- rather than two-dimensional objects. *See also* ELECTRON MICROSCOPE.

treponema

Although some of the diseases caused by this genus have been known to humans for centuries—syphilis, for example, was well described in medieval times—and even though *Treponema pallidum*, the syphilis agent, was visualized as far back as 1905, the bacteria resist close scientific scrutiny to this day. They are extremely fragile organisms with spiral or coiled bodies ranging in length from 3–20 μm and less than 0.15μm in diameter, and susceptible to any changes in their surroundings, including pH, temperature, salt concentrations (tonicity), moisture content, and the presence of even mild soaps and stains. Bacteria of this genus do not respond at all to the Gram stain and are best visualized using other microscopic techniques, such as using silver nitrate impregnation, negative staining, or via dark-field microscopy. This last technique reveals them to be capable of a corkscrew like motion, due to the rotation of the cells. Strictly anaerobic in nature, *Treponema* species have also proven remarkably resistant to cultivation, and scientists have not been successful in growing them either in media or tissue culture. There has been limited success with maintaining, but not growing, *T. pallidum,* in the lab under anaerobic conditions at 25°C. Examples of other identified species include *T. carateum,* which causes a skin disease called pinta; *T. pertenue,* the agent for yaws; *T. vincentii;* and a rabbit-spe-

cific species called *T. cuniculi. See also* BEJEL; PINTA; SYPHILIS; YAWS.

tricarboxylic acid (TCA) cycle. *See* KREBS CYCLE.

trichome

Row of cells that remain attached to one another following successive cell divisions. The main difference between a trichome—seen, for example in the cyanobacterium *Beggiatoa*—and a simple chain of bacteria, such as those of the streptococci, is the presence of pores between adjacent cells, which allows for cell-to-cell communication. Individual cells in a trichome may be externally demarcated by the presence of constrictions at the locations of the intracellular septa, or appear as a single linear filament, which may or may not be encased in a common sheath. *See also* CYANOBACTERIUM.

Trichomonas

A flagellate parasitic protozoan with a simple life cycle, different species of which are often found residing in various locations of the human body. The most common species, and also the only known pathogen, is *T. vaginalis*, found inhabiting the urino-genital tract; *T. tenax* may be found in the oral cavity and respiratory tract, but is generally not associated with any disease. The parasite is a globular cell with a prominent nucleus and a rigid structure termed as the "axostyle" that runs from one end to another. In addition, the cell consists of three to five flagella on one (anterior) end and an undulating membrane, which enable the organism to move with ease in their natural environments. These organisms multiply by binary fission and do not form cysts. They are transmitted from one individual to another via direct (typically sexual) contact, or fomites, including unclean bathroom facilities and infected clothes and towels.

T. vaginalis is the agent of a sexually transmitted disease that is usually called trichomoniasis. Usually this disease takes the form of vaginitis in women—the organism is one of the three main causes of this type of inflammation. Infections are confined to the vagina, vulva, cervix, and urethra, and rarely progress beyond. Common symptoms include a burning or itchy feeling and vaginal discharge, which may be thick and show traces of blood in heavy infections. Trichomoniasis is often asymptomatic in men, or may take the form of urethritis or prostatitis. Urination is typically painful and frequent, and small amounts of discharge may be noted in the urine. Diagnosis is confirmed by the microscopic observation of wet smears of clinical samples from the urino-genital tract. Metronidazole is the drug of choice in treating trichomoniasis.

trophozoite

Motile, vegetative form of a protozoan parasite that feeds, multiplies (asexually), and maintains the parasite population within its primary host. *See also* CYST; PROTOZOAN.

Trypanosoma

Genus of flagellate protozoa that live in the blood and tissues of their human hosts and cause human diseases such as African sleeping sickness and Chagas disease. These parasites have a two-host life cycle, alternating between humans and insect vectors, such as the tsetse fly and the *Triatoma* bug, with no

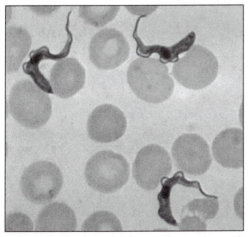

Trypanosoma forms in blood smear from patient with African trypanosomiasis. *CDC/Dr. Myron G. Schultz.*

sexual stages in either host. The two main groups of trypanosomes that may be differentiated on the basis of their geographic distribution, mode of transmission from insect to human, and the human tissues that they tend to infect. The two groups are discussed in separate entries.

Trypanosoma brucei

This trypanosome species is the causative agent of African sleeping sickness and appears to be confined to the African continent. These organisms are transmitted directly into the site of an insect bite when a tsetse fly (*Glossina* species) takes its blood meal. The species consists of two subtypes, *T. brucei rhodesiense* and *T. brucei gambiense*, which are indistinguishable morphologically, but follow a different course of infection. The *rhodesiense* type causes a much faster progressing and fulminant disease (East African sickness), with much higher levels of parasitemia. Geographically, this disease is less widespread and appears to be confined to a portion of central East Africa. Animals, not humans, are the primary reservoirs for the *rhodesiense* subtype. In contrast, the Gambian or West African parasite resides primarily in humans and is widespread in East *and* parts of Central Africa. A unique feature of the African trypanosomes is their ability to change the surface coat of their outer membrane in an attempt to evade the hosts' immune surveillance mechanisms.

When a tsetse fly deposits the infective parasites into the human, the cells multiply in peripheral blood for a few generations and then migrate towards cells in the lymph nodes and central nervous system. The parasites in this stage are highly pleomorphic, although the prevalent form during initial infection is a long, slender, nucleated cell with a single flagellum extending from the undulating membrane that runs along the length of the cell. They are capable of rapid movement in the body and multiply by longitudinal binary fission. In later cycles of multiplication, the parasites also begin to form short, stumpy forms with no flagella that are infective for the insect. In the fly, these parasites multiply in the lumen of the mid gut and hind gut of the flies to form motile, flagellated parasites. After about 2 weeks, the trypanosomes migrate to the salivary glands via the salivary ducts, and are ready for injection into the host when the fly takes a blood meal. Once infected, the tsetse fly is infective for life. *See also* AFRICAN SLEEPING SICKNESS.

Trypanosoma cruzi

The causative agent of Chagas disease, this New World (predominantly South American) trypanosome is different from all the other species in its mode of insect-to-human transmission via the feces of the vector. The main insect host for this species is *Triatoma,* which is also called the kissing bug. Rather than enter through the saliva, *T. cruzi* is deposited on the host's skin surface with the insect's feces, and rubbed or scratched into the bite wound or mucosal surfaces. While this is the main mode of infection, *T. cruzi* many also enter the body of a new host via contaminated blood transfusions, laboratory cultures or via the congenital route. Once they enter the wound, the parasites do not multiply in blood but invade the cells of local and distant tissues, where they lose their external structures—flagella and undulating membrane— and undergo binary fission to give rise to the nonflagellated forms. Further multiplication produces new flagellated trypanosomes, ready for infecting the insect host. The flagellated form of *T. cruzi* is a long, spindle-shaped cell, which curves around to give characteristic C or U shapes in blood smears. It has a large nucleus in the center of the cell and large, oval-shaped kinetoplast in the posterior end, with a basal body that gives rise to a flagellum that runs along the outer rim of the undulating membrane and emerges free at the anterior end. After ingestion into the insect along with the blood meal, the parasite transforms into a noninfective multiplicative phase that multiplies in the mid gut of the bug and after 8–10 days develops into the infective parasites that are then passed out with the feces. *See also* CHAGAS DISEASE.

Trypanosoma rangeli

Parasitic species, which like *T. cruzi,* are found only in the New World, but are transmitted via saliva and not feces. These organisms have thus far not been explicitly linked to any human disease.

tuberculosis

From the consumptives of the past to the modern day AIDS patients who suffer from new and frightening forms of the disease, tuberculosis has acquired a permanent place in human history. Over the centuries it has claimed the lives of many well-known figures—Chopin, Moliere, and Keats were among its victims. The very nature of the illness, namely the gradual decline in strength and vitality of the victim to an almost ethereal state, has provided fodder for many a work of literature, including Thomas Mann's celebrated *Magic Mountain.*

In medical terms, tuberculosis is a chronic bacterial disease caused primarily by an infection with *Mycobacterium tuberculosis.* Related mycobacterial species, notably *M. bovis* and *M. africanum* are also known to cause the disease. Although the bacteria may

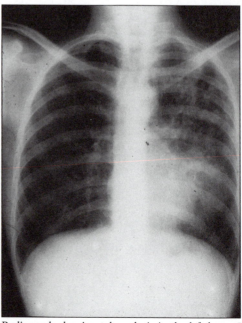

Radiograph showing tuberculosis in the left lung. © *Science Source/Photo Researchers.*

infect virtually any site or organ in the body, the most common location, as well as the best-recognized form of the infection occurs in the lungs, i.e., pulmonary tuberculosis. This is because by far the most common mode of infection, is by the inhalation of the bacteria. The organisms may spread from the lungs to the regional lymph nodes and from there to the bloodstream and other parts of the body, causing a more dispersed form of the disease.

The major symptoms of tuberculosis, as well as the progression of the disease, are largely a function of the immune system of the infected individual rather than the virulence of the organism *per se*. This involvement of the immune system is also reflected in the demographics of the disease—susceptible populations include very young children, whose immune systems have not yet had time to develop, and the very old. AIDS-infected and other immuno-compromised individuals also show a greater tendency to suffer more severe symptoms.

Soon after the bacteria invade the lungs, they induce a local delayed type hypersensitivity reaction, which results in the accumulation of immune cells, particularly ma—crophages, at the site of infection. In a person with a healthy immune system, the bacteria are destroyed and the infection may pass altogether unnoticed. However, if the immune status of the infected person is low, or if the initial bacterial load is high, the macrophages are unable to kill the mycobacteria. However, they stimulate the accumulation of other immune cells which results in the formation of a chronic granulomatous lesion at the site, due to the destruction of the host's own cells and tissue by the immune cells. Over time, the lesion becomes calcified to form a compartment (also known as a tubercle), which sequesters the bacteria away from the destructive influences in the body, and allows them to multiply. These lesions may be detected by X-rays because of the calcium content. As the disease progresses, the patient will suffer from signs of pulmonary distress, including coughing and difficulty in breathing, and will gradually grow weaker and

die. If the calcified lesion ruptures, bacteria may be released into the bloodstream and be carried to other locations, where they induce tubercular lesions and damage tissues.

Untreated, tuberculosis can have fatal consequences, but if detected in time, it is completely curable with appropriate drugs or antibiotics. A dreaded killer before the era of antibiotics, tuberculosis was thought nearly conquered toward the middle of the past century, but reemerged, stronger and more lethal than ever, when AIDS burst upon the scene. The disease is diagnosed with the help of multiple tests: a chest X-ray to detect calcified tubercles, the examination of sputum and other clinical samples for the presence of the tubercular bacilli, and a skin test with antigen from the bacteria (called tuberculin) to determine the exposure of the individual to the mycobacteria. The first vaccines against tuberculosis were developed in 1921 using a live avirulent bacillus developed by Calmette and Guerin (the BCG vaccine). This vaccine is used more frequently in developing countries where the disease incidence is high. However, it is not recommended for wide use in the United States where incidence is lower except among HIV-infected individuals, in whom the vaccine would not be effective anyway. Vaccinated individuals test positive for the tuberculin test. *See also* AIDS; EMERGING DISEASE/PATHOGEN; *Mycobacterium tuberculosis*.

tuberculosis

Also known as "rabbit fever" from the source of infection for the first identified cases of this disease, tularemia is a highly contagious, zoonotic infection of humans caused by the bacterium *Francisella tularensis*. The most common mode of infection of this organism is via ticks and deerflies, which transmit the organism to humans from reservoirs such as rabbits, rodents, deer, dogs, and cats. However, organisms may also infect humans via open skin lesions, the ingestion of infected foods or water, or by breathing in the bacteria via aerosols. The disease is marked by a sudden onset—within 2–5 days of bacterial exposure—of flu-like symptoms, including fever, chills, aches and pains, and coughing. If tick-borne, the systemic symptoms are preceded by a skin ulcer and lymphatic involvement at or near the site of the bite. Diarrhea and vomiting frequently accompany the intestinal infections, while inhalation might lead to the development of a pneumonia-like disease.

Although the disease may originate from any number of different small animals, rabbits appear to be the predominant reservoir for human infections. For instance, tularemia outbreaks are often seen to occur around hunting season during which time hunters are exposed to large amounts of bacteria from skinning the rabbits. A high rate of incidence also occurs during the summer months when deerflies and ticks are in great abundance. The organism may be easily identified by its requirement for cysteine for growth in laboratory media, although this test is discouraged except in laboratories with special biohazard containment facilities. The treatment of choice for tularemia appears to be the broad-spectrum antibiotic streptomycin. *See also* FRANCISELLA TULARENSIS.

tumor

A cancerous growth in solid tissues. *See also* CANCER.

tyndallization

Method of food preservation developed by John Tyndall during the late nineteenth century. The process involves intermittent heating at relatively low temperatures—i.e., not boiling temperatures—interspersed with cooling-down phases. The idea behind this treatment was to kill vegetative cells during the heating phase, which would also stimulate the germination of spores that would grow during the cooling phase. Numerous repetitions would increase the chances of eliminating more spores from the food, while the use of lower temperatures enables the food to remain largely unaltered in taste and texture. *See also* FOOD PRESERVATION.

typhoid fever

Systemic bacterial disease, with predominant symptoms of sustained fever and gastrointestinal disturbances, caused by infection with *Salmonella typhi*. A similar but milder disease known as paratyphoid fever is caused by a closely related but distinct species of the same genus, called *S. paratyphi*. Worldwide, the annual incidence of typhoid is about 17 million with a death toll of nearly 600,000. The vast majority of these cases occur in Asia, the Middle East, and Latin America.

Because *Salmonella* is an enteric bacterium, the primary site for infection is the gastrointestinal tract. Bacteria typically gain entry into the body through contaminated food or water sources. Raw fish from sewage-contaminated waters as well as fruits and vegetables, which have been watered with contaminated water, are common sources of infection. Both typhoid and paratyphoid are food-borne infections and not intoxications, and the development of disease depends on the ability of the bacteria to penetrate the intestinal lining and establish colonies. Within the intestines, the bacteria are taken up by the phagocytic cells, where they resist destruction by the host enzymes and are transported to the regional lymph nodes and bloodstream and eventually to the rest of the body. Often the bacteria establish themselves in the gall bladder, and continue to multiply, thereby providing an almost constant source of bacteria both to the blood as well as to the intestinal tract.

The onset of typhoid fever is insidious, occurring as a gradually increasing fever with a persistent headache and generalized weakness and aches during the first week of infection. The nonspecific nature of these symptoms may often be left untreated and may develop into more serious manifestations of sustained high fevers (104°F) accompanied by abdominal pains and gastrointestinal disorders that are as likely to run to constipation as diarrhea. This phase of the disease corresponds to the presence of the bacteria in the bloodstream. Often times patients, particularly light-skinned individuals, may develop rose-colored spots or a rash on the abdomen. At this stage, the disease has the potential to resolve itself in several ways, depending on the virulence of the infecting strain and the immune status of the individual. Complications, which arise from the multiplication of bacteria in different organs, can take the form of abscesses in tissues, perforation of the intestinal wall, or pneumonia (when the lungs are invaded). In many instances, the body's defenses are successful in clearing the organism from the body altogether. A third possible outcome is the development of the carrier state, either in convalescent individuals or in people in whom there was no apparent disease.

The nonspecific nature of the initial symptoms makes typhoid a difficult disease to diagnose, especially in the West where incidence is relatively low. Serological tests are only of limited value because sensitivity to these tests is low. Diagnosis is best achieved by the isolation and identification of the salmonellae from stools and blood. Pinpointing the exact source of the bacteria is also useful, especially from a public health standpoint. Antibiotic therapy has proven the most effective treatment, although the emergence of new resistant strains has seriously limited the choices in the past few decades. Preventive measures include ensuring proper sanitation and sewage disposal; proper storage, handling, and treatment of foods (see entry on food-borne diseases); and screening of potential carriers and limiting their contact with food to be distributed in public places. Vaccines are available and recommended for general use in areas of the world where typhoid is prevalent. In the United States, the vaccine is not administered routinely, but given to people who are deemed to be at risk for exposure either due to their profession or because of travel. *See also* SALMONELLA TYPHI; TYPHOID MARY.

Typhoid Mary

This nickname is now practically synonymous with persons who are sources of a specific infection or disease without exhibiting any apparent signs of the disease themselves. It originates from an article that appeared in *The Journal of the American Medical Association* in 1908, describing an asymptomatic carrier from New York City named Mary Mallon. An Irish immigrant who worked as a cook, Mallon was identified by the New York City Health Department as the source of infection for at least 28 cases of typhoid fever that had occurred in homes that she had cooked for between 1886 and 1906. After they traced the infections to her, the health department had Mallon arrested and admitted to a city hospital for infectious diseases. She was subsequently released because she showed no signs of the disease, but under the condition that she not cook for public consumption again. Unwilling to accept the medical community's verdict that she was the source of a disease of which she had no experience and that cut off her only means of livelihood, Mallon disregarded the injunction to stop cooking and continued to work as a cook for about five years while avoiding the authorities. Eventually, they tracked her down again and proceeded to hold her in isolated custody for 23 years until she died in 1938. *See also* CARRIER; SALMONELLA TYPHI; TYPHOID FEVER.

U

ulcer (also peptic or gastric ulcer)

Sores or necrotic lesions in the lining of the stomach and duodenum. Although lesions of this nature have been recognized and treated symptomatically for well over a century, it has only been in the past two decades that their real cause has come to light. Until then, physicians had attributed the formation of ulcers to a host of causes, including mental stress, smoking, and poor food habits. While it is true that any or all of these factors may serve to trigger or exacerbate the condition, the root cause of an ulcer is a *Helicobacter pylori* infection.

The organism colonizes the stomach and duodenal walls where it breaks down the protective mucus layer and exposes the epidermis to gastric juices, thereby triggering an inflammation of the area, i.e., gastritis. Prolonged gastritis will lead to further erosion of the stomach wall and eventually the formation of an ulcer. Left unchecked, an ulcer can result in the perforation of the stomach or duodenum and a leakage of the contents into the internal cavities of the body. In some cases, chronic gastritis or ulcers may serve as sites for the development of stomach or duodenal cancers.

Exactly how *Helicobacter* enters the stomach is not known for sure, although the epidemiological patterns suggest an oral-fecal route. Infections are prevalent worldwide, although there seems to be a link between infection and living conditions involving poor sanitation, close contact, and overcrowding. A telltale sign of infection is development of "urea breath" or the odor of ammonia in one's breath. This only occurs during an active infection. The definitive test for *H. pylori* infections is the ability to detect and culture these organisms from the patient. The material for these tests is obtained by a procedure called endoscopy in which a piece of the stomach lining is obtained through a tube to the stomach via the mouth. In addition, a blood test detects the presence of specific antibodies to this organism in the bloodstream.

H. pylori infections are only treated when they become symptomatic, i.e., when a patient develops visible gastritis or ulcers. The treatment of choice is a combination of antibiotics for 2–3 weeks and some common antacid such as Peptobismol. Successful elimination of the organism can only be confirmed by endoscopy because antibodies can persist in the bloodstream for several weeks after the infection is cured. *See also* HELICOBACTER PYLORI.

ultracentrifugation

Centrifugation that is performed under extremely high speed and g-forces, and usually used in microbiology and biochemistry in various preparatory and analytical procedures. An example is density gradient centrifugation, a technique in which cesium

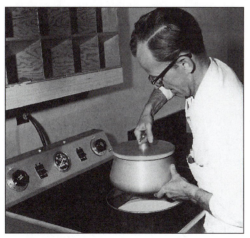

Centrifuging. *CDC.*

chloride is used to separate DNA molecules of varying sizes from different sources within a cell. *See also* CENTRIFUGATION.

ultrahigh temperature treatment (UHT)

Commercial process for treating milk and milk products so as to achieve antisepsis and increase shelf life. The UHT process involves the sterilization of the dairy food at extremely high temperatures, between 140–150°C for a few seconds, followed by aseptic packing in airtight containers. UHT-processed milk may be stored at room temperature for several weeks without danger of spoilage or change in quality. In general, UHT-treated foods are considered to have a higher consumer acceptability than ordinarily pasteurized products. *See also* FOOD PRESERVATION.

undulating membrane

Structure found in certain protozoan cells, e.g., *Trypanosoma*, consisting of a fin-like extension of the membrane on the outer edge of the organism's body. This membrane is capable of a wave-like motion that aids in the mobility of the parasites in the blood and interstitial spaces of tissues. *See also* TRYPANOSOMA.

urease

Enzyme that catalyzes the conversion—hydrolysis—of urea into carbon dioxide and ammonia. Urease is produced by many intestinal bacteria, which often help the host animal to clear urea from the body. The production of this enzyme by urinary tract pathogens such as *Proteus* and *Klebsiella* allows these organisms to colonize and thrive in the urinary bladder and surrounding areas.

UV microscope

A microscope that uses ultraviolet light to produce magnified images of objects. UV microscopes have higher magnification powers than ordinary light microscopes, which means that they resolve smaller distances. Since glass is opaque to UV rays, the lenses of these instruments are made of quartz. Images formed by these microscopes cannot be viewed with the naked eye because UV rays are harmful to the eyes. Images may be viewed directly using special UV-protective eyewear, but are usually captured on special photographic plates or by using UV-sensitive video systems. UV microscopes are useful for visualizing objects similar in size to those seen by optical microscopes but with a higher level of detail. They are also the instruments used in fluorescent microscopy. *See also* MICROSCOPE.

V

vaccine

A vaccine is any substance that is administered to an individual as a protective or preventive measure against disease. The name harks back to Jenner's use of the cowpox (vaccinia) material to confer immunity against smallpox. Basically a vaccine acts by either boosting or replacing the activities of the body's immune system, so that the individual is able to deal with the pathogen when he or she comes in contact with it. Vaccines against different diseases are either administered as a matter of course at childhood or given to special groups of individuals who are deemed as being at high risk for contracting a particular disease. *See also* IMMUNIZATION.

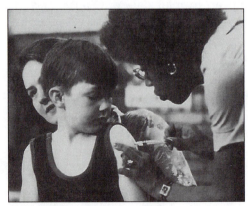

Child receiving a vaccination. *CDC.*

vaccinia virus

Member of the poxvirus family, which is used for immunizing against smallpox, and which was ultimately responsible for eradicating the disease. This virus is antigenically similar to the smallpox—variola—virus, but is safer to use because it is either less virulent (causing only localized lesions) or altogether non-pathogenic in humans. *See also* POXVIRUS; VACCINE.

vacuole

Membrane-bound compartment found in the cytoplasm of cells. Depending on the method of formation, the contents and their intended fate, vacuoles may be secretory, digestive, or merely performing a storage function in the cell. *See also* GAS VACUOLE.

Varicella-Zoster virus (VZV)

The causative agent of both chickenpox and shingles in humans, the Varicella-Zoster virus (VZV) is a large enveloped virus with a double-stranded DNA genome of 125 kb, belonging to the family of herpesviruses. Humans are the only known natural hosts. One of the most notable features of VZV is its dual role in infected hosts, which arises from its ability to exist within the host in a latent form for several years. The exact mechanism of viral latency is not fully understood, although genetic analysis suggests that during this period VZV probably resides in the trigeminal nerve ganglion, where it enervates parts of the face and neck. *See also* CHICKENPOX; SHINGLES.

Variola virus

This member of the poxvirus family is the causative agent of smallpox, a disease that has time and again wreaked devastation upon the human race throughout history. The viruses are divided into two main groups, variola major, which has fatality rates of 25–30 percent, and variola minor, which causes the same symptoms but far fewer deaths. Unlike most other viruses in this family, variola viruses can infect human hosts only. Another characteristic feature is their ability to cause acute infections, without the possibility of developing into chronic (e.g., hepatitis) or latent (e.g., as in chicken pox/shingles) states. Viral antigens induce long-term immunity to the disease, and indeed the development of the science of vaccines (vaccionology) has firm roots in the smallpox virus/immunity model.

Because the variola viruses have no natural hosts besides humans, the only known sources of virus to remain since the disease was eradicated in 1979 are frozen stocks stored in the facilities of the CDC, and its Russian counterpart. Over the past two decades, international committees have met on several occasions to try and decide about the fate of these stocks, but the debate rages on with different scientists of international repute offering diametrically opposite views on the subject. Scientists who advocate destruction warn against the high virulence and fatality rates of the disease, should it accidentally ever infect humans again, as well as the potential hazards of biological warfare, if the stocks were to fall into the wrong hands. The dissenters to this view argue that destruction of the virus is detrimental for research purposes, particularly as it pertains to the development of vaccines. Although the cur-

rent vaccines are in wide use and have succeeded in keeping smallpox at bay for nearly 20 years, some argue that there are still dangers associated with this particular preparation, and that research into new types of vaccines would require the original viruses. Study of other important and unique properties of these viruses, such as their host specificity and virulence, would require these virus stocks rather than avirulent relatives, such as the vaccinia virus. Proponents of destruction counter these with arguments that the entire viral genome has been cloned and is preserved in a vector library. At the time this book went to print, the debate had not yet been settled and both countries were retaining their smallpox virus stockpile. *See also* SMALLPOX.

vector

A general term for any entity that is used as a vehicle for transferring something from one place to another. For instance, in the context of public health and infectious diseases, any non-human animal that transmits pathogens from one person to another is a vector for disease. A vector may transmit the disease from one infected human to another, e.g., malaria by the *Anopheles* mosquito, or from an animal to human hosts, e.g., Lyme disease, which is carried over from animal reservoirs by ticks. In genetic engineering, viruses and plasmids are used as the vectors for transferring genes from one cell to another. *See also* DISEASE TRANSMISSION; HOST; PLASMID.

Ixodes scapularis, tick vector for Lyme disease. *CDC.*

venereal disease

Infectious diseases whose primary mode of transmission is via sexual contact, e.g., gonorrhea, syphilis, genital herpes, and chlamydial and HIV infections. The act of sex itself is not the cause of any of these diseases, and serves only as the vehicle of spread, usually because of the transfer of bodily fluids—e.g., saliva or semen—which contain the infectious agent. Most venereal diseases may also be spread by other means, such as blood transfusions and shared needles (in the case of blood-borne infections), and even, on occasion, by simple contact. *See also* SEXUALLY TRANSMITTED DISEASE.

verruga peruana

A cutaneous form or phase of an infection with the bacterium *Bartonella bacilliformis*, seen almost exclusively in the mountain valley regions of Peru (hence the label "peruana" to describe the disease), Ecuador, and southwest Colombia. This type of bartonellosis typically occurs some 1–3 weeks after primary infection, although the disease may manifest itself as late as 3–4 months later. The initial infection, which occurs due to the injection of the bacteria into the body via the bite of a sandfly, may be either asymptomatic or manifest as Oroya fever. The cutaneous disease is characterized by the eruption of lesions called verrugas, on different parts of the body. These may be widely distributed small, rash-like eruptions or nodular lesions that are fewer in number but more deep seated in character. Verruga lesions, particularly near the joints, may develop into tumor-like masses with ulcerated surfaces. The skin eruptions are often accompanied by severe, shifting pain in the muscles or joints. This type of bartonellosis is rarely, if ever, fatal. The disease is diagnosed by the demonstration of the infectious bacteria in material from the lesions during the eruptive stage or by serological tests. The bacteria are sensitive to a wide range of antibiotics and the disease may be treated with penicillin, streptomycin, chloramphenicol, or tetracycline. *See also* BARTONELLA.

vesicular stomatitis virus (VSV)

Enveloped, single-stranded RNA virus of the rhabdovirus family, found primarily in farm animals such as horses and cattle. VSV-induced human infections are zoonotic in origin, and are believed to be transmitted to humans from the vesicular fluids or tissues of infected animals or via sandflies. The febrile illness resembles an acute case of influenza with fever, chills, and muscle aches that usually resolve within 7–10 days without further complications. *See also* RHABDOVIRUS.

Vibrio

Named for the characteristic comma or curvy shape of the bacteria when isolated from their natural habitats, this genus of gram-negative bacteria is the only genus within its family known to be associated with human disease. The best-known species of this genus is the causative agent of cholera. Vibrios are facultative anaerobes but show a marked preference for aerobic conditions. Organisms can use a variety of sugars, including glucose, sucrose, galactose, and mannitol, which they metabolize to produce acid but no gas—i.e., fermentation does not proceed to completion. They reduce nitrates and are strongly proteolytic. The vibrios survive adequately outside the body, especially in warm climates and near water, but are easily killed by heating at high temperatures or with chemicals. Organisms are motile due to the presence of usually a single polar flagellum, although some

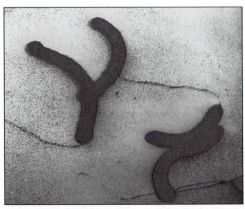

Micrograph of *Vibrio cholerae;* note the single terminal flagella. © *Science Source/Photo Researchers.*

species may contain a tuft of flagella at the poles. Because the organisms often lose their curved shape upon serial passages in laboratory culture, this feature should not be used as an index for identification.

Vibrio alginolyticus

A normal inhabitant of sea water, *V. alginolyticus* has been found associated with external wound or ear infections that occur after marine or sea-related activities. The multiplication of the organisms at the site of infection may often lead to formation of pus which may require drainage for proper healing, but infections with this species are amenable to antibiotic therapy.

Vibrio cholerae

This species is the causative agent of Asiatic cholera in humans. The establishment of disease by this organism depends on the attachment and multiplication of the bacteria in the intestinal epithelium as well as the production of a powerful enterotoxin called choleragen. The presence of flagella plays an important role in the colonizing abilities of the bacteria and mutants of *V. cholerae* that lack these structures are rendered avirulent. The cholera toxin is responsible for the diarrhea and vomiting typical of this disease. Structurally, the toxin is made up of two protein subunits, A and B, which act in concert to induce changes in the permeability of the host cells and cause the cells to discharge massive quantities of water and electrolytes into the intestinal lumen. *See also* CHOLERA.

Vibrio fetus

This organism shares the principal structural and biochemical properties of *V. cholerae* but prefers to infect farm animals over humans, and causes infectious abortions.

Vibrio parahaemolyticus

This species of the genus *Vibrio* is a normal inhabitant of coastal waters, where it is frequently found associated with zooplankton and marine animals such as crabs. Like *V. cholerae,* it may be ingested with improperly cooked seafood and cause human infections. The most common type of disease is a self-limited gastrointestinal condition marked by watery diarrhea and abdominal cramps. Sometimes the disease may be more severe and resemble cholera in the manifestations of dehydration and dysentery. *V. para-haemolyticus* has also been isolated from external wound infections, which can be treated with antibiotics. Pathogenic strains may be distinguished from nonpathogens by their ability to lyse human blood cells when grown in a special medium.

vinegar

Edible acetic acid product made by the action of special bacterial species such as *Acetobacter* and *Acetmonas* on alcoholic beverages. The name is derived from the French phrase *vin aigre,* meaning sour wine, and wine is perhaps the most common starting material, although virtually any alcoholic beverage might be used. Other common examples include cider, beer (malt vinegar), and saki (rice wine vinegar). The bacteria grow spontaneously in wine when it is exposed to air. The chemical reaction in making vinegar is simply the oxidation of alcohol to acid. The commercial process involves trickling wine down a vinegar tower composed of birch or other twigs lined with films of the acetic acid bacteria and other organisms (exact composition is unknown). This process exposes the wine to air, and allows the bacteria to grow aerobically, and convert ethanol to acetic acid. Specialty vinegars, such as balsamic vinegar, are made by adding special ingredients to the product either before or after fermentation, and usually do not involve any microbial activity.

Acetic acid or vinegar may also be made in a cheaper, purely chemical process in which methanol rather than ethanol is the starting product. This vinegar is for industrial and laboratory use rather than consumption. *See also* ACETOBACTER.

Virchow, Rudolph (1821–1902)

German pathologist who set the foundations for describing human diseases in terms of the anatomical location of the lesions. In his major work, *Die Cellularpathologie* (1858), he demonstrated the applicability of the cell theory to healthy as well as diseased tissues. *SEE ALSO* SMALLPOX.

viremia

Disease state characterized by the presence of viruses in the bloodstream.

virion

An intact virus particle. *See also* CAPSID; VIRUS.

virus

Term used to represent the group of sub-microscopic infectious agents made up of a single type of nucleic acid enclosed in a protein envelope. The viruses occupy a unique niche in our world, defying description within the parameters that scientists have set for defining living organisms, and yet undeniably in existence, as causative agents for several, often devastating diseases. The term, which means "poisonous fluid" in Latin, was first used to specify this group of entities when it became obvious that there were several infectious diseases that could not be attributed to pathogens like bacteria, fungi, or protozoa, but which were caused by some smaller, subcellular agent that remained infective even after being filtered through materials capable of screening out all cellular material. Too small to be seen under even the most powerful light microscopes, the viruses were first seen only after the development of electron microscopes. Many scientists have described viruses as infectious agents at the "threshold of life" because virus particles, in and of themselves, are inert and incapable of performing any of the activities commonly associated with life, such as metabolism, growth, and replication. And yet, when it enters a living cell of another organism, a virus is capable of diverting that cell's machinery to its own ends,

replicating itself, and ultimately producing several new virus particles, in short, behaving like a living organism. Thus, another apt description for them is "obligate intracellular parasites"—namely, entities that require other living cells to carry out even the most basic functions. The organism whose cells are infected by the virus is called the viral host.

All viruses consist of a nucleic acid molecule called the core, surrounded by a protein coat known as the capsid. This basic structure containing the core and capsid is called the nucleocapsid. In addition, some viruses possess an outer envelope made of phospholipids or protein-lipid complexes. The viral core contains the genetic material of the virus and encodes the instructions for the protein coat as well as for the enzymes needed for its own replication. It is made up of either DNA or RNA but never both. Either species of nucleic acid may be present in a double- or single-stranded form, thus dividing the viruses into four main groups (See Table 10, the virus classification table with this entry). To express and replicate its genes, the virus requires protein synthetic machinery—ribosomes and polymerases—which can only be found inside whole cells. Thus, the need for a living host becomes apparent. The capsid of a virus protects the nuclear material from the outer environment. Typically, the capsid does not enter the host cell but plays a key role in recognizing the cell and initiating the infective process. The individual protein units of the capsid are called the capsomeres. The number, arrangement, and symmetry of the capsomeres around the core are characteristic for a given group of viruses. When present, the outer envelope also plays a role in protection and host recognition. Unlike the capsid, this component is not encoded in the viral genome but is derived from the membranes of the host cell when the virus emerges from the cell. Enveloped viruses thus tend to have a more complex structure than the naked viruses.

Although submicroscopic, the viruses exhibit a wide range of sizes, depending on the structure and organization of the nucleocapsid and the presence or absence of the outer envelope. The smallest known viruses may be no larger than some proteins, e.g., the yellow fever virus, a naked, single-stranded RNA virus, is only 40 nanometers (nm) in diameter, which is on the same scale as a protein such as egg albumin at 10 nm. On the other hand, the enveloped, double-stranded, DNA-containing smallpox virus at 300 nm approaches the size of a small bacterium.

A distinguishing characteristic of viruses is often the range of hosts whose cells they infect—they have a predilection for animals or plants or bacteria, but typically do not infect more than one group. This host specificity seems to be independent of the nucleic acid content; for instance, the smallpox and yellow fever viruses both cause human disease but have different types of nucleic acids.

Because of viruses' unique mode of existence, the conventional systems of taxonomy and nomenclature used to classify and identify bacteria and higher living organisms cannot be applied to them. No formal scheme exists for naming and grouping these entities. Different viruses are named for diseases they cause, the hosts they infect, or the people who discovered them, e.g., poliovirus, the avian leukosis virus (which causes the disease leukosis in birds), or the Rous sarcoma virus (discovered by Peyton Rous). As mentioned earlier, viruses are often grouped according to their structure and composition, or host range, but these schemes are strictly functional and often overlap. For example, while both the poliovirus and tobacco mosaic virus (TMV) contain RNA, the former is an animal virus while TMV is specific for plants. And while poxviruses may have the same host specificity as poliovirus, the former contain DNA and the latter RNA.

Viruses do not grow and reproduce in the typical manner of other intracellular parasites like the rickettsiae but rather hijack the host cell and take over the cell's metabolic machinery for their own purposes. The exact mechanism by which this takeover is accomplished varies according to the structure and composition of the infecting virus.

The first step, common to all viruses, is the adsorption of the virus particle on the surface of the host cell and subsequent penetration into the cell. The first step occurs when specific attachment sites called receptors on the membranes of the host cell recognize the viral capsid or envelope. When these receptors are blocked or absent, the virus cannot attach and infect the cells. Consequently, the study of viral receptors is of great interest in the pharmaceutical industry as potential targets for drug therapy in viral infections. Following attachment, the virus enters the cytoplasm of the host cell. Bacterial viruses typically inject only their nucleic acid into the cell, leaving the outer coats behind. However, many animal and plant viruses enter the host cell by a form of phagocytosis—engulfment by the host cell. Most plant viruses typically enter their host with the aid of an insect vector that breaks through the cell walls and allows viruses to come in contact with the cell membrane or cytoplasm. Once inside the cell, the viral envelope and protein coats are removed and the virus is ready for the next step.

After the virus enters the cell, it begins to replicate its genome. The specific pattern of replication depends on whether the genome is DNA or RNA, single- or double-stranded. The simplest mode of replication happens in cases of double-stranded DNA viruses, such as the T4 bacteriophage and the vaccinia (cowpox) virus. They simply harness such cellular components as ribosomes, t-RNAs, enzymes—DNA and RNA polymerases and the protein synthesizing enzymes—as well as the amino acid and nucleotide building blocks of proteins and nucleic acids to synthesize multiple copies of the viral coats and genomes. These components are then assembled to yield intact virus particles. This latter process is known as viral maturation.

The specific steps in the replication of the different viruses are somewhat different. For instance, some DNA viruses do not enter the replicative cycle immediately but integrate into the host's DNA. This cycle is called the lysogeny. Single stranded DNA must first be converted to a double-stranded form before the genome can be replicated. RNA viruses also have more than one mode of replication.

Table 10. Classification of the Virus Families of Human/Public Health Relevance

Note: This chart outlines the general scheme of virus classification. The major criteria used for classification into different families include: (i) the type of nucleic acid genome; (ii) the nature of the nucleic acid—namely, number of strands, polarity or "sense" of molecule (if single-stranded), and its shape; (iii) the presence or absence of an envelope; and (iv) nucleocapsid symmetry. Subdivisions within a family—into genera and species—is based on different, less uniform criteria that vary from family to family.

DNA Viruses:

Single-Stranded DNA Viruses:

Parvovirus   Linear genome with negative sense, i.e. cannot be used directly as template for synthesis;   Icosahedral nucleocapsid;   Lacks envelope;   Smallest known DNA viruses with a genome size of 5 kilobases and total diameter of 20 nm.

  e.g. Virus B19   "Causative agent of "slapped-cheek disease."
        Adeno-associated virus

Double-Stranded DNA Viruses:

Naked viruses (i.e. lacking envelope):

Adenovirus   Linear genome;   Icosahedral nucleocapsid;   Both human and avian species isolated;   Implicated in respiratory infections.

Papovavirus   Circular genome;   Icosahedral capsid;   Oncogenic, i.e. induce tumors in host organisms.

  Papilloma viruses   Isolated from human warts.

    e.g. Shope's papilloma virus

  Polyoma virus   Cause sold tumors at multiple sites simultaneously.

    e.g. BK virus;   JC virus;   SV40 (simian virus 40)

Hepadnavirus   Circular genome;   Icosahedral capsid;   Encodes a reverse transcriptase enzyme;[1]

  e.g. Hepatitis B virus   Causes infectious or serum hepatitis in humans.

Enveloped viruses:

Herpesvirus   Linear genomes ranging in size from 125-230 kilobases;   Icosohedral nucleocapsid;   Associated with wide spectrum of human diseases.

  e.g. Herpes simplex virus;   Varicella-Zoster (Chickenpox) virus;   Cytomegalovirus;   Epstein-Barr virus.

Poxvirus   Linear genomes ranging in size from 130-250 kilobases;   Nucleocapsid possesses complex symmetry;   Envelope, which endows viruses with large brick-like or ovoid shapes is not essential for viral infectivity;   Viral transcriptase is encoded in genome.

  e.g. Smallpox (Variola) virus;   Vaccinia virus.

RNA Viruses:

Single-Stranded RNA Viruses:[2]

Naked viruses (lacking envelope):

Picornavirus   Positive sense genome;   Icosahedral capsid;   Individual genera differentiated on basis of host and tissue preferences.

  e.g. Poliovirus, enteroviruses, cardiovirus, rhinovirus, Hepatitis A virus and foot-and-mouth-disease virus.

Astrovirus   Positive sense genome;   Icosahedral nucleocapsid;   with characteristic star shape visible under electron microscope;   Family resembles picornaviruses closely but has antigenically distinct capsid proteins;   Viruses cause mild diarrhea in infected individuals.

Calicivirus   Positive sense genome;   Icosahedral capsid with cup-like depressions on surface;   Causes gastrointestinal infections.

  e.g. Norwalk virus, Hepatitis E virus.

Deltavirus   Negative sense, circular genome that encodes protein component of virion called delta antigen;   Virion particles exhibit icosahedral symmetry;   Viruses associated with hepatitis D;

Enveloped viruses:

Togavirus Positive sense genome; Icosahedral symmetry; Envelope lacks a matrix protein; Mostly arboviruses that cause variety of infections in humans.

Alphavirus Causative agents of encephalitis in animals (horses) and humans; Naturally maintained in a rodent/bird to insect (mosquito) cycle.

Rubivirus Sole non-arthropod-borne member of the togavirus family; May acquire envelope from endoplasmic reticulum rather than plasma membrane of the host cell.

e.g. Rubella virus

Flavivirus Positive sense genome; Icosahedral capsid; Envelope lacks a matrix protein; Mostly arboviruses (originally classified as Group B arboviruses) transmitted by mosquitoes and ticks; Different individuals isolated from a number of viral hemorrhagic fevers and encephalitides.

e.g. Dengue fever virus, Yellow fever virus, Japanese encephalitis fever virus etc.

Hepatitis C virus Non-arthropod borne member of flavivirus family; Leading cause of non-A non-B hepatitis in humans; Transmitted sexually or via blood.

Retrovirus Positive sense, diploid genome; Usually found in cells in the form of DNA integrated into the host cell genome; Icosahedral nucleocapsid; Virus family encodes own reverse transcriptase.

e.g. Human immunodeficiency virus (HIV), Human T-cell leukemia virus (HTLV)

Coronavirus Positive sense genome; Helical capsid; Envelope lacks matrix protein; Viruses are best known as one of the frequent causes of the common cold in humans although some isolates are seen in feces;

Paramyxovirus Negative sense genome; Helical nucleocapsid; Encodes a DNA transcriptase enzyme; Family consists of several genera associated with different human diseases.

e.g. Parainfluenza virus, measles virus, mumps virus, and respiratory syncytial virus.

Rhabdovirus Negative sense genome; Bullet-shaped viruses enclosing a helical nucleocapsid; Viruses mature at site near host cell membrane and form characteristic inclusion bodies called negri bodies; Genome encodes a viral transcriptase.

e.g. Rabies virus and Vesicular stomatitis virus (VSV).

Filovirus Negative sense genome; Helical nucleocapsids; Viral transcriptase encoded in genome; These viruses resemble the rhabdoviruses in many respects but often assume very long, thread-like shapes; Known to cause hemorrhagic fevers in humans.

e.g. Ebola virus, Marburg virus and Reston virus.

Orthomyxovirus Negative sense genome consisting of 7-8 segments with a total size of about 13.5 kilobases; Helical capsids are surrounded by a lipid envelopes to form spherical particles of 80-120 nm diameter; Encode own transcriptase enzyme.

e.g. Influenza viruses types A, B and C.

Arenavirus Mostly negative sense genome (actually ambisense due to the presence of certain positive sense portions) consisting of two segments held together by hydrogen bonds in an apparently circular configuration; Helical nucleocapsid enclosed along with host ribosomes within virus envelope; Encode viral transcriptase; Cause certain hemorrhagic fevers in humans.

e.g. Lassa virus, Lymphocytic choriomeningitis virus (LCM virus), Machupo virus.

Bunyavirus Negative sense genome consisting of 3 segments held in a circular shape by hydrogen bonds; A single genus contains ambisense RNA, i.e. portions of the molecule are either positive or negative sense; Helical or tubular nucleocapsids enclosed in trios within a single lipid envelope derived from the membrane of the Golgi apparatus of the host cell; Encode own transcriptase in one of the segments; Several genera of mostly arboviruses associated with a number of hemorrhagic fevers and some encephalitides in humans.

e.g. Sandfly fever virus, La Crosse virus, Crimean-Congo hemorrhagic fever virus.

Double- Stranded RNA viruses

Reovirus Segmented genome with 10-12 pieces and a total size ranging from 15-20 kilobases; Icosahedral nucleocapsid; Lacks envelope.

Rotavirus Characteristic "wheel"-like appearance under electron microscope; Most common cause of gastrointestinal infections in infants.

Orbivirus Classified with other arboviruses; Mostly animal pathogens causing such diseases as bluetongue in sheep and African horse sickness.

Coltivirus Arbovirus that causes Colorado tick fever in humans.

Notes:
[1]Unless otherwise indicated viruses do not produce any transcriptase enzymes and rely completely on host enzymes for nucleic-acid synthesis.
[2]All RNA viruses except for Deltavirus possess linear genomes.

While some genomes are capable of acting directly as the messengers for translation, others can only act as templates (antisense RNA) and still others must first be converted to DNA—by an enzyme called reverse transcriptase—before individual components can be synthesized.

Once the new viral particles are ready, they must be released from the host cell. Some viruses, called the lytic viruses, emerge from the host by causing the cell membrane to lyse or rupture. This is typical of the naked or unenveloped viruses. Other viruses may emerge through a process of budding or "pinching off" from the cell surface. Such particles will contain an outer envelope whose makeup reflects the composition of the host's cell membrane. This budding process does not lyse the cell and continues indefinitely until the host cell dies from severe damage to its metabolism.

Their intracellular mode of replication presents a special set of problems for growing and studying viruses in the laboratory. Unlike bacteria, they cannot be grown on simple media, and require some form of living cells in which to grow. Viruses of different types may be grown in the laboratory by introducing particles into their natural hosts—by injection if they happen to be plants or animals or by simple absorption in the case of bacteriophages. Both plant and animal viruses may also be grown in tissue cultures of appropriate origins. In addition, animal viruses may be propagated by injection into eggs, i.e., chick embryos.

Viruses are a significant cause of a variety of infectious diseases, and have been implicated in some of the most sweeping epidemics in history. They are responsible for many common childhood diseases, including measles, mumps, and chicken pox, as well as sexually transmitted diseases, such hepatitis and AIDS. Even certain forms of cancer and tumors are attributed to these parasites. The eventual outcome of a viral infection is cellular damage and death, although the mechanisms by which different viruses cause damage are different for different viruses. These mechanisms are reflected in the changes in the cell or the cytopathic effects of the virus. For instance, viruses with lytic life cycles will cause the host cells to lyse. Other changes that viruses may induce include changes in host-cell membrane permeability, the loss of normal functions of the host cell due to damage or repression of the host's genes, and changes in the surface properties of the infected cells that result in either fusion or clumping. Some viruses aggregate inside the cell to form inclusion bodies that are visible even under an ordinary light microscope. The changes caused by viral infection are often difficult to detect in the early stages of infection because they affect only a few cells. Clinical symptoms only become apparent after a significant number of cells become infected. Depending on the original site of infection as well as the nature of the virus, the symptoms may be either localized or generalized. A significant portion of this book is devoted to describing various types of viruses and viral diseases. *See also* CAPSID; VIRION.

vitamin

Collective term for a number of organic compounds, soluble in either water or fat, that are necessary in trace amounts for normal metabolic functions in most living organisms. Normal intestinal organisms, such as nonpathogenic strains of *Escherichia coli*, synthesize vitamins like B-complex and Vitamin K and release them into the gut from where they are absorbed.

W

Waksman, Selman A. (1888–1973)

Selman Waksman discovered streptomycin, the first effective antibiotic to be developed for use against tuberculosis. In contrast to Alexander Fleming's "accidental" discovery of penicillin, Waksman found streptomycin by meticulous searching and screening among the actinomycetes for antibiotic substances. He received the Nobel Prize in Medicine for his efforts in 1952. His other contributions include studies on soil microorgansims and various actinomycetes. *See also* STREPTOMY-CIN.

wart

A usually benign growth or tumor in the epidermis of the skin and mucus membranes caused by a human papilloma virus. A single wart usually represents the transformation of a single cell. Different types of warts are classified according to the appearance and location of the lesions. The simplest and most common type of wart is a small painless papule with a well-defined border and rough texture, often found in the extremities. Laryngeal papillomas are warts that form on the vocal cords and epiglottis of children. Flat warts are smooth, painless, slightly elevated warts that usually appear at multiple foci at once. Plantar warts appear on the feet and are similar in appearance but painful. Warts that appear in the moist areas of the skin around the genitalia and anus are called condylomas.

Although warts are for the most part benign in nature, laryngeal and genital papillomas may become malignant. Condylomas are frequently spread by sexual contact. Some evidence indicates the transmission of common and flat warts by direct contact. The most effective treatment appears to be surgical excision. *See also* PAPILLOMA VIRUS.

water activity

Numerical value or parameter that describes the moisture content of an environment. It is typically used in food microbiology to describe the amount of water that a particular food might make available to a microbe. In mathematical terms, the water activity of a substance may be defined as the ratio of the vapor pressures of the food substance to pure water. Thus pure water has an a_w of 1.000, while a 22-percent salt solution has an a_w of approximately 0.86. Most fresh foods (such as fruits and milk) have an a_w greater than 0.99. Bacteria can grow over a range of values, although the optimal values are typically over 0.92. The lowest level of tolerance among the prokaryotes is demonstrated by the extreme halophiles, which can grow at values as low as 0.75. *See also* HALOPHILE.

water-borne infection

This term is frequently used to denote outbreaks of microbial diseases that may be

Cryptosporidium parvum oocysts seen on stool smear, acid-fast stain.

Micrograph of *Cryptosporidium* and photograph of *Cryptosporidium*-contaminated water. *CDC/NDIC.*

traced to a contaminated water source. As in the case of food-borne diseases, outbreaks are characterized by the occurrence of a disease type among people exposed to a common source. Water-borne infections are a serious public health concern because of the infrastructure of water supplies and waste water treatment. Contamination at even a single source can thus mean the dissemination of infection to large populations at once. However, the same factors that contribute to dangers in infection also help to institute damage control more easily; once the source of the problem is identified, it can be dealt with effectively and efficiently. In less developed countries, water-borne infections pose different, and perhaps even larger threats. Due to inadequate facilities for sewage treatment and disposal, drinking water supplies frequently contaminated through seepage serve as reservoirs for a number of infections. Furthermore, because many of these countries are in tropical areas, the natural environment provides highly conducive conditions for the proliferation of microbes. Consequently, many epidemic diseases—cholera is a particularly notorious example—may be traced to water sources in these parts of the world.

Watson, James D. (1928–)

One of the two scientists who proposed the double-helical model for the structure of DNA, with specific base pairing between A and T or G and C to hold the opposite strands together. The complementary nature of the molecules, as indicated by this model, had implications for replication. Watson, an American, shared the 1962 Nobel Prize in Physiology with his colleague, Francis Crick, as well as with Maurice Wilkins, who provided pictorial evidence in the form of X-ray crystallographs of DNA molecules. *See also* CRICK, FRANCIS; DNA.

whooping cough

Bacterial disease caused by a respiratory infection with *Bordetella pertussis* (and, occasionally, in a milder form by *B. parapertussis*) that is usually associated with violent paroxysms of coughing followed by inspiratory whooping noises. The disease begins with the symptoms of an ordinary cold as organisms colonize the upper respiratory tract. During this initial period, the pertussis bacilli multiply at a rapid rate and are present in large numbers in the saliva and nasal discharges. The infection is spread rapidly via the droplet route. The characteristic cough is part of the inflammatory response of the trachea and bronchi to the toxins produced by the bacteria. It may persist for several weeks, well after the infectious stage of the disease has ceased. Although whooping is observed mostly in pertussis patients, it is not due to any specific action of the bacteria and may also occur in other respiratory infections such as those by adenoviruses.

Perhaps even more serious than the actual disease are such complications as bronchitis and pneumonia that arise from *Bordetella* infections. These conditions are especially serious among small children, and are responsible for most of the fatalities associated with whooping cough. Diagnosis of *B. pertussis* may be confirmed by culturing clinical samples—sputum or nasal secretions—on plates containing a medium made of glycerine, potato, and blood (called Bordet-Gengou medium) along with penicillin to inhibit the growth of other organisms. Characteristic mercury droplet-like colonies will begin to appear on the plates 2–3 days after incubation at 37°C. Antibiotics appear to have little effect in the outcome of whooping cough, except perhaps as a preventive measure against complications, secondary infections, and the spreading of the disease. General precautions observed in the case of respiratory diseases—i.e., limiting exposure of infected people and containing aerosols—should be observed. The DTP vaccine has been used successfully to control the onset of disease. *See also BORDETELLA*; DTP VACCINE.

wine

Alcoholic beverage produced by the fermentation of fruit juices by yeasts. While grapes are by far the most popular and widespread fruit used to make wines, a number of other fruits, such as blackberries, apples, pears, and plums, and even certain flowers, such as dandelion blossoms, may be used to make wines.

The basic chemistry in winemaking is the breakdown of fruit sugars to alcohol. Grapes are crushed and the juice—called must—is collected and fermented in wooden barrels

with wild yeasts, usually *Saccharomyces cerevisiae* and a closely related species that wine specialists call *S. ellipsoideus*. Bacterial growth is controlled during this initial fermentation by treating with sulfur dioxide. At this stage, the wine may be bottled. After bottling, certain wines are subjected to a maturation process, during which time interactions take place between the alcohol and fruit acids, and certain compounds precipitate out of solution. Microbes typically play little or no part in the maturation of wines.

The eventual taste and appearance of a wine depends on many factors, including the specific type of grape used; the initial composition of the must—the amount and type of sugars present, acidity, and the presence of fruit solids; the nature of the cultures; and, when relevant, the maturation process. For instance, when fermentation is carried out in the presence of the skin, pips, and stalks of the grapes, the final product is a red wine, which will have a high content of tannins. The quality of many red wines is greatly improved by maturation. White wines, produced by separating the must from the skin and seeds before fermentation, typically show little improvement after they have been bottled. The light, blush colored *rosé* wines are made by briefly exposing the must to the grape solids during the early stages of fermentation only, and *not* by mixing red and white wines.

In addition to the basic fermentation by yeasts, some wines are acted upon by other organisms. The must of Burgundy grapes, for example, has a high content of malic acid, which is broken down into lactic acid by a species of *Lactobacillus* during the fermentative process. Sauternes, a sweet white wine, is made from grapes that have already been infected and partially dehydrated by a fungus called *Botrytis cinerea*. Instead of maturation, sherries are made by passing the fermented must through a sequence of as many as 10 casks over a period of several years. Each cask develops a floating scum of a yeast called *S. beticus*, a relative of *S. cerevisiae*, which does not change the alcohol content of the wine, but imparts certain special flavors and aromas. Before bottling, the sherry is fortified with some fresh wine.

Champagnes and sparkling wines make use of a special double fermentation process. Blends of white wine are mixed with syrup and put in reinforced bottles with bolted-on corks and subjected to a second fermentation. The bottles are left on special racks and over a period of several months turned over slowly, so that the yeast and sediments settle down on the inside of the cork. The plug of sediment is carefully removed from the bottles and replaced by an equal volume of a syrup-brandy mixture. Because the second fermentation takes place in sealed containers, the gases produced are trapped within the bottle, resulting in an effervescent drink that can retain its sparkle long after opening. Only those sparkling wines made with grapes from the Champagne region in France may carry that name. *See also* ALCOHOLIC FERMENTATION; FERMENTATION; YEAST.

World Health Organization (WHO)

An international organization whose main mission is to work for the betterment of human health all over the world. The WHO was founded in 1948 under the aegis of the United Nations and consists of nearly 200 member nations, whose representatives include a diverse group of health-related professionals, including physicians, epidemiologists, health administrators, and researchers in various basic and applied scientific disciplines. The major functions of the organization are to set worldwide standards and offer advice about the field of health, to cooperate with various local governments in administering various national and international health programs (including preventive measures, therapies, and general procedures for maintaining health), and to aid in developing various technologies and implementing efficient transfer of these technologies to areas where they are most needed. The WHO is perhaps best known for its efforts and successes in the area of infectious diseases, e.g., it was the driving force behind the eradication of smallpox, and is well known for its role in administering various vaccines, especially in developing nations. However, its functions are not limited to this area of health, but extend into a much wider arena, including mental health, social medicine, and non-communicable diseases such as cancer and heart disease. Headquartered in Geneva, Switzerland, the WHO conducts its various activities and programs out of laboratories, offices, and field stations all over the world.

X

X-ray microscope

A type of microscope that uses X-rays to create enlarged images of objects. X-rays pose a special problem in microscopy because materials such as glass and quartz, which can bend visible light, are not refractive for X-rays. Furthermore, these rays are not charged like electron beams and thus cannot be redirected using electromagnetic fields. It is therefore difficult to devise an adequate apparatus to focus X-rays to form images. One approach is to use mirrors rather than lenses; this technique produces images by reflection rather than refraction of the X-rays. Alternatively, the enlarged images are produced with the help of zone plates, which rely on similar image-resolving principles as the phase contrast microscope. Both approaches have proved only partially successful and the use of X-ray microscopes is limited in comparison to optical and electron microscopes. *See also* MICROSCOPE.

xenodiagnosis

Diagnostic technique that makes use of the arthropod vectors or other animal hosts of a disease as the primary indicator of infection. The procedure is most commonly used for the diagnoses of trypanosomal infections such as Chagas disease. Uninfected insects—which have been raised in the laboratory—are allowed to feed on the blood of suspected patients of Chagas disease. After 30–60 days, the feces of these insects are examined regularly, over a period of three months, for the presence of developmental stages of *Trypanosoma cruzi*. This method is used widely in field experiments, especially in South America. In a xenodiagnostic procedure for *Trichinella* infections, muscle tissue from suspected patients is fed to uninfected rats, which are then checked periodically for the appearance of the larval stage of the parasite. This is a relatively rare procedure and is not available in most standard laboratories. *See also* CHAGAS DISEASE; *TRYPANOSOMA CRUZI*.

xerophile

Term used to describe an organism that requires little water for growth. Fungi, especially molds, tend to be the most xerophilic microbes. For example, species such as *Xeromyces bisporus*, a mold associated with food spoilage, can tolerate environments with a water activity of 0.61. *See also* WATER ACTIVITY.

Y

yaws

A severely mutilating, chronic disease caused by the spirochete *Treponema pertenue*, yaws occurs almost exclusively in tropical regions of the world. The clinical progression of yaws closely parallels that of syphilis—beginning with a primary lesion at the site of infection; progressing to a secondary stage of generalized spread and inflammation in various tissues; and ending with a chronic, destructive tertiary phase during which organisms continue to multiply and destroy tissues. Unlike syphilis, however, yaws is not primarily a venereal disease and person-to-person contact appears to be the primary mode of transmission. A primary lesion or "mother yaw" usually develops on the arms or legs and then progresses into more generalized superficial lesions, which become increasingly destructive. Yaw lesions have a characteristic raspberry-like appearance for which reason the disease has also been called framboesia (from the French word *frambois* for raspberry). The lesions appear to be limited to the skin, bones, and cartilage and there is no involvement of the nervous tissue, arteries, or internal organs, as in syphilis. *See also* SYPHILIS; TREPONEMA.

yeast

Yeast is a general, non-scientific term for fungi characterized by a single-celled structure and reproduction by budding. Common examples include species of *Candida*—a common cause of vaginal yeast infections—and *Saccharomyces*, commonly known as baker's or brewer's yeast. Yeasts will typically appear to stain gram positive but this property is not a reflection of the peptidoglycan content of their cell walls (they do not contain any). They may be distinguished from bacteria by their markedly large size as well as by their eukaryotic nature. It should be noted that while the unicellular mode of existence appears to be the norm, many yeasts are nevertheless capable of a dimorphic way of life and may form short hyphae under special conditions. This tendency is especially marked in various opportunistic pathogens discussed throughout this book but even species of *Saccharomyces* have occasionally exhibited such properties. Certain species of yeast, most notably *S. cereviseae* and *Candida utilis*, are used in a dry form as dietary supplements for providing protein and B complex vitamins. *See also* BUDDING; FUNGI.

yellow fever

An acute, communicable, viral hemorrhagic fever in humans caused by a member of the flavivirus family, yellow fever is an ancient but somewhat confusing figure in the history of human disease. At least part of the confusion lies in the name of this disease—historians have identified at least 150 different names that this same disease has been called in different parts of the world through the centuries. Some other common names include jungle fever and patriotic fever. In addition, the term yellow fever has also been used in the past to label other diseases such as malaria. Also, yellow fever remains less amenable to accurate diagnosis than most other infectious diseases, and is often mistaken for malaria, typhoid fever, Weil disease (leptospirosis), or dengue fever, because of the similarity of many of its symptoms to these other diseases, especially at the outset of infection.

Epidemiologically, yellow fever presents an interesting case because of its apparent geographic containment within the tropics. It is endemic in parts of Africa and South America within regions where temperatures do not fall below 71°C, and has never been known to occur in Asia. This latter fact is particularly puzzling because the primary means of transmission is the domestic mosquito, *Aedes aegypti*, which is common the world over. Once infected, a mosquito remains so for the duration of its life. Because of the prevalence of these mosquitoes in urban settings, yellow fever is today an urban disease, despite its possible origins with forest primates.

Following its entry into the bloodstream via a mosquito bite, the virus enters blood cells and subsequently liver cells and multiplies, causing the death of the host cells with every round of replication. After an incubation period of about three to six days, the patient will suffer a sudden onset of fever, chills, muscle aches and pains, and nausea and vomiting. The patient's vomit is often black due to the massive shedding of bile. A fairly common and characteristic symptom is the development of jaundice, which is moderate during the early stages of yellow fever and intensifies later, particularly if the interventions are inadequate. Jaundice is also one of the major causes of death in these infections; while the overall fatality rate of yellow fever is less than 5 percent, about 20–50 percent of the patients who develop jaundice die from the disease.

As mentioned earlier, yellow fever is frequently mistaken for several other infectious diseases of both viral and nonviral origin. The infection is virtually impossible to detect during the acute phase of infection and special immunological tests are needed to detect viral antigens in blood and liver tissues. A vaccine made of live, inactivated virus has been developed and is recommended for people traveling to areas of the world where the disease is endemic. This vaccine confers effective immunity for about 10 years, beyond which a booster shot is recommended. The inclusion of yellow fever vaccines as part of the routine childhood immunizations is on the rise in countries where the disease is endemic. Control measures, particularly among urban settings in endemic areas, is targeted at controlling mosquito populations and exposure to mosquitoes. *See also* FLAVIVIRUS; HEMORRHAGIC FEVER.

Yersinia

This genus in the Enterobacteriaceae family of gram-negative, non-spore-forming bacilli consists of mostly non-lactose-fermenting organisms, often found as parasites or pathogens in humans and other animals. Most species may exhibit some motility at temperatures below 30°C, but, with few exceptions, are non-motile at 37°C (i.e., body temperature). Pathogenic species of interest are discussed in individual entries. *See also* CYST; PROTOZOAN.

Yersinia enterocolitica

A natural pathogen of deer, pigs, birds, cattle, and other animals, this organism may cause a number of different gastrointestinal infections in humans. The primary mode of infection is via the consumption of con–taminated water or improperly processed meats and milk. Because *Y. enterocolitica* can grow at 4°C, it can pose problems even in refrigerated foods.

Upon their entry into the gut, these organisms infect the lymphoid tissues where they cause chronic infections. The nature and severity of disease depends largely on the age and general health of the affected individual. For instance, infants and young children are more likely to develop enterocolitis, while ileitis—typically accompanied by an acute inflammation of the lymph nodes in the abdominal region—occurs more frequently in older children and adults. The former is characterized by a prolonged diarrhea lasting at least two to several weeks, accompanied by abdominal pains and possible fever. The latter conditions resemble a case of acute appendicitis. *Y. enterocolitica* does not usually cause any systemic infections, although septicemia may be seen as a complication in elderly or weak patients. Diagnosis is based on the isolation and identification of organisms from feces. This organism can be differentiated from other *Yersinia* species on the basis of biochemical tests, susceptibility to phages, and motility and growth at different temperatures. Aggressive antibiotic therapy is required, and there are no vaccines.

Yersinia pestis

This organism is the cause of the dreaded bubonic plague, and normally resides in animal reservoirs such as rats. The bacteria may be transmitted via vectors (rat fleas) or through aerosols. Freshly isolated cultures produce copious amounts of slime due to the production of a capsule. This capsule is quickly lost when the organisms are grown *in vitro* or in the insect vectors. Virulent strains produce the capsule as well as other antigens, and show a high degree of pathogenicity for laboratory rats and guinea pigs. Other characteristics that distinguish *Y. pestis* from other species include the non-motility of this organism at any temperature and its sensitivity to bacteriophage. *See also* PLAGUE.

Yersinia pseudotuberculosis

This species, a natural pathogen of rodents and birds, can infect humans and cause severe enterocolitis, with lesions resembling those of intestinal tuberculosis (hence the name). It may also be implicated in local abscesses. Infections require aggressive chemotherapy with broad-spectrum antibiotics *Y. pseudotuberculosis* is distinguished from other *Yersinia* species by its motility at 25°C.

yogurt

Fermented food product produced by the action on milk of bacteria—mainly *Lactobacillus bulgaricus* in conjunction with *Streptococcus thermophilus*. Yogurt is prepared by inoculating milk with a starter consisting of these organisms and incubating the mixture at 35–45°C for several hours. The primary action of the two bacterial species is to ferment lactose (milk sugars) into lactic acid, which curdles the milk proteins and forms the semi-solid end product that we call yogurt. The combined growth of these two bacteria results in acid production at a rate faster than either organism by itself. *S. thermophilus*, which multiplies first, produces formic acid that accelerates the growth of the lactobacilli and stimulates them to produce compounds such as acetaldehyde (acetoin), which is responsible for the characteristic creamy taste of the end product.

Yogurt is a wholesome food not only because it is high in protein and low in fat, but also because the growth of lactobacilli creates an environment that is unfavorable for the growth of such pathogens as *Salmonella*, which would otherwise proliferate in milk. It is one of the fermented foods that is easily and frequently made at home in small batches, as well as on a commercial scale. The practice of making yogurt at home is especially prevalent in places where milk tends to spoil more easily because of a warm climate and lack of standardized pasteurization; the conversion to yogurt extends the shelf life of the milk by several days. *See also* LACTIC ACID FERMENTATION; PROBIOTIC.

Z

Zoogloea ramigera

Widely used in sewage treatment plants for the treatment of liquid wastes, *Zoogloea ramigera* is a gram-negative, motile, rod-shaped bacterium that is typically found as masses in a mixture of polysaccharides in polluted freshwater sources (such as lakes or ponds). The organisms are aerobic, heterotrophic, and produce both catalase and oxidase enzymes.

zoonosis

A general term for an infectious disease that is transmitted to humans from vertebrate host animals for which the organism is normally pathogenic (i.e., disease causing). Common sources of zoonotic infections include rodents, cattle, and swine. Common examples of zoonotic diseases of humans include anthrax, brucellosis, and the plague. *See also* ANTHRAX; *BRUCELLA*; PLAGUE; *YERSINIA*.

zooplankton

Minute floating animal life, e.g., protozoa, which may be found in virtually all different bodies of water on earth.

zygote

Diploid cell produced as a result of fusion or fertilization between two gametes or sex cells. *See also* DIPLOID; GAMETE.

APPENDIX 1
A Chronology of Epidemics in History

430–427 B.C.	The "Plague of Athens," also known as the "Plague of Thucydides," is said to have wiped out at least one-third of the city's population. Despite Thucydides' detailed eye-witness accounts, the identity of the disease remains a mystery to modern historians. Speculations include **smallpox**, the **plague**, an equivalent of acute **measles**, and **typhus**.
A.D. 161–166	An epidemic disease, possibly **smallpox**, which decimated the Roman army during their Parthian war.
452	A **smallpox** outbreak in the south of France; Nicaise, Bishop of Rheims (who became the patron saint for this disease), was a survivor of the epidemic.
541–544	Generally referred to as the "Plague of Justinian," this is the first recorded pandemic of the **bubonic plague**. It appears to have originated in Asia Minor then swept through Africa, Europe, and Turkey. Various merchant ships and troops spread the disease throughout the Western world. The death toll was enormous, with a total exceeding 300,000 in Constantinople in the first year alone.
664	The first recorded epidemic in Britain, known today as the "Yellow Plague," although this was not a name used when the disease first struck. Historians have tentatively identified the disease as an unusual variant of the **bubonic plague,** possibly complicated by **jaundice**. Some have suggested **relapsing fever** as the possible culprit, and still others speculate on the possibility of **smallpox**.
1173	An **influenza** epidemic recorded for the first time in Italy, Germany, and England.
1347–52	Popularly known as the "Black Death" or "Black Plague," this widespread epidemic of **bubonic plague** devastated large areas of Asia and Europe. It originated in central Asia and killed an estimated 25 million people in China, India, and neighboring regions, entering Constantinople and spreading upward and westward into all of Europe. About 25 million more people succumbed to the disease in Europe between 1347 and 1351, including half the population of London, four-fifths of the population of Marseilles, and over one-third of Italy's people.
1494–95	Outbreak of **syphilis** within the French army during the Siege of Naples. A hitherto unknown disease, syphilis was not defined in medical terms until 35 years later.
1496–1500	**Syphilis** spreads through Europe.
1557–58	The first known epidemic of **malaria** in Europe.
1617	The introduction of **smallpox** to North America by Europeans. This epidemic wiped out about 90 percent of the population of the Massachusetts Bay Indians, who had never encountered this disease before and were hence extremely susceptible.
1633	The first **smallpox** epidemic among whites in North America breaks out in Boston.
1647	The first recorded epidemic of **yellow fever** in the New World begins in Barbados. The disease, which was called the "Barbados distemper," killed more than 5,000 people within a few months.

Appendix 1: A Chronology of Epidemics in History

1664–65   The "Great Plague of London." This outbreak of the **bubonic plague** killed 75,000–100,000 people in the city of London, wiping out nearly 20 percent of its population in less than a year.

1678–82   One of the largest epidemics of **malaria** to hit Europe, killing thousands of people, especially in areas of England, Belgium, and Holland, and becoming endemic in France.

1721   A **smallpox** epidemic in Boston and elsewhere in New England that has been characterized as "particularly vicious." About 6,000 of a total population of 11,000 were infected, and nearly 850 people died in this outbreak.

1751–53   Possibly the most severe epidemic of **smallpox** to afflict England until that point in history,

1781–82   A major outbreak of **influenza** in Europe, considered by many historians to be one of the most significant medical epidemics of that time. Although yet a matter of controversy, many believe it originated in China. By the time it disappeared in 1782, it had spread over extensive areas of Europe and claimed many lives.

1793   A **yellow fever** epidemic in Philadelphia. Described in detail by one of the leading physicians of the time, Benjamin Rush, who attributed the disease to a pile of rotting coffee on the wharf. The outbreak killed nearly one-eighth of the population of the city—an estimated 11,000 people out of the 50,000 contracted the disease, of whom 4,000 died.

1817   An epidemic of **cholera** in the town of Jessore in the Indian province of Bengal, believed to be the source of the pandemic that was to sweep the world over the next several years.

1817–23   This pandemic was the first of many **cholera** pandemics to sweep the world in the nineteenth century. It is particularly interesting because it was the first time that this disease was introduced into the Western world, although it had been endemic in India and surrounding areas for centuries. It traveled into the Middle East and Russia, but an exceptionally severe winter in 1823–24 checked its spread into Europe.

1826–37   This second **cholera** pandemic also originated in Bengal, India, and covered even more terrain than the first outbreak, reaching not only Europe but also the United States by 1832.

1824–29   A **smallpox** epidemic in Europe.

1837–40   A **smallpox** epidemic in Europe.

1837–63   The third **cholera** pandemic. Beginning with a series of outbreaks in and around India in 1837, this epidemic suddenly exploded in 1846 and proceeded to spread throughout the world.

1847–48   A widespread outbreak of **influenza** though Europe and the Mediterranean. Most likely, this epidemic originated in Russia.

1863–67   The fourth **cholera** pandemic, and possibly the most widespread outbreak in history. Like all others before it, the pandemic originated in India and spread via various land and sea routes to different parts of the world.

1870–74   A **smallpox** epidemic in Europe.

1878   A **yellow fever** epidemic in the eastern and southern United States. It ultimately spread to eight states, but hit Memphis, Tennessee, with the greatest intensity, affecting 17,500 people within three months and killing some 5,150.

1881–96   The fifth **cholera** pandemic, marked by Koch's discovery of the causative agent, *Vibrio cholerae*.

1889–90   An **influenza** pandemic that appears to have begun almost simultaneously in Siberia and in western Canada in May 1889. By the end of summer, Greenland reported a severe outbreak, followed by reports from England in October and New York City in December. According to various estimates, an hefty 40 percent of the population was attacked in Massachusetts, and 20–30 percent in England. The epidemic recurred in waves, twice in 1891 and again in 1892. The final outbreak is reported to have been more severe in the United States than the original outbreak.

1894   The first relatively large **poliomyelitis** outbreak described in the United States, believed to have originated in Rutland, Vermont. Some 132 cases were described within a few months in and around that area, which is significant given the relatively sparse population of the region.

1899   The sixth **cholera** pandemic begins.

1904	A **typhoid** and **paratyphoid** outbreak in the American army during the Spanish-American War. Records from the time report that about one-fifth of the total army population was afflicted.
1905	The last recorded outbreak (in epidemic proportions) of **yellow fever** in the United States, in Galveston, Texas, and New Orleans, Louisiana.
1906–14	A series of outbreaks of **tuberculosis** in South Africa that claimed more than 5,000 lives during this period and continued to remain a problem until the introduction of streptomycin in the early 1950s.
1916	A **poliomyelitis** epidemic in the United States, affecting some 26 states, although incidence predominated in the Northeast. At least 29,000 cases and 7,000 deaths were recorded. New York City was particularly hard hit and accounted for some 9,000 cases and nearly 2,500 deaths.
1917–19	An **influenza** pandemic, popularly known as the "Spanish Influenza Epidemic of 1917–19." This was one of the deadliest and most widespread pandemics recorded in history, killing about 15–50 percent of the affected populations in various countries, with a total death toll estimated at 20 million. In the United States alone, this figure was estimated at around .5 million. The mortality rate was significantly higher than preceding epidemics and the disease appeared to strike very young and very old populations. Arriving as it did during World War I, it aroused less panic than other great epidemics, despite the fact that it claimed many more lives than the war. It was dubbed the "Spanish Influenza," because Spain was the first place where its effects were seriously felt.
1919–31	An epidemic of **encephalitis** (presumably of viral origin) that claimed several thousand lives in England, Scotland, and Wales over its 12-year duration. An unknown disease at the time of the outbreak, it was at first mistaken for botulism by many physicians.
1921–22	The first epidemic outbreak of louse-borne **relapsing fever** reported in Mali and neighboring areas of West Africa. The disease affected an estimated 108,000 people and killed close to 15,000.
1924	The first major outbreak of **Japanese encephalitis** reported in Japan, with over 6,000 cases and close to 3,500 deaths (a mortality rate of 62 percent was recorded).
1926–40	Outbreaks of **pneumonia**, primarily streptococcal pneumonia, among miners in South Africa. Over 41,000 people were infected during this time and nearly 5,000 deaths were recorded.
1931	A second epidemic of **poliomyelitis** hits the United States. Centered in the northeast, and especially in New York City, this outbreak killed over 4,000 people.
1938–40	A devastating epidemic of **malaria** in the northeast coastal region of Brazil that infected as many as 290,000 people and killed almost one-tenth of this population within two years. Children under the age of five were particularly susceptible, and the use of quinine met with only limited success.
1942–53	The most devastating of the **poliomyelitis** epidemics in the United States, peaking in 1952 with about 60,000 cases reported from all over the country. The disease also received more media attention than ever before because of President Franklin Roosevelt's personal involvement in the cause.
1951–54	Incidence of a strange new disease, characterized by symptoms of hemorrhagic fever compounded with renal failure, that broke out among the American troops in Korea. The disease agent was found to be a member of the **hantavirus** family called haantaan virus.
1955	One of the largest recorded epidemics of **infectious hepatitis** in Delhi, India. Long considered endemic to that region, the 1955 outbreak was a massive explosion of the disease attributable to the contamination of one of the city's main water sources. Although it lasted for only a few weeks, there were over 97,000 cases, one-fourth of which showed symptoms of jaundice.
1957	An **influenza** pandemic that originated in southwest China and within six months had spread to most parts of the globe. Anywhere from 10 to 35 percent of the population was afflicted in different parts of the world, but the overall mortality rate was relatively low at 0.25 percent.

Appendix 1: A Chronology of Epidemics in History

1964	One of the largest recorded epidemics of **rubella** in the United States, which infected 12.5 million people and resulted in 30,000 terminated pregnancies and the birth of close to 20,000 congenitally handicapped babies. The epidemic is estimated to have cost $1–$2 billion.
1967	Outbreak of the **Marburg virus** hemorrhagic fever in Germany and Yugoslavia. This outbreak was largely limited to the group of people who worked in laboratories handling African green monkeys, all of which had been shipped from a single region in Uganda.
1968	An **influenza** pandemic, better known as the "Hong Kong" pandemic, for the place where the causative agent was pinpointed. Like the epidemic of a decade earlier, this too originated somewhere in southwest China but had far more devastating consequences in terms of severity of the disease and mortality.
1969–71	Epidemic of an acute hemorrhagic type of **conjunctivitis**, which began in Africa and spread to Asia and parts of Europe. The causative agent was found to be a new strain of enterovirus (in the same family as the poliovirus).
1974	A **malaria** epidemic in India. Believed to have been brought in 1974 by refugees from Sri Lanka and Pakistan, malaria experienced a dramatic resurgence during the 1970s, with the development of drug-resistance on multiple fronts: the mosquitoes' resistance to DDT and the malarial parasites' resistance to quinine. According to reports, about 5 million people were infected in different parts of the country in 1975, and 6 million in the following year. By 1977, nearly 30 million people in the country were infected.
1976	Outbreaks of hemorrhagic fevers, due to the new **Ebola virus**, occured in Zaire and Sudan. The virus killed over 200 people in a region of Zaire near the Ebola River between mid-August and September, and subsequently an equivalent number in a remote village in Sudan.
1976	The first outbreak of **Legionnaire's disease** in Philadelphia among the attendees at a convention of the American Legion. The disease made headlines because of the bizarre nature of the disease and its apparent mode of spread via the air conditioning system. In reality, however, it had a relatively low impact and claimed only about 26 lives of the 260 who contracted the disease.
1977–78	An **influenza** pandemic, better known as the "Russian" or "Red" pandemic, even though it originated in China, because the disease only began to grow in importance to the public after the Russian outbreak occurred. Of moderate intensity, this epidemic was widely dispersed over the world but moved relatively slowly in comparison to previous influenza pandemics.
early 1980s	**AIDS** first makes an appearance in the United States in apparently epidemic proportions.
1987	Worldwide spread of **AIDS**.
1992–93	A **diphtheria** outbreak in Russia, with reports of 4,000 cases and 106 deaths (60 percent in adults), spurred in part by a lack of proper immunization or booster shots against the disease.
1993	**Hantavirus pulmonary syndrome** caused by the Sin Nombre Virus in the United States.
1994	Unexpected outbreak of the **pneumonic plague** in India, originating in the town of Surat, which sent residents of the city into a panic and spurred a mass migration of nearly 200,000 people out of the city in less than a week.

Sources

Note: This list is for quick reference only. For the complete citations of these references, please consult the bibliography at the end of the book.

Ackerknecht, E. H. *History and Geography of the Most Important Diseases.*
Kohn, George C., ed. *Encyclopedia of Plague and Pestilence.*
Marks, Geoffrey, and William K. Beatty. *Epidemics.*
Porter, Roy, ed. *The Cambridge Illustrated History of Medicine.*
Rhodes, Philip. *An Outline History of Medicine.*
Top, Franklin H., ed. *The History of American Epidemiology.*

APPENDIX 2
Large Infectious Disease Outbreaks

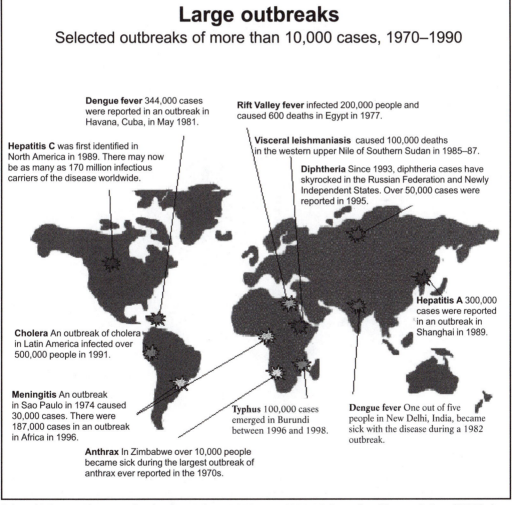

Large outbreaks
Selected outbreaks of more than 10,000 cases, 1970–1990

Dengue fever 344,000 cases were reported in an outbreak in Havana, Cuba, in May 1981.

Rift Valley fever infected 200,000 people and caused 600 deaths in Egypt in 1977.

Hepatitis C was first identified in North America in 1989. There may now be as many as 170 million infectious carriers of the disease worldwide.

Visceral leishmaniasis caused 100,000 deaths in the western upper Nile of Southern Sudan in 1985–87.

Diphtheria Since 1993, diphtheria cases have skyrocked in the Russian Federation and Newly Independent States. Over 50,000 cases were reported in 1995.

Hepatitis A 300,000 cases were reported in an outbreak in Shanghai in 1989.

Cholera An outbreak of cholera in Latin America infected over 500,000 people in 1991.

Meningitis An outbreak in Sao Paulo in 1974 caused 30,000 cases. There were 187,000 cases in an outbreak in Africa in 1996.

Typhus 100,000 cases emerged in Burundi between 1996 and 1998.

Dengue fever One out of five people in New Delhi, India, became sick with the disease during a 1982 outbreak.

Anthrax In Zimbabwe over 10,000 people became sick during the largest outbreak of anthrax ever reported in the 1970s.

Selected infectious disease outbreaks of more than 10,000 cases, 1970–90. Reproduced by permission of WHO, from *Report on Infectious Diseases 1999.* © World Health Organization 1999.

APPENDIX 3
Leading Causes of Death in the World

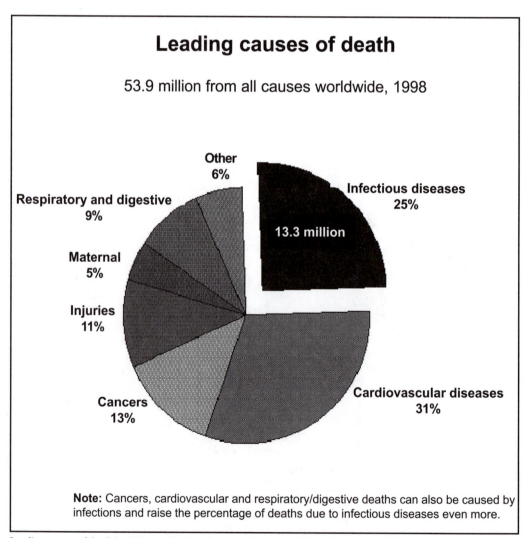

Leading causes of death in 1998, worldwide, all causes. Reproduced by permission of WHO, from *Report on Infectious Diseases 1999.* © World Health Organization 1999.

APPENDIX 4
Leading Infectious Disease Killers

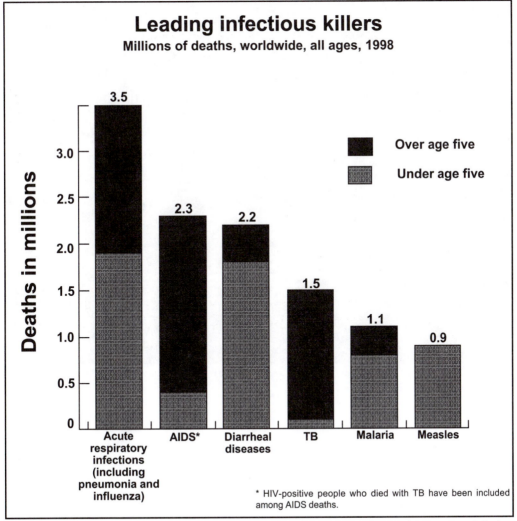

Leading causes of death in 1998, worldwide, all causes. Reproduced by permission of WHO, from *Report on Infectious Diseases 1999.* © World Health Organization 1999.

APPENDIX 5
Death Rates from Leading Causes of
Death in the United States

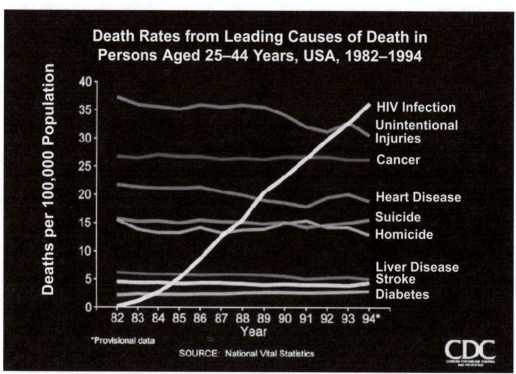

Death rates from leading causes of death in persons aged 25–44 years, United States, 1982–94. Courtesy of CDC/NCHS/National Viral Statistics.

BIBLIOGRAPHY

As the reader will learn upon a visit to the library, any one of the volumes listed below could be easily substituted for a dozen or so equally informative ones that lie next to it on the shelf. While I have attempted to include a diversity of references from the materials that I found and frequently used while doing research for this book, I heartily recommend that the interested reader use this list as a starting point only.

Books

Ackerknecht, E. H. *History and Geography of the Most Important Diseases*. New York: Hafner Publishing Company Inc., 1965.

Benenson, Abram S., ed. *Control of Communicable Diseases Manual*, 17th ed. Washington, D.C.: American Public Health Association, 2000.

Bogitsh, Burton J., and Thomas C. Cheng. *Human Parasitology*, 2nd ed. San Diego: Academic Press, 1998.

Boyd, Robert F. *Basic Medical Microbiology*, 5th ed. Boston: Little, Brown, 1995.

Collier, Leslie, and Albert Balows, eds. *Topley and Wilson's Microbiology and Microbial Infections*, 9th ed. 6 volumes. New York: Arnold, 1998.

De Kruif, Paul. *Microbe Hunters*. New York: Harcourt, Brace and Company, 1926.

DeLaat, Adrian N. C. *Microbiology for the Allied Health Professions*, 3rd ed. Philadelphia: Lea and Febiger, 1984.

Dimmock, N. J., and S. B. Primrose. *Introduction to Modern Virology*, 4th ed. Boston: Blackwell Science, 1994.

Dixon, Bernard. *Power Unseen: How Microbes Rule the World*. New York: W. H. Freeman, 1994.

Doyle, Michael P., Larry R. Beuchat, and Thomas J. Montville. *Food Microbiology: Fundamentals and Frontiers*. Washington, D.C.: ASM Press, 1997.

Dusenbery, David B. *Life at Small Scale: The Behavior of Microbes*. New York: Scientific American Library, 1996.

Fuerst, Robert. *Frobisher and Fuerst's Microbiology in Health and Disease*, 15th ed. Philadelphia: W. B. Saunders Company, 1983.

Garrett, Laurie. *The Coming Plague: Newly Emerging Diseases in a World Out of Balance*. New York: Farrar, Straus and Giroux, 1994.

————. *Microbes versus Mankind: The Coming Plague*. New York: Foreign Policy Association, 1996.

Gest, Howard. *The World of Microbes*. Madison, WI: Science Tech Publishers, 1987.

Gravé, Eric V. *Discover the Invisible*. Upper Saddle, NJ: Prentice-Hall Inc., 1984.

Kohn, George C.. ed. *Encyclopedia of Plague and Pestilence*. New York: Facts on File, Inc., 1995.

Kolata, Gina B. *Flu: The Story of the Great Influenza Pandemic of 1918 and the*

Search for the Virus That Caused It. New York: Farrar, Straus and Giroux, 1999.

Koprowski, Hilary, and Michael B. A. Oldstone, eds. *Microbe Hunters, Then and Now*. Bloomington, IL: Medi-Ed Press, 1996.

Lederberg, Joshua. *The Encyclopedia of Microbiology*. 4 volumes. San Diego: Academic Press, 2000.

Marks, Geoffrey, and William K. Beatty. *Epidemics*. New York: Charles Scribner's Sons, 1976.

Pitot, Henry C. *Fundamentals of Oncology*, 3rd ed. New York: Dekker, 1986.

Porter, Roy, ed. *The Cambridge Illustrated History of Medicine*. New York: Cambridge University Press, 1996.

Rhodes, Philip. *An Outline History of Medicine*. Boston: Butterworths, 1985.

Salminen, Seppo, and Atte von Wright, eds. *Lactic Acid Bacteria*. New York: Dekker, 1993.

Shilts, Randy. *And the Band Played On: Politics, People, and the AIDS Epidemic*. New York: St. Martin's Press, 1987.

Singleton, Paul. *Bacteria in Biology, Biotechnology, and Medicine*, 5th ed. New York: John Wiley, 1999.

Slack, John M., and Mary Ann Gerencser. *Actinomyces, Filamentous Bacteria: Biology and Pathogenicity*. Minneapolis: Burgess Publishing Company, 1975.

Slayter, Elizabeth M., and Henry S. Slayter. *Light and Electron Microscopy*. New York: Cambridge University Press, 1992.

Sugar, Alan M., and Caron A. Lyman. *A Practical Guide to Medically Important Fungi and the Diseases They Cause*. Philadelphia: Lippincott-Raven, 1997.

Top, Franklin H., ed. *The History of American Epidemiology*. St. Louis: The C.V. Mosby Company, 1952.

Varmus, Harold, and Robert A. Weinberg. *Genes and the Biology of Cancer*. New York: Scientific American Library, 1993.

Wagner, Edward K., and Martinez J. Hewlett. *Basic Virology*. Malden, MA: Blackwell Science, 1999.

White, David O., and Frank J. Fenner. *Medical Virology*, 4th ed. San Diego: Academic Press, 1994.

Useful Web Sites on Microbes and Public Health

American Public Health Association (APHA)
http://www.apha.org
American Society for Microbiology (ASM)
http://www.asm.org
American Type Culture Collection (ATCC)
http://www.atcc.org
Centers for Disease Control (CDC)
http://www.cdc.gov
FDA's "Bad Bugs" webpage
http://vm.cfsan.fda.gov/~mow/intro.html
Medical Subjects Headings
http://www.nlm.nih.gov/mesh/meshhome.html
Web site with lists of bacteria for investigation
www.eibe.reading.ac.uk/NCBE/SAFETY/bacteria.html
World Health Organization (WHO)
http://www.who.org

INDEX

by Dottie M. Jahoda

Figures are designated by *f* and tables by *t*.

Acetmonas, 255
Acetobacter, 1
    A. aceti, 1
    A. diazotrophicus, 1
    A. xylinum, 1
    vinegar production by, 255
Acid fast stain, 1
Acinetobacter, 1–2
    A. calcoaceticus, 2
    in sludge, 6
Acquired immunodeficiency syndrome
    (AIDS), 2
    Burkitt lymphoma in, 43
    cancers associated with, 3
    cases of, in United States, 2*f*
    Coccidioides immitis infection in, 66
    cocktail therapy for, 131
    cytomegalovirus infection in, 77
    definition of, 4
    diseases associated with, 2–3
    epidemics of, in history, 274
    first recognition of, 3
    HIV virus causing, 2–3
    incidence of, and deaths in United States, 3*f*
    infection process in, 4
    infections associated with, 3
    mortality in United States due to, 278*f*
    Pneumocystis carinii infection in, 199
    progressive multifocal leukoencephalopathy
        in, 203
    risk factors for, 4
    transmission of, 3–4
    treatment of, 4
    tuberculosis in, 247
Actinobacillus, 5
    A. actinomycetemcomitans, 5
    A. lignieresii, 5

    associated with *Actinomyces,* 6
Actinomyces, 5
    A. bovis, 6, 56
    A. eriksonii, 6
    A. israelii, 6, 156
    A. naeslundii, 6, 82
    A. odontolyticus, 6
    A. viscosus, 6, 82
    Arachnia and, 20
    associated with *Actinobacillus,* 6
    characteristics of, 5
    colony of, 5*f*
    in composting, 69
    hypersensitivity reactions to, 5–6
    importance of, 5
    infections caused by, 6
    Nocardia associated with, 6
Activated sludge, 6
Adenosine triphosphate. *See* ATP
Adenovirus
    characteristics of, 7
    discovery of, 6–7
    infections caused by, 7
ADP (adenosine diphosphate), 23
Aerobe, 7
    facultative, 103
Aerococcus viridans, 7
Aeromonas, 7
    A. caviae, 7
    A. hydrophila, 7
    A. salmonicida, 7
    characteristics of, 8
    infections caused by, 7–8
Aerosol, 8
Aflatoxin(s), 8
    produced by *Aspergillus,* 8, 22

African sleeping sickness, 8
    diagnosis of, 9
    microorganism causing, 8–9
    stages of, 9
    treatment of, 9
Agar, 9
Agrobacterium, 9
    A. rhizogenes, 10
    A. rubi, 10
    A. tumifaciens, 9, 73
    plant diseases caused by, 9–10
AIDS. *See* Acquired immunodeficiency
    syndrome
Akinete, 10
Alcaligenes, 10
    A. eutrophus, 10
    A. faecalis, 10
    in sludge, 6
Alcohol dehydrogenase, 10–11
Alcohol fermentation, 10
    microbes used in, 10–11
Alexandrium catanella, toxic blooms caused by,
    40
Algae, 11
    blooms caused by, 39–40, 212
    blue-green, 40
    compounds produced by, 11
    pigments in, 11
Allergen, 11
Allergic reaction, 11–12. *See also* Hypersensi-
    tivity reaction(s)
Alper's syndrome, 202
Alphavirus(es), 240
Ameba, 12
Amebiasis, 12
    diagnosis and treatment of, 13
    microorganism causing, 12–13
Ameboma, 13
American trypanosomiasis, 55
American Type Culture Collection (ATCC), 13
Ames test, 13
Amino acid, 13–14
Amino group, 13
Aminoglycosides, 14
Ammonia assimilation, 14
Ammonification, 14
Anabaena, 76
Anaerobe, 14
    facultative, 103
Anaerobic respiration, 15
Anemia, 15
Animal virulence test, 15
Anthrax
    forms of, in humans, 15–16
    microbe causing, 15
Antibiogram, 17

Antibiotic(s), 16. *See also specific antibiotic*
    aminoglycosides as, 14
    broad-spectrum, 16
    cephalosporin, 55
    chloramphenicol, 62
    cyclosporins, 76–77
    gentamycin, 14
    kanamycin, 14
    β-lactam antibiotics, 149
    neomycin, 14
    novobiocin, 179
    penicillin, 190
    polymyxin, 200
    rifamycin, 217
    selective toxicity of, 16
    sensitivity to, testing, 17–18
    streptomycin, 14
    sulfonamides, 235
    tetracycline, 238
Antibiotics, resistance to
    increase in, 17f
    microbes with, 16
    prescribing antibiotics and, 17
Antibody(ies), 18
Antigen, 18
    Australia, 24
Antigenic drift, 18–19
Antigenic shift, 19
Antisepsis, 19
Antiserum, 19
Antitoxin, 19
Antiviral agents, 19–20, 19t
Aphthovirus, 20
Arachnia, 20
    A. propionica, 20
    and *Actinomyces,* 20
Arbovirus(es), 20
Archaebacteria, 20–21
Arenavirus(es), 21
Asepsis, 21
Aseptic technique, 21–22
Aspergillus
    A. flavus, 8, 22
    A. fumigatus, 22
    A. niger, 22
    A. parasiticus, 8
    aflatoxins produced by, 8, 22
    diseases caused by, 22
Assimilation, 22
Astrovirus(es), 22–23
ATCC. *See* American Type Culture Collection
Athlete's foot, 23
ATP (adenosine triphosphate), 23. *See also*
    Glycolysis; Krebs cycle; Respiration
    phosphorylation of, 191–192
Attack rate, 23

Attenuated vaccine, 23–24
Australia antigen, 24
Autoclave, 24
    walk-in, 24*f*
Autoradiography, 25
Autotroph, 25
Auxotroph, 25
Avery, Oswald T., 25
Avian sarcoma virus(es), 25–26
Axenic, 26
Azomonas, 26
    nitrogen fixation by, 176
Azotobacter, 26
    A. chromococcum, 26
    A. vinelandii, 26
    nitrogen fixation by, 176

B cell, 27
B cell lymphoma, associated with AIDS, 3
B19 virus, 188
Babesia, 27
    B. microti, 27
    infections caused by, 27
    life cycle of, 27
    symptoms and treatment of infection with,
        27–28
Bacillus, 28–29
    B. anthracis, 15–16, 28, 51
    B. brevis, 30*f*
    B. cereus, 28–29
    B. israeliensis, 29, 36
    B. licheniformis, 15, 37
    B. polymyxa, 29, 200
    B. stearothermophilus, 29
    B. thuringiensis, 29, 36
    in composting, 69
    denitrification by, 81
Bacteremia, 29
Bacteria, 30. *See also specific bacterium; specific*
    disease
    autotrophic, 58
    capsule of, 51
    characteristics of, 30–31, 30*f*
    chemoautotrophic, 58
    chemotrophic, 59
    coenocytic, 67
    coliform, 67
    colony of, 67
    conjugation of, 69–70
    diseases caused by (*see specific bacterium;*
        specific disease)
    endotoxins of, 96
    exotoxins of, 102
    fimbria of, 104–105
    flagella of, 105
    glutamic dehydrogenase in, 14

    hemolysins produced by, 124
    heterocystic cells of, 130
    human dependence on, 31
    L forms of, 149
    metabolic enzymes of, 31
    nitrifying, 175
    outer membrane of, 185
    pilus of, 194
    plasmids in, 196
    pure culture of, 206–207
    R strains of, 51
    resistance of, to antibiotics, 16–17
    spherical, 67
    spore formation by, 230
    strains of, 232
    transduction in, 243
    transformation in, 243
    wall of, 54, 191
Bacteriocin, 31
Bacteriophage, 31–32
    bacterial infection with, 196
Bacteroides, 32
    B. fragilis, 32
Baculovirus(es), 32
Balantidium coli, 32–33
Baltimore, David, 33
Bartonella, 33
    B. bacilliformis, 33, 184, 254
    B. henselae, 33, 52
    B. quintana, 33
Bassi, Agostino, 33
Beer, production of, 33–34
Behring, Emil A. von, 34
Bejel, 34
Bergey, David H., 34
Bergey's Manual of Systematic Bacteriology, 34
β-lactam antibiotics, 149
β-lactamase, 149
Binary fission, 34–35
Binomial nomenclature, 35
Biochemical pathway, 92
Biofilm(s), 35
    commercial use of, 35
    formation of, 35
Biological aerated filter, 36
Biological oxygen demand (BOD), 36
Biological warfare, 36–37
Bioluminescence, 37
Biosensor, 37–38
Biosphere, 38
Biotechnology, 38
BK virus, 200
Blastomyces dermatitidis, 38
    diagnosis and treatment of infections caused
        by, 38–39
    infections caused by, 38

Blood-borne pathogens, 39
Bloom(s)
    algal, in Lake Michigan, 39f
    environmental impact of, 39
    microbes causing, 39
    red tide, 212
    toxins produced by, 39–40
Blue-green algae, 40
Blumberg, Baruch S., 24, 40
BOD (biological oxygen demand), 36
Bordatella, 40
    B. bronciseptica, 40
    B. parapertussis, 40, 262
    B. pertussis, 40, 262–263
    infections caused by, 40–41
Bordet, Jules, 40
Borrelia, 41
    B. burgdorferi, 41, 157
    B. hermsii, 41, 212
    B. parkerii, 41, 212
    B. recurrentis, 41, 212
Botrytis cinerea, 108, 263
Botulism, 41
Bread, production of, 42
Broad-spectrum antibiotics, 16
Brucella, 42
    B. abortus, 42
    B. canis, 42
    B. melitensis, 42
    B. neotomae, 42
    B. ovis, 42
    B. suis, 42
Bubonic plague, epidemics of, in history, 271, 272
Buchanan, R. E., 34
Budding, 42
Bunyavirus(es), 20, 42
    diseases caused by, 43
    encephalitis caused by, 94, 95t
Burkitt lymphoma, 43
Butanediol fermentation, 43
Butter, production of, 43
Buttermilk, production of, 43–44

Calicivirus, 45
California encephalitis, 43
Campylobacter, 45–46
    C. jejuni, 45
Cancer, 46
    abnormal cell behavior in, 46–47
    benign and malignant, 47
    biopsy for, 48
    cell growth and causes of, 46, 47–48
    diagnosis of, 48
    DNA mutations in, 50
    immune system reaction to, 48
    metastasis in, 47

neoplasms in, 48
    screening for, 48
    staging systems for, 47
    treatment of, 48, 49
    viruses and, 46, 49–50
Candida, 50–51
    C. albicans, 51
    C. glabarata, 51
    C. krusei, 51
    C. parapsilosis, 51
    C. tropicalis, 51
    C. utilis, 51
    infections caused by, 239
Cane gall, 10
Canning, 51
Capsid, 51
Capsule, 51
Carbolfuchsin stain, 1
Carbon cycle, 51–52
Carboxyl group, 13
Cardiovirus(es), 52
Caries, dental, 82
Carrier, 52
Case fatality rate, 52
Cat scratch disease, 33, 52–53
Catalase, 53
Caulobacter, 53
CDC (Centers for Disease Control and Prevention), 55
Cell(s), 53
    cytoplasm of, 77–78
    death of, 173
    division of, 53–54
    endocytosis in, 96
    endoplasmic reticulum of, 96
    exocytosis in, 102
    Golgi apparatus in, 118
    inclusion bodies in, 139
    membrane of, 185
    metabolism of, 163
    mitochondria in, 165
    motility of, 167
    mutant, 168
    nucleus in, 179
    phagocytosis in, 191
    respiration of, 148, 214
    ribosomes in, 216–217
    signal transduction pathways of, 47–48
    spheroplast, 229–230
    transformation of, 243
    vacuoles in, 113, 252
    wall of, 54, 190
Cellulase, 54
Cellulomonas, 54–55
C. flavigena, 55
Centers for Disease Control and Prevention (CDC), 55

Centrifugation, 55
    density gradient, 82
    ultra-, 250–251
Cephalosporin, 55
Chagas disease, 55
Chagoma, 56
Chain, Ernst Boris, 56
Cheese, production of, 57*f*
    butanediol fermentation in, 43
    microorganisms used in, 56
Chemoautotroph, 58
Chemoorganotroph, 58
Chemotaxis, 58
Chemotherapy, 58
Chemotroph, 59
Chickenpox, 59
Chlamydia, 59
C. pneumoniae, 59, 61
C. psittaci, 59, 61
C. trachomatis, 59, 61–62. *See also Chlamydia
    trachomatis*
    characteristics of, 59–60
    comparison of, to *Rickettsia* and viruses, 60*t*
    infections caused by, 60
Chlamydia trachomatis, 59, 61–62
    infection with, 157–158, 242
Chloramphenicol, 62
Chlorophyll, 62
    in algae, 11
Chloroplast, 62
Cholera, 62
    diagnosis of, 62–63
    epidemics of, in history, 272
    epidemiology of, 63
    microorganism causing, 62
    prevention of, 63
    symptoms of, 62
    treatment of, 63
Chromomycosis, 63
Chromosome(s), 63
Chronic fatigue syndrome, 64
Ciguatera fish poisoning, 39–40
Cilia, 64
Cloning, 64
Clostridium, 64
    anaerobic respiration by, 15
    C. acetobutylicum, 65
    C. botulinum, 41, 65
    C. difficile, 65
    C. perfringens, 65–66, 140
    C. ramosum, 64
    C. tetani, 66, 237–238
    gangrene caused by, 112–113
    nitrogen fixation by, 176
Coagulase, 66

Coccidioides immitis, 66
    diagnosis and treatment of infection by, 66–67
    infection by, in AIDS syndrome, 66
    symptoms of infection by, 66
Coccus, 67
Cocktail therapy, 131
Coenocyte, 67
Colicin, 67
Coliform, 67
Colony, 67
Colorado tick fever, 67–68
Coltivirus(es), 68
    causing Colorado tick fever, 67
Commensalism, 68
Communicable disease control, 140
Complement, 68
    activation of, 68
    importance of, in immunity, 68–69
Compost, 69
Conjugation, 69–70
Conjunctivitis, epidemics of, in history, 274
Contact microradiography, 70
Contact tracing, 70
Contaminant, 70
Contrast, 70
Coronavirus(es), 71, 71*f*
Corynebacterium
    C. diphtheriae, 71
    C. pseudodiphtheriticum, 86
    C. renale, 71–72
    C. xerosis, 86
Corynebacterium diphtheriae, 71
    antitoxin to, 85*f*
    characteristics of, 71
    toxins produced by, 71–72, 84–85
Cowpox, 72
Coxiella, 217
    C. burnetii, 72, 208
Coxsackie virus(es), 72
    infections caused by, 122–123
Creutsfeld-Jacob disease, 72–73
Crick, Francis, 73
Crimean-Congo hemorrhagic fever, 124
Crown gall disease, 73
    microorganism causing, 9–10, 73
Cryptococcus, 73
    C. neoformans, 73–74
    characteristics of, 73–74
    infection by, 73
    symptoms and treatment of infection by, 74
Cryptosporidium, 74
    C. parvum, 74
    diagnosis and treatment of infection by, 75
    infection by, in AIDS syndrome, 3
    infections caused by, 74
    life cycle of, 74–75, 262*f*

Curing food, 75
Cutaneous leishmaniasis, 75
   diagnosis and treatment of, 76
   microorganism causing, 75
   symptoms of, 75
Cyanobacteria, 76
Cyclosporins, 76–77
   microorganisms producing, 76
Cylindocarpon lucidum, 76–77
Cyst, 77
Cytomegalovirus (CMV), 77
Cytopathic effect (CPE), 77
Cytoplasm, 77–78

Darkfield microscopy, 79
Darwin, Charles, 79
Defective virus, 79
Dehydration, 79
Delayed type hypersensitivity, 80
Delbrück, Max, 80, 130
Dengue fever, 80
   diagnosis and treatment of, 81
   epidemiology of, 80–81, 81*f*
   microorganism causing, 80
   symptoms of, 80–81
   transmission of, 80
Denitrification, 81–82
   microorganisms performing, 81
Density gradient centrifugation, 82
Dental caries, 82
Dental plaque, 82–83
Dermatophytes, 83
Desiccation, 84
Desulfococcus, anaerobic respiration in, 15
Desulfomonas, 84
Desulfovibrio, anaerobic respiration in, 15
Desulfuromonas, anaerobic respiration in, 15
Diagnosis, 84
Diagnostic microbiology
   American Type Culture Collection in, 13
   Ames test, 13
   animal virulence test, 15
   antibiotic sensitivity testing, 17–18
   autoradiography in, 25
   Bergey's Manual of Systematic Bacteriology
      in, 34
   centrifugation in, 55, 82, 250–251
   colonies of bacteria in, 67
   contact microradiography in, 70
   contact tracing in, 70
   contaminants in, 70
   culture media in, 9, 162–163
   dilution test, 18
   disc diffusion test, 18
   electrophoresis in, 92
   Elek diffusion test, 93

   filtration in, 104
   IMViC testing, 139
   inoculation of culture media in, 142–143
   laminar flow hood in, 22
   lyophilization in, 158
   metal decorating in, 163
   microscopy in (*see* Microscopy)
   microtome in, 165
   minimum inhibitory concentration in, 18
   quality control in, 208
   Schick test, 225
   selective toxicity of antibiotics in, 16
   sensitivity discs in, 18*f*
   staining in, 1, 70, 84, 119, 173, 231
   sterile technique in, 232
   sterilization in, 232
   streptococcal classification in, 150, 233
   tissue culture in, 239
   xenodiagnosis in, 265
Diarrhea, 84
   infantile, 101
   shellfish poisoning, 39–40
   traveler's, 101
Diatom, 84
Differential staining, 84
Dilution test, 18
Diphtheria
   control of, 86
   diagnosis of, 85
   epidemics of, in history, 274
   microorganism causing, 71, 84
   symptoms and signs of, 84–85
   transmission of, 86
   treatment of, 85–86
Diphtheroid, 86
Diploid, 86
Disc diffusion test, 18
Disease, infectious. *See also* Infection
   clustering of cases of, 87
   control of communicable, 140
   diagnosis of, 84
   endemic, 94, 96
   enzootic, 98
   epidemic, 98
   epizootic, 98
   food-borne, 106–107
   hosts in transmission of, 131–132
   incidence of, 87
   notifiable, 87, 178–179, 178*t*
   outbreaks of, 184, 275*f*
   pandemic, 186
   prevalence of, 87
   reservoirs in transmission of, 213
   transmission of, 87
   vectors in transmission of, 253
   venereal, 254

worldwide, as major cause of death, 276f, 277f
xenodiagnosis of, 265
Disinfection, 87
DNA, 87–88
    fingerprinting technique, 88
    hybridization of, 88
    recombinant, 211–212
    recombination of, 212
    replication of, 88
Domagk, Gerhard, 88–89
Domoic acid poisoning (DAP), 39–40
Droplet infection, 89
Drug resistance, 89
DTP vaccine, 89
Dulbecco, Renato, 33, 89
Dumdum fever. See Kala-azar
Dysentery, 89
    caused by E. coli strain, 101

E. coli. See Escherichia coli
Ebola virus, 90, 104, 124
    epidemics of, in history, 274
Ecologic processes, 90
    ammonia assimilation, 14
    ammonification, 14
    biochemical pathways, 92
    biofilms, 35
    biological oxygen demand, 36
    blooms, 39–40
    carbon cycle, 51–52
    composting, 69
    denitrification, 81–82
    eutrophication, 102
    food chain/food web, 107
    nitrogen cycle, 175
    nitrogen fixation, 26, 175, 176
    oxygen cycle, 185
    pest control, 36
    photosynthesis, 192–193
    putrefaction, 207
    respiration, 15, 214
Ehrlichia, 217
Electron microscope, 90–92, 91f
Electron transfer chain, 92
Electrophoresis, 92
Elek diffusion test, 93
Embden-Meyerhof-Parnas pathway. See Glycolysis
Emerging pathogen/infection, 93
    definition of, 93
    examples of, 93t
    identification of, 93–94
EMP pathway. See Glycolysis
Encephalitis, 94
    alphaviruses causing, 240
    epidemics of, in history, 273

equine, 240
Japanese, epidemics of, in history, 273
Encephalopathy, spongiform, 230
    bovine, 159
    prions associated with, 201–202
Endemic, 94, 96
Enders, John, 96
Endocytosis, 96
Endoplasmic reticulum, 96
Endosymbiont, 96
Endosymbiont hypothesis, 96
Endotoxin, 96
Entamoeba, 96–97
    E. gingivalis, 97
    E. hartmanni, 97
    E. histolytica, 12–13, 97
    E. polecki, 97
Enterobacter
    butanediol fermentation by, 43
    E. aerogenes, 97
Enterotoxin, 97
Enterovirus(es), 97–98
Enzootic disease, 98
Enzyme, 98
Epidemic(s), 98
    in history, 271–274
Epidemiology, 98
Epidermophyton, 83
    E. floccosum, 23
Epizootic disease, 98
Epstein-Barr virus (EBV), 99
    Burkitt lymphoma associated with, 43
    infection with, 120, 140
Erwinia, 99
    butanediol fermentation by, 43
    E. carotovora, 108
Erysipelothrix, 99
    E. rhusiopathiae, 99
    E. tonsillarum, 99
Escherichia coli, 99, 100f
    among indigenous microflora, 139
    anaerobic respiration in, 15
    antigens of, 100
    characteristics of, 100
    enterovirulent strains of, 100, 101
    in food-borne disease, 106
    genome of, 99–100
    infection with pathogenic, diagnosis and treatment of, 100–101
    in intestinal wall, 140f
    mixed acid fermentation in, 165
    pathogenic, 100
    role of, in intestinal tract, 100
    toxins produced by, 100
Eukaryote, 101–102
Eutrophication, 102
Evans, Alice C., 102

Exocytosis, 102
Exotoxin, 102
Eyach virus, 68

F factor, 70
Facultative aerobe, 103
Facultative anaerobe, 103
Fatal family insomnia, 103
Fermentation, 103–104
    alcohol, 10–11
    in beer production, 33–34
    in bread production, 42
    lactic acid, 150, 202, 268
    mixed acid, 165, 202
    in wine production, 263–264
Filovirus(es), 104, 104f
Filter, biological aerated, 36
Filtration, 104
Fimbria, 104–105
Fission, binary, 34–35
Flagella, 105
Flavivirus(es), 20, 105
    causing Dengue fever, 80
    causing yellow fever, 266–267
    encephalitis caused by, 94, 95t
Flavobacteria, denitrification by, 81
Fleming, Alexander, 105
Florey, Howard Walter, 105–106
Fluorescent microscopy, 106
Fomite, 106
Food
    inspection of, 107
    irradiation of, 107
    organoleptic, 182–183
    preparation of (see Food preparation/
        production)
    preservation of, 107–108
    spoilage of, 108
Food-borne disease, 106–107
Food chain/food web, 107
Food poisoning. See Food-borne disease
Food preparation/production
    alcohol, 10–11
    beer, 33–34
    botulism, 41
    bread, 42
    butter, 43
    buttermilk, 43–44
    canning, 51
    cheese, 56, 57f (see also Penicillium)
    contaminants in, 70
    curing in, 75
    fermentation in, 103–104 (see also Fermen-
        tation)
    pasteurization in, 189
    probiotics in, 202
    quality control in, 208

tyndallization, 248
ultrahigh temperature treatment, 251
vinegar, 255
water activity in, 261
wines, 263–264
yogurt, 268
Foot-and-mouth disease, 108–109
    virus causing, 20, 109
Fournier's disease, 109
Fracastoro, Girolamo, 109, 115
Francisella tularensis, 109
    infection with, 247–248
Freeze-drying, 158
Fumigation, 109
Fungus (fungi), 109–110
    characteristics of, 110
    hyphae of, 110, 135
    infections caused by (see Aspergillus;
        Blastomyces; Candida; Dermatophytes;
        Histoplasma capsulatum; Mucormyco-
        sis; Mycetoma; Paracoccidioides
        brasiliensis; Piedra; Pityriasis versicolor;
        Saccharomyces; Yeast)
    replication/reproduction of, 110–111
Fungus ball, 111

Gadjusek, Carleton, 112
Gallo, Robert C., 4, 112
Gamete, 112
Gametocyte, 112
Gas gangrene, 112–113
Gas vacuole, 113
GC ratio, 113
Gene(s), 113–114
    cloning, 64
    lux, 156
    mutation in, 16–17, 50, 168
    Nif, 26, 174
    operon of, 180
Gene expression, 114
Genetic code, 114
Genetic engineering, 114
Gengou, Octave, 40
Genital herpes, 128–129
Genome, 115
Genotype, 115
Gentamycin, 14
Genus, 115
Germ Theory, 115
German measles, 115–116
Gerstmann-Straussler-Scheinker syndrome,
    116
Giardia, 116
    characteristics of, 116
    cysts of, 116f
    diagnosis of infection with, 117
    life cycle of, 116–117

symptoms and treatment of infection by, 117
trophozoites of, 116*f*
Global warming, 117
Gloeothece, 76
Glucose, 117
Glycolysis, 117–118
in fermentation, 103
Glycoprotein(s), 118
Golgi apparatus, 118
Gonorrhea, 118
GPT (guanosine triphosphate), 120
Gram stain, 1, 119
Granuloma, 119
Granulosis virus particles, 32
Greenhouse effect, 119
Griffith, Fred, 243
GSS syndrome. *See* Gerstmann-Straussler-
Scheinker syndrome
Guillain-Barré syndrome, 120
Gymnodinium breve, 212

Haemophilus, 121
H. aegyptius, 121
H. ducreyi, 121
H. haemolyticus, 121
H. influenzae, 121–122
Hairy root, 10
Haldane, J. B. S., 183
Halobacterium, 113
H. salinarium, 122
Haloduric, 122
Halophile, 122
Hand, foot, and mouth disease, 122–123
Hansen's bacillus, 169
Hantavirus, 123
epidemics of infections, in history, 273, 274
Haploidy, 123
HeLa cell line, 239
Helicobacter
H. cinaidi, 123
H. fennelliae, 123
H. pylori, 123–124, 250
Hemadsorption, 124
Hemagglutinin, 124
Hemolysin(s), 124
Hemorrhagic colitis, caused by *E. coli* strain,
101
Hemorrhagic fever(s), 124–125, 125*t*
caused by bunyaviruses, 43
caused by filoviruses, 104
caused by flaviviruses, 266–267
South American, 21
Hepatitis
etiology of, 125–126
infectious, epidemics of, in history, 273
serum, 126–127

viral causes of, 126 (*see also specific hepatitis
virus*)
Hepatitis A virus (HAV), 127
in food-borne disease, 106
infections caused by, 126
Hepatitis B virus (HBV), 127
infections caused by, 126–127
Hepatitis C virus (HCV), 127–128
Hepatitis E virus, 45, 128
Herpes simplex virus (HSV), 128
characteristics of, 128
infections caused by, 128–129
skin lesions caused by, 128–129, 129*f*
treatment of infections caused by, 129
types of, 128
Herpes zoster virus, post infection herpetic
neuralgia, 200
Herpesvirus(es), 129
alpha group of, 129–130
beta group of, 130
characteristics of, 129
gamma group of, 130
Hershey, Alfred D., 130
Heterocyst, 130
Heterotroph, 130
Histoplasma capsulatum, 110*f,* 130
characteristics of, 130
symptoms of infection with, 130–131
treatment of infection with, 131
types of, 130
Ho, David, 131
Hook, Robert, 131
Host, 131–132
HTLV. *See* Human T cell leukemia virus
Human immunodeficiency virus (HIV), 132
characteristics of, 132
infection of T cells by, 132–133
infection with, in AIDS, 2–3, 133
T cells infected by, 132*f*
transmission of, in humans, 133
Human T cell leukemia virus (HTLV), 133
diagnosis and control of infection with, 134
discovery of, 133
symptoms of infection with, 133–134
transmission of, in humans, 133
Human T cell lymphotropic virus. *See* Human T
cell leukemia virus
Hydrolysis, 134
Hyperbaric oxygen therapy, 134
Hypersensitivity reaction(s), 134–135
delayed-type, 80
Type I, 12, 134
Type II, 134
Type III, 134
Type IV, 80, 134
Hypha(e), 110, 135

Immune response, 136
Immune system, 136–137
  and cancer, 48
Immunity, 137. *See also* Vaccine(s)
Immunization, 137–138. *See also* Vaccine(s)
  childhood schedule of, 138*f*
Immunoglobulin. *See* Antibody(ies)
Immunology, 138–139
Impetigo, 139
IMViC tests, 139
Inclusion body, 139
Incubation period, 139
Indigenous microflora, 139–140
  Acinetobacter, 1
  Escherichia coli, 139
  Lactobacillus acidophilus, 139
  Micrococcus, 139
  Staphylococcus epidermidis, 139
Infantile diarrhea, 101
Infection
  droplet, 89
  nosocomial, 178
  opportunistic, 181*t*
  waterborne, 261–262
Infection control, 140
Infectious mononucleosis, 140
Inflammation, 140–141
Influenza, 141
  epidemics of, in history, 271–274
  viruses causing, 141–142
Influenza virus(es), 141
  characteristics of, 141–142
  diseases caused by, 142
  replication of, 142
Inoculating loop, 142–143
  use of, 142*f*
Inoculation, 143
Interferon, 143
Isolation, patient, 190

Jaundice, epidemics of, in history, 271
JC virus, 200, 203
Jenner, Edward, 144, 144*f*

Kala-azar, 145
Kanamycin, 14
Kaposi's sarcoma, 145–146
  associated with AIDS, 3, 145
Khurthia, 148
Kinetoplast, 146
Klebs-Lodffler bacillus, 71
Klebsiella
  butanediol fermentation by, 43
  K. ozenae, 146
  K. pneumoniae, 146
  K. rhinoscleromatis, 146

nitrogen fixation by, 176
  urease production by, 251
Koch, Robert, 115, 146–147
Koch's postulates, 147
Krebs, Hans A., 147
Krebs cycle, 148
Kuru, 112, 148

L forms, 149
β-lactam antibiotics, 149
β-lactamase, 149
Lactic acid bacteria, 149
Lactic acid fermentation, 150
Lactobacillus, 150
  beneficial effect of, in humans, 202
  in bread production, 42
  L. acidophilus, 44, 139, 150
  L. bulgaricus, 43–44, 150, 268
  L. casei, 150
  L. delbruckei, 150
  L. sanfrancisco, 150
  lactic acid production by, 149
  in wine production, 263
Lactococcus, 150
  L. lactis, 150
Laminar flow hood, 22
Lancefield, Rebecca, 150
Lassa fever, virus causing, 21
Laveran, C. L. A., 151
Lederberg, Joshua, 151
Leeuwenhoek, Antoni van, 115, 151, 183
  discovery of *Giardia* by, 116
Legionella pneumophila, 151
  characteristics of, 151
  disease caused by, 152
  epidemiology of infection with, 151–152
  infection process of, 152
Legionnaire's disease, 152
  epidemics of, in history, 274
Leishmania, 152
  infections caused by, 75, 75*f*, 145
  L. aethiopica, 75
  L. braziliensis, 75, 152
  L. chagasi, 145
  L. donovani, 145, 152
  L. infantum, 145
  L. major, 75
  L. mexicana, 75, 152
  L. tropica, 75, 145, 152
  life cycle of, 152–153
Leprosy, 153
  diagnosis and treatment of, 153–154
  epidemiology of, 153
Leptospira, 154
  L. biflexa, 154
  L. hollandia, 154

L. interrogans, 30*f,* 154
L. noguchii, 154
Leuconostoc, 154
in butter production, 43
L. cremoris, 43
L. mesentroides, 154
lactic acid production by, 149
Leukocyte(s), 155
Lichens, 155, 168
Life, origin of, 183–184
Linnaeus, Carl, 155
Lipmann, Fritz A., 155
Lipopolysaccharide (LPS), 155
in outer membrane of gram-negative
bacteria, 185
Lister, Joseph, 155
Listeria, 155–156
infections caused by, 156
L. monocytogenes, 156
Lithotroph, 156
Lockjaw, 156
Lumpy jaw, 156
Luria, Salvador E., 130, 156
lux genes, 156
Lyme disease, 157
microorganism causing, 41
Lymphocyte, 157
Lymphocytic choriomeningitis virus (LCMV),
21
Lymphogranuloma venereum, 62, 157–158
Lyophilization, 158
Lysozyme, 158

MacLeod, Colin, 159
Mad cow disease, 159
Madura foot, 169
Magnifying power, 159
Malachite green stain, 1
Malaria
caused by *Plasmodium falciparum,* 198
caused by *Plasmodium malariae,* 198
caused by *Plasmodium ovale,* 198
caused by *Plasmodium vivax,* 198
diagnosis of, 160
epidemics of, in history, 271, 272, 273, 274
epidemiology of, 159–160
malignant, 198
microorganisms causing, 159, 160
ovale, 198
quartan, 198
symptoms of, 160
transmission of, 160
treatment of, 160–161
Malassezia furfur, infection with, 194
Mallon, Mary, 52, 249
Marburg virus, 104, 104*f,* 124, 161
epidemics of, in history, 274

Marshall, Barry, 123
McCarty, Maclyn, 161
Measles, 161
complications of, 161
epidemics of, in history, 271, 274
infection process in, 161
prevention of, 161–162
virus causing, 161
Medium (media), 162
basic nutrient, 162
definition of, 162
selective, 162–163
Mendel, Gregor, 163
Metabolism, cell, 163
ATP in, 23
glycolysis in, 103, 117–118
phosphorylation of ATP in, 191–192
respiration in, 15, 214
signal transduction pathway and, 47–48
Metal decorating, 163
Metchnikoff, Elie, 163
Methanobacterium thermoautotrophicum, 163–
164
Methanococcus jannaschii, 164
Methanogen, 164
Methylophilus methylotrophus, 164
Methylotroph, 164
MIC (minimum inhibitory concentration), 18
Microbiology, 164
Micrococcus, 139
Microcyctis, 76
Microradiography, contact, 70
Microscopy, 164–165
contrast, 70
darkfield, 79
electron, 90–92
fluorescent, 106
magnifying power in, 159
metal decorating technique, 163
optical, 181–182, 182*f*
phase-contrast, 191
resolution in, 213
scanning electron, 224
scanning transmission electron, 224
transmission electron, 244
UV, 251
X-ray, 265
Microsporum, 83
Microtome, 165
Miescher, Friedrich, 165
Miller, Stanley, 184
Minimum inhibitory concentration (MIC), 18
Mitochondrion, 165
Mixed acid fermentation, 165
MMR vaccine, 165–166
Monkeypox virus, 166
Mononucleosis, infectious, 140

Montagnier, Luc, 3, 166
Montagu, Mary Wortley, 229
Morbidity and Mortality Weekly Report, 166
Morbidity rate, 166
Mortality, worldwide, due to infectious disease, 276f, 277f
Mortality rate, 166–167
Motility, cell, 167
MOTT bacilli, 169
Mucor, infections caused by, 167
Mucormycosis, 167
Muller, Hermann, 167
Mumps, 167
    diagnosis of, 167–168
    infection process in, 167
    prevention of, 168
Mumps virus, 167–168
Mutagen, 168
Mutant, 168
Mutation(s), 168
    and bacterial resistance to antibiotics, 16–17
    of DNA in cancer, 50
Mutualism, 168
Mycetoma, 169
Mycobacteria, acid fast staining of, 1
Mycobacterium, 169
    M. africanum, 246
    M. avium, 169
    M. intracellulare, 169
    M. leprae, 153–154, 169
    M. microti, 169
    M. tuberculosis, 80, 169–170, 246–247
Mycolic acids, 1
Mycoplasma, 170
    antibiotic resistance of, 16
    M. hominis, 170
    M. hyopneumoniae, 171
    M. pneumoniae, 170, 171
    M. urealyticum, 170
Mycoplasma(s), 170–171
Mycosis, 171

Nalidixic acid, 172
Nanobe, 172
Necrosis, 173
Negative staining, 173
Negri bodies, 173, 211
Neisseria, 173
    characteristics of, 173
    diagnosis of infection by, 173
    N. animalis, 173
    N. gonorrhoeae, 118, 173
    N. meningitidis, 173, 230
    N. mucosa, 173
    N. ovis, 173
    N. pharyngis, 173
Neomycin, 14

Neoplasms, 46–47
    and cancer, 48
Neurotoxic shellfish poisoning (NSP), 39–40
Nicolle, C. J. H., 173–174
Nif genes, 174
    of *Azomonas,* 26
Nitrate assimilation, 174
Nitrate respiration, 174
Nitrification, 174
Nitrifying bacteria, 175
Nitrobacter, 175
    N. hamburgensis, 175
    N. vulgaris, 175
    N. winogradskyi, 175
Nitrococcus mobilis, 175
Nitrogen cycle, 175
Nitrogen fixation, 175
    microorganisms that cause, 26, 176
Nitrogenase, 176
Nobel, Alfred, 176
Nocardia, 176
    associated with *Actinomyces,* 6
    characteristics of, 176–177
    diagnosis of infection with, 177
    infections caused by, 177
    N. asteroides, 177
    N. brasiliensis, 177
    N. caviae, 177
Nomenclature, binomial, 35
Non-Hodgkin's lymphoma, in AIDS, 3
Nongonococcal urethritis (NGU), 61
Norwalk virus, 45, 177–178
Nosocomial infection, 178
Nostoc commune, 76
Notifiable disease, 87, 178–179
    in United States, 178t
Novobiocin, 179
Nuclear polyhedrosis viruses (NPV), 32
Nucleus, 179

Oncogenic virus, 180
Oparin, Alexander, 180, 183
Operon, 180
Opportunistic infections, 181t
Opportunistic pathogen, 180
Opsonin, 181
Optical microscope, 181–182
Orbivirus(es), 182
Organoleptic food, 182–183
Organotroph, 183
Origin of life, 183–184
Oroya fever, 33, 184
Otitis media, 121
Outbreak(s), of infectious disease, 184, 275f
Outer membrane, 185
Oxygen cycle, 185

Pandemic, 186
Papillomavirus(es), 186
    warts caused by, 261
Papovavirus(es), 186–187
Paracoccidioides brasiliensis, 187
Parainfluenza virus, 187
Paralytic shellfish poisoning (PSP), 39–40
Paramyxovirus(es), 187
    characteristics of, 187–188
    diseases caused by, 188
Parasitism, 188
Paratyphoid, epidemics of, in history, 273
Parvovirus(es), 188
Pasteur, Louis, 115, 183, 188–189, 189*f*
Pasteurization, 1, 189
Pathogen(s), 189–190
    blood-borne, 39
    emerging, 93–94
    opportunistic, 180
Patient isolation, 190
Pediococcus, lactic acid production by, 149
Pelvic inflammatory disease (PID), 61
Penicillin, 190
Penicillium, 190
    P. camembertis, 190
    P. notatum, 105, 190
    P. roqueforti, 58, 190
Peptic ulcers, *Helicobacter pylori* in, 123
Peptide bond(s), 14, 190
Peptidoglycan, 190
Peptococcus niger, 191
Pest control, biological, 36
Phage. *See* Bacteriophage
Phagocytosis, 191
Phase-contrast microscope, 191
Phenotype, 191
Phosphorylation, 191–192
Photobacterium, 192
    luminescence of, 37
    lux genes of, 156
    P. fisheri, 192
    P. phosphoreum, 192
Photolithotroph, 192
Photoorganotroph, 192
Photosynthesis, 192–193
Phototroph, 193
Phytoplankton, 193
Picornavirus(es), 20, 193
Piedra, 193–194
Piedraia hortae, infection with, 194
Pilus, 194
Pinta, 194
Pityriasis versicolor, 194
Pityrosporon orbiculare, infection with, 194
Plague, 194
    diagnosis and treatment of, 195
    epidemics of, in history, 271–272

history of, 195–196
    infection process in, 194–195
    microorganism causing, 194
    prevention of transmission of, 195
    symptoms of, 195
Plaque, 196
    dental, 82–83
Plasma membrane. *See* Cell(s), membrane of
Plasmid, 196
Plasmodium, 196
    immature schizont stage of, 197*f*
    life cycle of, 196
    life cycle of, in humans, 196–198
    malarial disease caused by, 159–161
    mature schizont stage of, 197*f*
    P. falciparum, 160, 161, 198
    P. malariae, 198
    P. ovale, 197*f*, 198
    P. vivax, 160, 161, 197*f*, 198
    ring nucleus in trophozoite stage of, 197*f*
    trophozoite stage of, 197*f*
Plesiomonas shigelloides, 198–199
Pneumocystis carinii, 3, 199
Pneumonia, epidemics of, in history, 273
Pneumonic plague, epidemics of, in history, 274
Poliomyelitis, 199
    epidemics of, in history, 272, 273
Poliovirus, 200, 200*f*
Polyhydroxybutyrate (PHB), 10
Polymyxin, 200
Polyoma virus(es), 200
Postherpetic neuralgia (PHN), 200
Poxvirus(es), 201
Prion(s), 201–202
    in bovine spongiform encephalitis, 159
    in Creutsfeld-Jacob disease, 72–73
    in fatal family insomnia, 103
    in Gerstmann-Straussler-Scheinker syn-
        drome, 116
    in kuru, 148
    in scrapie, 225
Probiotic, 202
Prognosis, 202–203
Progressive multifocal leukoencephalopathy
    (PML), 200, 203
Prokaryote, 203
Propionibacterium, 203
    P. freudenreichii, 203
    P. shermanii, 58, 203
Protein(s), 203
Protein A, 203
Protein M, 203–204
Proteus, 204
    P. mirabilis, 204
    P. myxofaciens, 204
    P. vulgaris, 204
    urease production by, 251

Protista, 204
Prototroph, 204
Protozoa, 204–205
Prusiner, Stanley, 201–202, 205
Pseudomonad, 205
Pseudomonas, 205–206
    denitrification by, 81
    P. aeruginosa, 30*f,* 206
    P. marginalis, 108
    P. stutzeri, 15, 206
Pseudopod, 206
Public health. *See also* Disease; Pathogen(s);
    Vaccine(s)
    CDC (Centers for Disease Control and
        Prevention), 55
    communicable disease control in, 140
    contaminants in, 70
    disease carriers and, 52
    drug resistance of microorganisms and, 89
    emerging pathogens and, 93–94
    epidemiology in, 98
    fomites and, 106
    fumigation and, 109
    Germ Theory and, 115
    global warming and, 117
    immunization in, 137–138
    infection control and, 140 (*see also* Infec-
        tion)
    isolation of patient in, 190
    large infectious disease outbreaks, 275*f*
    leading causes of death worldwide, 276*f*
    morbidity and mortality rate reports in, 166–
        167
    notifiable diseases and, 87, 178–179, 178*t*
    quarantine in, 208–209
    risk assessment in, 219
    risk ratio in, 219
    sanitation in, 224
    sexually transmitted disease and, 226–227
    surveillance in, 235
    World Health Organization in, 264
Pure culture, of bacteria, 206–207
Putrefaction, 207

Q fever, 72, 208
Quality control, 208
Quarantine, 208–209

Rabbit fever, 109
Rabies virus, 210
    diagnosis and treatment of infection with,
        211
    infection with, 173, 210
Rat bite fever, 211
Recombinant DNA, 211–212
Red blood cells, hemadsorption by, 124

Red tide, 212
Redi, Francisco, 183, 212
Reed, Walter, 212
Relapsing fever, 212–213
    epidemics of, in history, 271, 273
Reovirus(es), 213
Reservoir, 213
Resolution limit, 213
Respiration, 214
    anaerobic, 15
Respiratory syncytial virus, 214
Retrovirus(es), 214–215
Reverse transcriptase, 215
Rhabdovirus(es), 215
Rheumatic fever, 216
Rhinovirus(es), 216
Rhizobium, 168
    nitrogen fixation by, 176
    R. leguminosarum, 216
Rhizopus, infections caused by, 167
Ribosome, 216–217
Rickettsiae, 217
    diseases caused by, in humans, 218*t*
Rickettsiae (genus), 217
Rifamycin, 217
Rift Valley fever, 43, 104, 124
Ringworm lesion, 83*f*
Risk assessment, 219
RNA, 219
Rochalimaea, 217
    R. quintana, 217
Ross, Ronald, 219
Rotavirus(es), 220
Rothia, 220
    R. dentacariosa, 220
Rous, Peyton, 26, 220
    and virus-cancer theory, 49
Rous sarcoma virus, 26, 49
Rubella, epidemics of, in history, 274
Rubella virus, 220
    characteristics of, 220
    infections caused by, 115–116, 221, 221*f*
Rubeola virus, causing measles, 161–162
Rubin, Harry, 50
Ruminococcus, 220

Sabin, Albert B., 222
Saccharomyces
    S. beticus, 263–264
    S. carlsbergensis, 33–34
    S. cerevisiae, 266
    S. ellipsoideus, 263
Saccharomyces cerevisiae, 222
    in alcoholic fermentation, 10–11
    in beer production, 33–34
    in bread-making, 42
    in wine production, 263–264

Salk, Jonas, 222
Salmonella, 223
    Ames test for, 13
    in food-borne disease, 106
    mixed acid fermentation in, 165
    S. enteritidis, 223
    S. paratyphi, 248
    S. typhi, 223, 248–249
    S. typhimurium, 13, 223
Sandfly fever, 43
Sanitation, 224
Saprophyte, 224
Sarcina ventriculi, 224
Scalded skin syndrome, 224
Scarlet fever, 224–225
Schick test, 225
Schizogony, 225
Scrapie, 225
Selective toxicity, of antibiotics, 16
SEM (scanning electron microscope), 224
Sensitivity discs, 18*f*
Septicemia, 225
Serotypes, of *Streptococcus*
    classification of, 233
    discovery of, 150
Serratia, 226
    S. liquifaciens, 226
Serum hepatitis, 126–127
Sewage disposal, 226
Sewage treatment, 6, 226, 269
    activated sludge, 6
    biofilms, 35
    biological aerated filter, 36
Sexually transmitted disease (STD), 226–227
Shigella
    in food-borne disease, 106
    S. dysenteriae, 227
Shingles, 227
Signal transduction pathways, 47–48
Sludge
    microorganisms in, 6
    treatment of, 226
Smallpox, 227
    epidemics of, in history, 271–272, 272
    immunization against, 228–229
    infection route in, 228
    lesions of, 228*f*
    prognosis in, 228
    symptoms of, 228
Snow, John, 229
South American hemorrhagic fevers, 21
Spallanzani, Lazzaro, 183, 229
Species, 229
Spheroplast, 229
Spinal meningitis, 230
Spirillum minus, 230
    infection with, 211

Spirulina, 76
Spongiform encephalopathy, 230
    bovine, 159
    prions associated with, 201–202
Spore, 230
Sporothrix schenckii, 230–231
Staining, 231
    acid fast, 1
    contrast, 70
    differential, 84
    Gram, 1, 119
    negative, 173
Staphylococcus, 231–232
    hemolysins produced by, 124
    S. aureus, 231
    S. epidermidis, 139, 231
    S. saprophyticus, 231
Staphylococcus aureus
    infection with, 224
    lysostaphin in, 158
    protein A in cell wall of, 203
    toxic shock syndrome caused by, 240
STD (sexually transmitted disease), 226–227
STEM (scanning transmission electron
    microscope), 224
Sterile technique, 232
Sterilization, 232
Strain (bacterial), 232
Streptobacillus moniliformis, 232
    infection with, 211
Streptococcus, 232
    in butter production, 43
    classification of, 233
    colony of, in feta cheese, 57*f*
    hemolysins produced by, 124, 124*f,* 233
    α-hemolytic, 233
    β-hemolytic, 233
    γ-hemolytic, 233
    lactic acid production by, 149
    protein M classification of, 233
    S. cremoris, 43
    S. diacetilactis, 43
    S. lactis, 43, 150
    S. mutans, 30*f*
    S. pneumoniae, 30*f,* 233
    S. pyogenes, 234
    S. thermophilus, 56, 58, 268
    serotypes of, classification of, 233
    serotypes of, discovery of, 150
    in yogurt production, 202, 268
Streptococcus pyogenes, 234
    infection with, 139
    protein M in, 203–204
    rheumatic fever caused by, 216
    scarlet fever caused by, 224–225
    toxic shock caused by, 241

Streptomyces, 234
  S. clavuligerus, 234
  S. coelicolor, 234
  S. griseus, 234
  S. niveus, 179
Streptomycin, 14, 234
Subacute sclerosing panencephalitis, 161, 234–235
Subtilisins, 37
Sulfonamides, 235
Surveillance, 235
SV40 virus, 200
Symbiosis, 235
Syncytium, 235
Syphilis, 235–236
  epidemics of, in history, 271

T cell(s), 237
  in development of AIDS, 2–3
  HIV infection of, 132*f*
  normal, 132*f*
Taxonomy, 237
TCA cycle. *See* Krebs cycle
TEM (transmission electron microscope), 244
Temin, Howard M., 33, 50, 237
Tetanus, 237–238
Tetracycline, 238
Theiler, Max, 238
Thermophile, 238
Thermotoga maritima, 238
Thiobacillus ferrooxidans, 238–239
Thylakoid, 239
Tinea pedia. See Athlete's foot
Tissue culture, 239
Togavirus(es), 20, 239
  characteristics of, 239–240
  encephalitis caused by, 94, 95*t*
  transmission cycle of, 240
Toxemia, 240
Toxic blooms, 39–40
Toxic shock syndrome (TSS), 240–241
Toxoid, 241
Toxoplasma
  infection with, in AIDS syndrome, 3
  T. gondii
Toxoplasma gondii
  characteristics of, 241
  diagnosis of infection with, 242
  infection by, in humans, 242
  life cycle of, 241
Toxoplasmosis, 242
Trachoma, 242
Transcription, 243
Transduction, 243
Transformation
  bacterial, 243
  cellular, 243

Translation, 243–244
Traveler's diarrhea, 101
Treponema, 244
  T. carateum, 194, 244
  T. cuniculi, 244
  T. pallidum, 34, 79*f*, 235–236, 244
  T. pertenue, 244, 266
  T. vincentii, 244
Trichoderma polysporum, 76–77
Trichome, 244
Trichomonas, 244–245
  T. vaginalis, 244
Trichophyton, 83
  infection with, 23
  T. mentagrophytes, 23
  T. rubrum, 23
Trichosporon
  infection with, 193–194
  T. beigelii, 193–194
Trophozoite, 245
Trypanosoma, 245
  T. brucei, 8–9, 245–246
  T. brucei gambiense, 9
  T. brucei rhodesiense, 8–9
  T. cruzi, 55–56, 246
  T. rangeli, 246
Tuberculosis
  microorganisms causing, 246–247
  prevention of, 247
  radiograph of, 246*f*
  symptoms and treatment of, 247
Tularemia, 109, 247–248
Tumor, 248
Tyndallization, 248
Typhoid fever
  epidemics of, in history, 273
  infection process in, 248–249
  microorganisms causing, 248
  prevention of, 249
  symptoms of, 248–249
Typhoid Mary, 52, 249
Typhus, epidemics of, in history, 271

Ulcers, peptic, or gastric, 250
Ultracentrifugation, 250–251
Ultrahigh temperature treatment, 251
Undulating membrane, 251
Ureaplasma, 170
Urease, 251
Urey, Harold, 183
UV microscope, 251
Uwins, Philippa, 172

Vaccine(s), 252
  attenuated, 23–24
  DTP, 89
  MMR, 165–166

Vaccinia virus, 252
Vacuole, 252
  gas, 113
Varicella-Zoster virus, 59, 227, 252–253
Variola virus, 253
Vector(s), 253, 253*f*
Venereal disease, 254
Verruga peruana, 254
Vesicular stomatitis virus, 254
Vibrio, 254–255
  V. alginolyticus, 255
  V. cholerae, 255
  V. fetus, 255
  V. parahaemolyticus, 255
Vibrio cholerae, 254*f,* 255
  in food-borne disease, 106
  infection with, 62–63
Vinegar, 255
Virchow, Rudolph, 256
Viremia, 256
Virion, 256
Virus(es), 256. *See also specific virus; specific disease*
  antigenic drift in, 18–19
  antigenic shift in, 19
  and cancer, 46, 49–50
  capsid of, 51
  characteristics of, 256
  classification of, 258*t*–259*t*
  defective, 79
  diseases caused by, 260 (*see also specific disease; specific virus*)
  genetics of cancer and, 49–50
  hemagglutinin on surfaces of, 124
  hemorrhagic fevers caused by, 124–125
  hosts of, 257
  infection process by, 257
  nomenclature of, 257
  oncogenic, 50, 180
  replication of, 256, 257, 260
  size of, 256–257
  study of, in research, 260
Vitamin, 260

Waksman, Selman A., 261
Warming, global, 117
Wart, 261
Washing powders, biological, 37
Water activity, 261
Waterborne infection, 261–262
Watson, James D., 262
WHO (World Health Organization), 264
Whooping cough, 262–263
Wiezmann, Chaim, 65
Wilkins, Maurice, 73
Wine, 263–264
World Health Organization (WHO), 264

X-ray microscope, 265
Xenodiagnosis, 265
Xerophile, 265

Yaws, 266
Yeast, 266. *See also Candida; Saccharomyces*
  budding in, 42
Yellow fever, 266–267
  epidemics of, in history, 271, 272, 273
Yersinia, 267
  Y. enterocolitica, 267–268
  Y. pestis, 194–196, 268
  Y. pseudotuberculosis, 268
Yogurt, 268

Ziehl-Neelsen stain, 1
Zoogloea
  in sludge, 6
  Z. ramigera, 269
Zoonosis, 269
Zooplankton, 269
Zygote, 269

Neeraja Sankaran is a science writer and a Ph.D. student in the history of medicine and science at Yale University. She holds degrees in both microbiology and science writing. Sankaran has contributed articles to a number of scientific publications including *The Scientist, The NCRR* (National Center for Research Resources) *Reporter, Annals of Internal Medicine,* and *Yale Medicine.*

3